Living and Dying Well

Lewis Petrinovich

University of California
Riverside, California

Plenum Press • New York and London

Library of Congress Cataloging-in-Publication Data

On file

ISBN 0-306-45171-9

A Division of Plenum Publishing Corporation
233 Spring Street, New York, N.Y. 10013

10 9 8 7 6 5 4 3 2 1

Printed in the United States of America

Preface

Problems of profound ethical and practical significance have arisen due to developments in biotechnology and the medical sciences. These problems make it necessary to reconsider many moral issues regarding life and death more carefully than has been done in the past. I believe that this reconsideration can profit if issues are viewed multidisciplinarily, so I attempt to frame moral problems in light of social, psychological, and biological realities and examine the implications in terms of economic factors. The issues perhaps are best exemplified in the rapid developments that have taken place in the science of human genetics—developments that have made it possible to utilize scientific advances to achieve practical ends, resulting in an explosion in the area of genetic engineering.

In *Human Evolution, Reproduction, and Morality*, I discussed evolutionary, ethical, and economic problems produced by developments in the technologies by which one can assist humans to reproduce when reproduction is difficult or impossible under normal conditions. In the present book, attention is devoted to technological developments that make it possible to routinely transplant organs from donors to patients who need them. Problems arise because it is necessary to obtain organs from cadavers as soon as possible after death in order that the organs be in a physical condition permitting successful transplant. The criteria by which death can be said to have occurred must be considered carefully because technological advances have made it possible to maintain organisms in a persistent vegetative state for extended periods.

Other economic and moral concerns are related to the development of technologies that extend the length of human life. It is possible to maintain organic functioning for years, even though the patient will never recover psychological functions. The ethical questions raised by all of these technological advances should be faced proactively by the political, medical, and ethical communities, and the discussions pursued here are intended to move toward understanding those psychological, moral, biological, and economic considerations. Finally, I pose some broad issues in medical ethics, consider the nature of the responsibilities of physicians regarding suicide and euthanasia, and discuss—as a case history in applied ethics—the problems encountered in attempts to devise an adequate health-care delivery system in the United States.

I express my appreciation to the staff at Plenum Press. With their able assistance, the production of this and the previous book was a pleasure for me. Once I provided the manuscript, everything flowed smoothly and proceeded ahead of schedule—a state of affairs that many of us do not always experience in our publishing endeavors. I appreciated the efficiency and professionalism of the production editors, Robin Cook and Robert Freire, and of Jeffrey Leventhal. I especially want to thank Executive Editor Eliot Werner and his editorial assistant Kathleen Lucadamo. Meeting with them is always a pleasant occasion, no matter how mundane the matters that might be our concern of the day. It was also pleasant to discover that Eliot and I are both enthusiastic Francophiles, which led to wistful exchanges regarding experiences in the wilds of France and the streets of the Left Bank in Paris.

I also thank Patricia O'Neill, who read some of the material in what turned out to be the penultimate draft and who led me to recognize some major organizational problems. Once these difficulties were pointed out, I was able to produce a final draft that was much more satisfactory than otherwise would have been the case.

I hope that these two Plenum books contribute positively to the discussion of issues facing contemporary society and focus attention on some data and perspectives that are often overlooked in treatments of these issues. Now that I have considered these issues regarding the permissible treatment of humans by humans, I will turn my attention to those concerning the permissible use of animals by humans.

Contents

CHAPTER 1

Objectives and Background Principles

In this book, evolutionary, developmental, and ethical principles are used to evaluate the quality of ongoing life, the ending of that life, and the merits of mapping the human genome and manipulating its structure and function. This discussion continues arguments developed in *Human Evolution, Reproduction, and Morality* (Petrinovich, 1995), in which an evolutionary view of human reproduction was presented. These evolutionary ideas apply to all sexually reproducing organisms, and aspects of human morality were interpreted in their light. It was argued that insights regarding the structure of human moral intuitions and existing systems of morality result when they are viewed from the perspective of actions that would be expected to perpetuate the genotypes of individuals.

After discussing evolutionary and moral principles, issues involving reproduction were considered: abortion, infanticide, and the reproductive technologies used to assist infertile couples to reproduce. The moral implications of these developments were discussed and recommendations made that could lead to reasonable and just social policies.

This chapter briefly reviews the basic evolutionary, philosophical, and cognitive principles argued in Part I of *Human Evolution, Reproduction, and Morality*. Only the major points of the argument are sketched, and those interested in the literature supporting them should consult that book.

BASIC EVOLUTIONARY PRINCIPLES

The web of ideas that make up the theory of evolution can be viewed as a broad network adequate to understand the origins, changes, and current states of organic systems. A basic axiom is that the interaction of elements in the organic universe can be viewed from the perspective of cost-benefit analyses. At the level of individual organisms, there must be a benefit in terms of passing on as many genes as possible to the next and succeeding generations, and avoiding the ultimate cost that would occur if a population of organisms becomes so narrowly specialized that the genetic line could face the risk of extinction in the face of changed pressures. To maintain a reasonable balance, it is necessary to forego the maximal level of propagation of genes that would be achieved by the mother cloning herself and thereby contributing all of her genes to the next generation. Such a cloning strategy is problematic, because some degree of variation must be maintained among the organisms that constitute the breeding population in order that some individuals will survive and reproduce whenever the characteristics of the environment change radically and quickly, or when some new players (competitors, cooperators, predators, or prey) suddenly appear. A second benefit from maintaining variability in individual genotypes is that a varying set of characteristics can, as Hamilton phrased it (Hamilton, Axelrod, & Tanese, 1990), present a continually moving target to more quickly reproducing parasites which, because of their rapid rate of reproduction and short generation time, could quickly come to specialize on the weaknesses of each generation of longer lived and more slowly reproducing organisms, until they drive the members of the host species to extinction. One way to defeat parasitization is to recombine individual genotypes at each mating, making it less easy for parasites to overcome.

The alternative strategy to cloning that has been adopted by many organisms is to sexually reproduce. The cost of sex is that each of the parents gives up one-half of its genes in each mating. The benefit is that genes are recombined at each reproduction, which maintains the desired variability.

Evolutionary change occurs when certain behavioral and physiological traits (called *phenotypes*) are produced by the underlying gene structures (called *genotypes*). This change takes place if the individuals displaying those traits are better able to reproduce and foster progeny than those not displaying those traits. Such individuals are considered to be better adapted, and they will become more numerous and successfully outreproduce competitors, providing that the phenotypes are heritable. When such differential reproduction occurs, the structure of the gene pool (the total number of different genes found in the population of individuals that make up the species) will change. The changes in the phenotypes produced must be *heritable*, which means that the offspring of such better adapted parents have some likelihood of inheriting those genes and gene combinations that produce physical characteristics conferring a selective advantage. These offspring, in turn, must be able to pass those genes on to their offspring.

Although this description of evolutionary change may seem straightforward, there is no simple relationship between genes and morphological, physiological, or psychological traits. There is no one-to-one relationship between genes and the phenotype. Different environments influence the range of phenotypes that can and will be expressed. Cumulative changes in the traits of successive generations can be produced through genetic inheritance, as well as through the perpetuation of environmental and cultural changes that influence the expression of genetic potential. This point is critical when applying genetic analyses to understand the functioning of human communities.

The process of natural selection is important because of its influence on the phenotypic variation presented by different organisms in the population, favoring some at the expense of others. Another factor is what is called *isolation*—that all genetic lines within a species do not interbreed freely because of geographic or behavioral separation. Such isolation often results in the appearance of characteristics that make one group of individuals so different (genotypically and phenotypically) from others in behavior or morphology that they are unable (or unwilling) to interbreed with those others. These changes can be produced by peculiarities of the different ecologies of some breeding communities, as well as by random genetic drift that can result in distinctive characteristics appearing in the individuals that define those communities. When such changes occur, the group of interbreeding organisms can be considered to be a new biological species, or at least on the way to specieshood.

One critical distinction is between the *proximate* and *ultimate* levels at which evolutionary processes can be viewed. At the proximate level, the questions involve how processes occur and the nature of the mechanisms within the organism and the environment that drive evolutionary processes. The ultimate level involves differential reproductive success, and the output at this level is reckoned in terms of the number of genes replaced in the gene pool relative to the performance of other individuals in the population. Maintaining the proximate–ultimate distinction helps to avoid considerable confusion that results when explanations are framed at one level (e.g., changes in physiological functions) and then are uncritically extended to explain events at the other level (e.g., reproductive success).

Sex and Reproduction

Because reproduction is an essential aspect of evolution, and sexuality has evolved as a primary strategy to further the reproductive end, one would expect evolutionary processes to have had the greatest influence on those characteristics (morphological, physiological, and psychological) that are involved in reproduction, and such is indeed the case. The basic games played in pursuit of human reproduction start at the moment there are sperm present and an egg to be fertilized (Baker & Bellis, 1989). The interests of the two parties, egg and sperm, lead them to a physiological contest. The male reproductive system has been tuned evolutionarily to maximize the likelihood of reproduction given the varying conditions surrounding the copulatory episode.

The human female, however, is not just a passive receptacle in which males play out their sperm-competition games, in which one individual's sperm competes to fertilize eggs and to block the attempts of others' sperm to succeed. Females can influence the outcome of the contest between sperm in several ways (Baker & Bellis, 1993; Bellis & Baker, 1990). The female orgasm can regulate the number of sperm retained at both the current and a succeeding copulation. Nocturnal, masturbatory, and copulatory orgasms are the primary mechanisms by which the female can influence retention of sperm, and intercopulatory orgasms are cryptic to the males. These various competitive strategies would have developed through the natural selection of characteristics enhancing the reproductive success of individuals who possess such heritable traits, and *no* conscious intent is implied. There is little doubt that, at these physiological levels, evolutionary arguments are reasonable. Are they reasonable when we consider interactions between the mother and the conceptus?

Trivers (1972) argued that the evolutionary interests of a mother and her offspring can be in conflict following conception, because the offspring benefits if it receives more parental care than the parent is prepared to give, with the mother accruing more interest in the offspring as her time of investment increases. Haig (1993) considered the interaction of the human mother and her fetus in considerable detail. The fetus benefits by extracting as many resources as possible from the mother, while the mother must strike a balance between nourishing the fetus and keeping some resources for herself, as well as providing for her existing and future children. Maternal genes pay the cost of fetal development throughout pregnancy to gain a future benefit (and natural selection "keeps tabs" on the relative benefit per unit cost). The conflict between what is best for the "mother's genes" and what is best for the "fetal genes" is marked by a high degree of interdependence—what Haig called a conflict of interest within a basically peaceful society. If the fetus has a genetic or developmental defect that will make it unlikely to survive, then it might be in the interest of the mother's overall reproductive success to miscarry and try to conceive again. The interest of the fetus, however, is to survive at all costs, so it should try to prevent the woman's body from causing a miscarriage, because that would result in total loss to the fetus.

An example of the working of such mutual coexistence has been provided by Profet (1992), who examined the phenomenon of pregnancy sickness from an adaptationist perspective. She argued that it evolved to protect the embryo against the *in utero* ingestion of toxins that are abundant in natural foods. Women who have moderate or severe pregnancy sickness have a higher pregnancy success rate than those who have mild or no pregnancy sickness. Pregnancy sickness is universal across human cultures, a fact that is compatible with the interpretation that the sickness conferred a selective advantage on ancestral humans.

This brief review supports the conclusion that the basic levels of reproduction can be understood in terms of cost–benefit analyses of the functional interests and strategies of the different players. Such basic strategies have been identified for most species, and their manner of operation has been studied carefully for a large number of species.

Misconceptions Regarding Development

A major misconception should be identified and never again be allowed to rear its ugly head. It is often assumed that if a trait appears as the result of genetic instructions, it is unmodifiable: that an individual has genes for some characteristic and that an inherited blueprint only has to unfold during development for the characteristic to be manifested. This misconception gives rise to concerns that any discussion of a universal human nature that might be encoded genetically is antithetical to a belief in fundamental human dignity, because it involves an undesirable deterministic view of human nature. Evolutionary psychologists (e.g., Cosmides & Tooby, 1987) pointed out that even a universal human nature would permit a large number of different traits, psychologies, and behaviors to manifest themselves between individuals and across cultures. Such variability is possible and likely because, even if there is such a common inherited psychology, it must operate under a variety of environmental conditions. The ability to respond to varying pressures is exactly what a successful gene pool must retain if the species defined by that gene pool is to successfully respond to a variety of pressures. It can be argued that humans, so far, have adapted successfully to an enormous variety of conditions.

The traits that appear early in the life of a developing organism are selected so that, given the almost inevitable circumstances that surround conception and birth, the organism will be provided with the necessary elements to sustain further physiological and behavioral development. It is critical that these early developmental processes run off in some appropriate sequence, and that alternative avenues of stimulation or input can be used if the most commonly occurring events are not encountered.

Although each organism begins with a distinct genotype, the differential events it encounters can influence its development in many different ways. However, if certain inputs are necessary for the development of essential characteristics of the organism, then there must be ways to buffer the organism's systems when those inputs do not occur. A useful way of considering the interaction of events and processes that are important in development is in terms of *experience-expectant* structures (Greenough, Black, & Wallace, 1987). These structures are designed to utilize environmental information that is so ubiquitous it is almost universal; it invariably occurs in the natural developmental history of individual organisms, and probably has throughout the evolutionary history of the species. One way these structures operate is through the production of an excess of connections among neurones early in the organism's developmental history. When certain experiential inputs occur, some of these neural cells and their connections will survive, and others will fade away. At the outset, the organism is primed genetically to be sensitive to these certain stimuli, and to respond to them when they occur.

When these early inputs are received, a second system becomes active, called the *experience-dependent* system. This second system stores information depending on the unique experiences of the individual organism, using it to generate new neural connections in response to the occurrence of a "to-be-remembered" event. Experience-expectant sensory systems make it possible for the organism to develop an extensive range of different performance capabilities, and predisposed to respond to those stimuli that would be available to all young animals of the species in the normal course of development. The genes need only outline the rough pattern of neural connectivity in a sensory system and determine the time when an experience-expectant system will be active and receptive, leaving the specific details to be established through the organism's interactions with its environment—which involves the experience-dependent system.

These two systems involve a great deal of general "prewiring," much of which is lost as development proceeds, with the successfully competing elements being those most actively utilized by the experience-dependent system. Even though the system was initially established structurally, there is a high degree of plasticity possible in the course of development. Such developmental plasticity in central neural representation has been documented for many species when one sensory modality suffers damage or when the organism is deprived of sensory input to a modality. Lacking the expected stimulation, the organism is able to use another sensory mode to continue the process of development.

A third system, called the *activity-dependent* system (Locke, 1993), utilizes the activity of the developing organism to produce sensory impressions that further the development of different sensory modalities. This system makes it possible to develop such things as speech and language, and provides the organism with the ability to actively generate the stimulation necessary for development to proceed.

The conception of how these three systems operate is quite different from the traditional one that considers development in terms of innate versus learned influences. Nowhere is one kind of process active for a time to be followed by another, with each controlling a certain percentage of the variance in development—there is a continuous interaction from the outset. This view emphasizes a continual dynamic interaction between biased perceptual and motor dispositions that are almost certain to be activated if the human infant is in the nurturant environment it must have to survive. Yet, if the usual array of stimuli is not available, because of a defective sensory system, for example, the developing organism can use stimuli from other modalities to continue along the path of development to generate sufficient stimulation that will provide the general activation required to sustain developmental processes.

If development is conceptualized in terms of these three systems, it is not meaningful to invoke the specter of genetic or biological "determinism." All organisms start out with experience-expectant systems that were established genetically, but phenotypic development is a function of a complex interplay between the experience-expectant, experience-dependent, and activity-dependent systems. Genes are only a piece of the story of development, because influences that occur, even in the uterine environment, can alter the expression of the innate programs inscribed in the developing organism's genes. There is a distinction between "biological" and "genetic" events: Environmental influences might produce biological differences between organisms in such things as receptor sensitivity, the structure of neural connections, and the levels of hormones. These biological differences, however, may not be attributable to genetic influences at all. Genetic differences may contribute to only one aspect of biological events that subsequently influence the nature of the organism.

Many of the initial insights regarding early development were investigated intensively, especially for birds and mammals, by those zoologists called *ethologists*, who study animals within the context of their natural environment. They identified a process they called "imprinting." Particularly important events occur at the time an organism is born. It has been found that adult behavior can be influenced crucially through exposure to stimuli that occur during specific times in development. For example, if a freshly hatched gosling is exposed to a moving object, and if this is the first thing the gosling has seen, it "imprints" on this object and behaves toward it as if the moving object was a parent (usually the mother)—which it almost always is in nature.

The general phenomenon of imprinting has been demonstrated in many species (including humans) for many stimuli, and for a wide range of behaviors. The first object experienced by human newborns is almost always the mother, the first sounds heard are usually those of the parents (especially the mother), and the first tastes experienced are those of food types in the immediate environment—either taken directly or through feeding of maternal milk. These early experiences drive the infant's development in certain directions; toward sensitivities and preferences for some classes of objects over others—such as a preference for the sound of the mother's face and voice, which leads it to respond to faces by smiling. There is genetic tuning of receptor systems; this tuning increases the likelihood that only stimuli with a restricted range of characteristics will be selected and responded to, and attention preferentially will be directed to those stimuli.

Evolution and the Human Condition

The next question concerns the relevance of evolutionary principles to help us understand the human condition. Wright (1995) discussed the views of evolutionary psychologists regarding the biology of violence, arguing strongly for the relevance of functional evolutionary analyses when considering the biochemical events involved in the kinds of violence that are prevalent in the inner cities of the United States. He concluded that there are evolved behavioral tendencies that are reactions to the loss of status, and that these tendencies represent adaptations that were useful in the environment of evolutionary adaptation

(EEA). The ideas of evolutionary psychologists are being represented accurately in the popular press and are finding their way into the thinking of the general public, as well as that of the medical community.

In *Human Evolution, Reproduction, and Morality*, the mechanisms involved in the development of human speech and in the initial stages of language development were discussed at length. Speech and language are accepted as two defining characteristics of humans. The developmental principles outlined earlier suggest that the early beginnings of language unfold in much the same way as other behavioral systems of both human and nonhuman animals. A number of social bonds, especially between the mother and neonate, are cemented almost from the moment of birth. Such bonds are important when considering issues regarding the moral status that should be accorded to the neonate, because they mark the entry of the organism as a player interacting with members of the social community. It is at this point that the neonate gains personhood and when the organism should be considered to be a moral patient—an individual that does not yet have the full moral standing of a moral agent. People have a predisposition to adopt moral principles as a result of early experience, and the specific principles adopted will reflect the coordinated interactions between early social experiences and an evolved genome that has been biased to enhance the developing organism's reproductive success.

Basic Evolved Processes. The question now concerns the extent to which evolutionary principles can be used to gain an understanding of the sensory, perceptual, cognitive, and social behaviors of humans. It is widely accepted that many fundamental processes are influenced by innate biasing mechanisms. One basic behavioral adaptation that almost all animals have is a tendency to orient toward a suddenly appearing stimulus, and if the stimulus is repeated over and again and is not followed by any particular consequence, the response to the stimulus wanes with repetition—a process called *habituation*. Few are distressed if it is suggested that such a complex mechanism is an evolved process.

Some organisms have receptor elements that react preferentially to selected types of stimulation and are difficult to habituate. Frogs have retinal cells that react preferentially to small, fast-moving objects ("bug detectors"), and there is little doubt, or disagreement, that this enhanced responsiveness to prospective food stimuli evolved as an adaptation to environmental conditions: Frogs that sensed insects more efficiently would be likely to be better nourished and enabled to reproduce more than those that lack the enhanced bug sensitivity (what has to be demon-

strated to complete the scenario is that the behavioral disposition is heritable and that reproductive success is enhanced). Animals of many species, including humans, have retinal cells that are stimulated or inhibited by such things as edges, and by lines with certain characteristics. It has been demonstrated that, without experience, figures of particular shapes are grouped together, objects moving in certain directions are detected as basic units, and depth of field is appreciated.

Humans have mechanisms that "automatically" process certain kinds of visual stimuli preattentively. Such stimuli are said to "pop out" of displays, and the time required to detect these stimuli is independent of the number of elements. Detection of certain other stimuli requires the use of an attentional mechanism that checks each item in the display in a serial fashion: The more elements to be checked, the longer the search time. Such simple sensory and perceptual mechanisms could have undergone strong selection and would have conferred a survival advantage in the EEA. When considering such simple levels of functioning, few reject the argument that these processes are evolved, content-specific information-processing adaptations. Many social scientists and humanists, however, become increasingly resistant to attempts to extend the arguments much beyond these simple behavioral processes.

There is an enormous body of evidence demonstrating that animals of many species have specialized detection systems, and that these systems adapt them to the demands of their environment. They have content-specific learning mechanisms that enhance the likelihood that certain kinds of events will be learned quickly. Some insects (e.g., wasps) learn the characteristics of the complex gestalt of stimuli surrounding nests containing the larvae they are provisioning, and do so with but one exposure. They can learn such characteristics better and more quickly than members of most avian and mammalian species, but can hardly be characterized as mental giants in other regards.

There is evidence for such content-specific learning mechanisms in a wide variety of situations that are critical to the survival of organisms. These mechanisms include species-specific defense reactions (Bolles, 1970), learned food aversions mediated by taste in rats and vision in birds (Garcia & Brett, 1977), and selective learning of certain sounds under certain conditions by birds (Petrinovich, 1990). When learning mechanisms are studied within the context of the ecology within which organisms exist and cope, one is led to the conclusion that searching for content-general "intellectual" mechanisms is not of much use to understand an animal's functioning, any more than trying to understand language development by studying the way humans process nonsense syllables can capture the essence of language acquisition.

Speech Development. When evolutionary mechanisms are invoked to understand more complex human behaviors, such as cognition, there is an increased resistance to accept them. One careful elucidation of the types of developmental processes that may be involved in cognition has been provided by Fernald (1992), who proposed a four-stage model characterizing the usefulness of intonation during infant speech development during the first year of life. This model is a concise summary of the communicative functions of infant-directed speech. A first level has the prelinguistic function of capitalizing on the infant's predisposition to respond differentially to certain prosodic characteristics (sound contour, including melody and rhythm) of infant-directed speech. These maternal vocalizations serve to alert (high-frequency sounds with a gradual rise-time in intensity) or to soothe (continuous, low-frequency sounds; especially white noise—such as shhh).

At a second level, the melodies of maternal speech become increasingly effective in directing infant attention and modulating arousal and emotion. Over the first six months of life, the infant's visual capabilities and motor coordination have improved; it can recognize individual faces and voices more quickly, and the social smile in response to voices and faces appears more frequently. The mother's speech not only captures attention, but also evokes emotional expression by the infant. By 5 months, infants from monolingual English-speaking families respond to vocalizations used in approval and prohibition when they are uttered using infant-directed speech only, and respond to vocalizations made in several languages. During the first six months, infants are more responsive to voices than to faces, but at about 7 months of age, they reliably recognize happy and angry facial expressions.

A third level is the communication of intention and emotion, with the vocal and facial expressions of the mother providing the infant initial access to the feelings and intentions of others. The infant begins to interpret the emotional states of others and to make predictions about the future actions of others, using signals based on vocal and facial expressions.

At a fourth level, prosodic elements are accepted as markers to help the infant identify linguistic units within the stream of speech. As Fernald so nicely phrased it, words begin to emerge from the melody. At 15 months, infants recognize familiar words better in infant-directed speech, but by 18 months have acquired the ability to identify familiar words equally well in adult-directed and infant-directed speech, although the exaggerated tone of infant-directed speech may still be important when acquiring new words. These events characterize the speech development of infants in all human societies that have been studied, indicating that there is the required cross-cultural universality that characterizes basic evolved processes.

Language Development. Can we move further along and understand a complex trait that is uniquely human? Studies of the development of natural language, done by cognitively oriented psycholinguists, provide us with major insights regarding the development both of speech and grammar. Pinker and Bloom (1991) made the heretical suggestion that the ability to use natural language belongs more to the study of human biology than to human culture.

Pinker (1994) argued that human language, including the underlying rules that produce grammar and syntax, evolved through the process of natural selection. The use of "motherese," the language mothers use when addressing infants, directs the attention of the infant in ways that would support emotional development, identify linguistic units within the stream of speech, and enhance the acquisition of new words. These functions assist caregivers to be more efficient, and make it more likely that offspring will survive to reproduce themselves. These language precursors precede and prepare the way for the development of human language. Bickerton (1990), Lieberman (1984), Locke (1993), and Pinker (1994) have all argued that the universality of critical elements in the development of human speech, grammar, and syntax support the idea that language has evolved.

Analysis of the human speech and language literature indicates that the development and use of human language, among the most cherished human abilities, has yielded many of its secrets to an evolutionary analysis. The human-language abilities that have been laboriously taught to members of nonhuman species are faint shadows of the complex grammar that spontaneously develops in all young human infants and whose basic aspects can be recognized at birth.

Evolution of Cognition

Development of Mathematics and Spatial Cognition. The developmental psychologist David Geary (1995) used an evolution-based framework to understand the development of the cognitive processes involved in mathematics. He identified numerical abilities that he considered to be universal, biologically primary abilities, many of which are shared with animals of other species, and which develop inevitably given normal experience. Among such biologically primary abilities Pinker (1994) noted that the brain organizes the world into discrete, bounded, and cohesive objects and arranges these objects into categories of the same kind. He argued that babies are designed to expect a language to contain words for kinds of objects and kinds of action, to display basic elements of a universal grammar, and to reflect the prosodic elements of their language community well before they have had any extensive exposure to a wide range of language exemplars.

Geary (1995) found that these tendencies to group objects of the same kind involve an implicit understanding of the number of objects in small arrays, and that it occurs as early as the first week of life. This sensitivity is intermodal, and by 18 months of age, infants show a sensitivity to ordinal relationships and engage in primitive counting behavior.

Geary (in press) suggested that the organisms of many species have an implicit understanding of some fundamental features of Euclidian geometry (e.g., that a line from one point to another is straight). He argued that the large sex differences that favor males in three-dimensional spatial abilities could be the result of a greater elaboration of the neuro-cognitive systems that support habitat navigation and representation. These differences are directly related to intramale competition and to males' courtship of females. In most preliterate societies, hunting is almost exclusively a male activity, and it has been shown that hunting success is directly related to the number of wives obtained in hunter–gatherer societies that permit polygyny. Characteristics such as the ability to mentally manipulate three-dimensional representations of information, to track and predict the trajectories of moving objects, and to navigate would enable men to hunt more successfully and to triumph in the small-scale warfare between kin-based groups that was likely to have prevailed in the EEA. There is little overlap in the distributions for males and females in throwing distance and velocity, even before males begin to engage in sports participation.

Geary argued that considerable confusion has occurred when sex differences are considered using broad categories, such as spatial abilities, verbal abilities, or mathematics, rather than considering the component features within these general categories. Petrinovich (1995) suggested that it is necessary to understand the characteristics of the EEA for humans in order to identify those behavioral tendencies that would have enhanced the reproductive success of those members of the society possessing them. One would expect males and females to be equivalent in their cognitive abilities, except for abilities that would affect the process of sexual selection, because there should be a direct benefit enhancing the performance of the differential tasks that could contribute to the success of each of the sexes. Men may have been selected to excel in hunting and fighting and women to excel in tasks that assess memory for objects and their location. Geary concluded that conceptualizing cognitive sex differences in terms of goal structure, and conceptual and procedural competencies in terms of sexual selection, will enable a more complete and satisfactory analysis of these differences than is possible by comparing scores on arbitrarily selected cognitive tasks. The strategies used to conceptualize language and mathematical development have provided insights into human nature, and these successes indicate that it might be possible to develop a descriptive base of actions that people engage in consistently enough that these actions can be considered to be universals.

Human Problem Solving. What about more complex human abilities, such as those involved in solving complex problems? Among social scientists, there has been a biophobia and intellectual isolationism that Tooby and Cosmides (1992) argued has become more extreme with time. What they call the Standard Social Science Model (SSSM) assumes that genetic variation cannot explain the purported fact that many behaviors are shared within groups of people, but not between such groups. The SSSM maintains that inputs are everywhere the same, although adults everywhere differ in behavioral and mental organization, and that these differences are produced by cultural events that are extragenetic. The generators of complex, meaningful organization in human life are considered to be emergent processes whose determinants are at the group level of sociocultural events. In this view, human nature is an empty vessel waiting to be filled, another *tabula rasa* on which the hand of experience can write. The SSSM allows that, although natural selection may have been involved at one time, human evolution has progressed to a point at which the influence of genetically determined systems of behavior has now been removed and replaced with general-purpose learning mechanisms involving content-independent cognitive processes. The arguments of the SSSM can be challenged using a database that ranges across phyla and other data that deal directly with uniquely human characteristics.

Tooby and Cosmides (1995) argued that the human brain would be expected to be organized functionally to construct information, make decisions, and generate behaviors that would tend to promote inclusive fitness in the ancestral environment and behavioral contexts of Pleistocene hunter–gatherers (the EEA). They make the case (as does Geary) that researchers might profitably spend their time looking for functional organizations that would be expected to have enhanced propagation in the EEA. Such a research program would use stimuli and tasks incorporating items that are representative of problems our hunter–gatherer ancestors would have encountered: faces, smiles, expressions of disgust, foods, the depiction of socially significant situations, sexual attractiveness, habitat quality cues, animals, navigational problems, cues of kinship, rage displays, cues of contagion, motivational cues, distressed children, species-typical body language, rigid object mechanics, plants, and predators.

Tooby and Cosmides (1992) and Gigerenzer and Hug (1992) conducted studies indicating that the standard, domain-general model of problem solving fails when the cognitive behaviors studied are representative of those that would be adaptive within the natural social ecology of humans. One of the most important processes used to explain the evolution of social cooperation and competition is that of inclusive fitness, on which is built the idea that one contributes one's genes to succeeding generations, not only by enhancing direct genetic contributions, but also by behaving in ways that enhance the genetic contribution of relatives, and even of members of the social community who might reciprocate any aid given them. Cognitions involving social exchanges, having undergone selection pressure for many thousands of years, should incorporate design features that are particularly appropriate to deal with such problems. Individuals should be especially adapted to reason efficiently when social contracts are involved, and should be attuned to detect cheating—a violation of a social contract. Gigerenzer and Hug (1992) found that people reasoned more efficiently (80% correct solutions) when solving problems requiring the detection of rule violations involving social contracts than when the same formal rule violations did not involve social contracts (30% correct solutions).

Social contract theory was the only one of several alternatives evaluated in these studies that could account for the consistently better performance on problems involving social contracts, although Davies, Fetzer, and Foster (1995) quarrel with the adequacy of this conclusion for one of the tasks used. People seem to have inference procedures that are applied specially to social contract problems: They are better able to detect cheaters, and are good at recognizing altruists, leading them to perform much better than they did on problems involving the same formal, logical steps but which did not involving social contracts. Gigerenzer and Hug (1992) found that people perform better on the same problem when it is posed in a perspective in which they are placed in a role requiring them to detect a cheater than if they are to search for information regarding the operative rule.

There certainly are domain-general mechanisms, such as those used in rote memory, a short-term memory load of seven plus or minus two, and in the attribution of cause. These mechanisms are deployed generally and are very useful aspects of people's cognitive abilities. There are, however, a multitude of domain-specific mechanisms that were selected to enhance the adaptation of organisms coping with evolutionarily significant problems.

The Evolved Human Social Condition

There is a large body of research that supports adaptational explanations of more complex human social behaviors. There are universal behavioral tendencies that influence patterns of homicide (Daly & Wilson, 1988), different patterns of jealousy shown by men and women (Wilson & Daly, 1992), sex differences in the age and status preferred for mates (Kenrick & Keefe, 1992), and in the different reproductive strategies employed by males and females in terms of a potential partner's reproductive potential (Buss, 1994). Buss and Schmitt (1993) proposed a contextual–evolutionary theory of mating, arguing that the adaptive logic of men and women should be different when they pursue short- versus long-term mating strategies, but both sexes should always be interested in a mate who would be a good parent. The pattern of human dispersal from natal communities to breeding communities was found by the anthropologists Clarke and Low (1992) to be consistent with predictions based on evolutionary mechanisms known to operate for many animal species.

There have been a number of studies of evolutionary mechanisms involved in the structure of legal principles (Wilson, 1987) and moral reasoning (Burnstein, Crandall, & Kitayama, 1994; Petrinovich & O'Neill, in press; Petrinovich, O'Neill, & Jorgensen, 1993; Wang, 1992). The evolutionary argument has been extended to understand problems involved in the practice of medicine (Maier, Watkins, & Fleshner, 1994; Nesse & Williams, 1995; Williams & Nesse, 1991). In *Human Evolution, Reproduction, and Morality*, the aforementioned literature was reviewed extensively to understand the nature of human moral systems and social policies, especially as applied to issues in reproduction.

Although a variety of specific rules and structures are found in different societies, there are general features that characterize all societies, and a large number of these features relate to reproduction and inheritance—of goods as well as genes. The traits of cooperation and communication provide the cohesive elements for the sexual partnerships on which society depends. These relatively permanent unions—based on economic and social cooperation—constitute the norm.

BASIC ISSUES IN MORALITY

In Chapter 6 of *Human Evolution, Reproduction, and Morality*, it was argued that when considering human moral systems from an evolutionary perspective, the most useful and applicable basic principles are those called consequentialist, which define what is good in terms of outcomes. The calculation of this good requires an estimate of the relative weight of the various values involved in order to do a cost–benefit analysis of the alternative outcomes. In terms of basic evolutionary theory, the ultimate benefit is reproductive success, and the costs are anything that diminishes that success in terms of reduced future reproduction or capacity.

Problems arise when attempting to calculate the relative weights that should be given to different values. How does one equate a given amount of pain and suffering to a given amount of pleasure and well-being? Is it necessary that some values always take precedence and cannot be allowed to fall below a minimal level? The question was phrased by Brandt (1980) in terms of how it can be decided that the value of speaking freely on political matters is stronger than the value of owning capital goods, and that both of these are weaker than the value of not being tortured.

Although these are difficult questions, values can be roughly ordered in terms of their essential importance. An absolute guarantee must be made that one should not be murdered, and this should take priority over not being subjected to unnecessary pain, which in turn would take precedence over enjoying a delightful meal. I think everyone's intuition would support relative orderings of these kinds. Another argument that most would accept in some version is John Rawls's (1971) maximin rule, which states that we should make decisions to maximize the outcome for the most disadvantaged in society, other things being equal. This rule makes good sense when combined with the idea that there is a minimal level below which no one should be allowed to fall. They must reach this level to be able to live a satisfactory life, and a just society must ensure that no one falls below that level on those values that are essential. Adequate health care is one such essential to which all should be entitled, and inequality in health care cannot be tolerated as long as the worst off have insufficient care.

Some Important Distinctions

There is an important distinction between moral agents and moral patients. Those with full standing in the moral community are considered moral agents; they have direct moral duties to one another and bear the load of all moral responsibilities and duties. Moral patients, on the other hand, lack the abilities that would make them accountable for the outcomes of their actions. Moral patients include such individuals as human mental defectives, the senile, very young children, fetuses, and most, if not all, nonhuman animals. A human who is incapable of reasoning or understanding abstract concepts cannot be held responsible for an act that injures another, because such moral patients are unable to understand the concepts of right or wrong, or sometimes even the causal relationship between their actions and the resulting injury to another.

Another important determination, when considering the beginning and termination of life, is at what point organisms attain the status of persons, when does the status of personhood begin and end. It was argued in Chapter 9 of *Human Evolution, Reproduction, and Morality* that personhood begins at the point of birth; that this is when a public human entity appears, and a social contract between the neonate and the members of the community comes into force. This contract makes infanticide impermissible, even though the neonate has not yet developed beyond the stage of a moral patient. This contract represents those aspects of humanity that provide the cohesive elements fostering the interests of the community and, as a result of the successful enforcement of its terms, serves to enhance the ultimate reproductive success of community members.

Justice

One basic element of the concept of justice is that all similar individuals should be treated similarly. The Kantian argument is that one should act in such a way that those principles used to regulate specific actions could be accepted as universal law. Rawls (1971) examined ideas regarding justice extensively and pointed out that happiness presupposes the enjoyment of primary human goods, such as health, a certain amount of wealth, and a respected place in a free society. He considered principles of justice in the light of fairness within the structure of society, and argued that the principles of justice should involve an agreement that free and rational persons accept, based on an original position of equality. It is within the bounds of agreements that basic rights and duties should be assigned and social benefits divided.

Rawls believes that the idea of a social contract is of primary importance in establishing justice. Feinberg (1989) construed Rawls's theory to fall within a social contract tradition, noting that Rawls emphasized that one has an obligation to do one's part whenever one accepts benefits and opportunities, in terms of goods provided by the society. It is also important that rules should not be changed in the middle of the game, because that would disappoint the honest expectations of those whose prior commitments and life plans were made with the assumption that the rules would be continued (Feinberg, 1989).

To establish basic principles we should, according to Rawls, invoke a difference principle: The position of the better-off is to be improved only if it concomitantly improves the position of the worst-off. Persons who are equals should not agree to a principle that would dictate lesser life prospects for some, simply for the sake of a greater *sum* of advantages enjoyed by others. This proposal is adopted to circumvent some of the problems encountered by a Utilitarian position that calculates morality by aggregating the total sum of welfare that different policies would produce.

Rawls (1971) argued for two principles: The first requires equality in the assignment of basic rights and duties, whereas the second holds that social and economic inequalities can be considered to be just only if they result in compensating benefits for everyone and, in particular, for the least advantaged members of society. There is no injustice in choices that produce greater benefits for the favored few, provided that the situation of the less fortunate persons is improved as well. Rawls refers primarily to our relations with other persons, and leaves out of the account how we are to conduct ourselves toward animals and the rest of nature. A contract theory such as his is compatible with principles that characterize evolutionary theory, and this line of thinking can be extended more broadly to include issues regarding animals and the environment with little difficulty.

In order for Rawls's theory to be applicable, there must be a way to settle questions regarding the priority of the plurality of principles involved. He considers the assignment of the weight of different principles to be an essential part of the conception of justice, and that rational discussion depends on an explanation of how these weights are to be determined. The basic principle of justice is that all primary goods—which he identifies as liberty and opportunity, income and wealth, and self-respect—must be distributed equally, unless an unequal distribution of them is to the advantage of the least favored.

Feinberg (1989) noted that the duty to uphold justice, as defined by the rules of established just practices and institutions, provides the sufficient principles to design social practices and institutions, and that these should be preferred over any principles that assume the actions of individuals will result in a benign society. Feinberg (1984), when considering how to apply the harm principle to the development of criminal law, made the suggestion that people have multiform interests that must be protected by uniform rules, and that the best way to proceed is by positing a "standard person." All standard persons share certain welfare interests, including an interest in continued life, health, economic sufficiency, and political liberty. The law must prevent harm from occurring to one's standard primary interests. The harm principle should take into account both the magnitude and probability of the harm: If there is a high probability of serious harm as a result of an action, then it should be forbidden by criminal law; if the probability is relatively low, then there is less reason to forbid it. He argued this position in the interest of respecting people's freedom, especially if the magnitude of the harm is small.

With this scheme for the standard welfare interest, it is necessary to consider what could be called secondary values and interests. These are relatively harmless, but are important to individual persons. These secondary values reflect differential experiences, abilities, and tastes, and include such things as possessing wealth, reputation, and applause, as well as enjoying friendship and comfort.

Kekes (1993) also argued that humans have universal, primary values. He identified a plurality of primary values: basic physiological needs (e.g., having food and not being tortured); psychological needs (e.g., love and freedom from humiliation); social needs (e.g., respect and freedom from exploitation). He considered the pluralism of secondary values in considerable depth and argued that the best way to encourage a personally satisfying and morally meritorious life is to adopt a moral pluralism that respects what we wish for ourselves. This pluralism emphasizes the possibilities whose realization may make our lives good. Kekes believes that pluralists must set their own conception of a good life as merely one possibility among a plurality of reasonable options, realizing that there is no absolute basis to prefer one particular set of secondary values over others. Individual differences are fostered, and experiments in different ways of living are encouraged, as long as the actions do not harm others or deny their interests and liberties. The assignment of weights to such incommensurate and incompatible secondary values is a serious problem when seeking moral principles to provide an appropriate basis on which to decide proper courses of action.

Rational Liberalism

In *Human Evolution, Reproduction, and Morality*, a philosophical position I called Rational Liberalism was discussed in order to illuminate some of the arguments that could be used when evolutionary and consequentialist considerations are less directly involved, especially when there is no concern with reproduction, or when utilitarian calculations are difficult to make. This position is based on John Stuart Mill's consequentialism, which emphasizes the role of intellectual pleasures to a greater extent than hedonistic ones. Feinberg (1986) accepted Mill's principle that society should prevent harm to individuals; that a person's (agent's) freedom should be absolute regarding things that affect the individual's self and body. He supplemented the harm principle with an *offense principle*, which he defined as hurt produced by deep revulsion toward an act of another. It might be reasonable to restrict the free choices of persons whenever it is reasonable to doubt that the choice is a rational, informed, and voluntary one, even though these choices might not violate either the harm or offense principles. A policy is morally justified when it is factually informed, rational, and voluntary. Society has an obligation to determine if acts meet this test, to permit choices that do meet the test, and to not permit them if they do not. Certain goods are essential and primary, and a degree of equality should be provided so that all members of society are guaranteed an adequate minimal level of these goods. The plurality of secondary moral interests and values should be respected as well, and the rules, regulations, and laws that society adopts should respect this pluralism. These rules must be within the permissible bounds of freedom of respect, freedom from harm, not give offense, and be consistent with the principle of liberty.

COGNITIVE SCIENCE

The ideas developed by cognitive scientists are important when considering a cognitive test for moral agency. The proper criteria for agency should rely on cognitive functioning and should be concerned with the quality of decisions the individual is capable of making. The essential characteristics that must be considered, as argued by Anderson (1990), include the goals the organism is pursuing, the structure of the environment that is relevant to the attainment of these goals, and the cost involved in using a cognitive process. Four major aspects of cognition were invoked by Anderson: Memory, Categorization, Causal Analysis, and Problem Solving. These concepts should be supplemented by an appeal to the essential quality of having an autobiographical sense, such that there is a continuing idea of a *self* that can be entertained by the organism. Any organism that can be shown to possess these qualities meets the test for moral agency and should be accorded the status of full personhood, with all its duties and responsibilities.

EVOLUTION AND MORALITY

In *Human Evolution, Reproduction, and Morality*, it was argued that understanding the nature and structure of people's moral intuitions might lead to a better appreciation of the kinds of moral imperatives that have developed, and that this understanding does not involve committing a naturalistic fallacy. Rather, it permits an understanding of patterns of moral beliefs that exist, which makes it possible to appreciate basic problems that might be encountered whenever society attempts to develop a set of *oughts* that conflict with people's deep-seated intuitions about realty—intuitions that could have evolved over thousands of years of human existence.

To better appreciate the structure of human moral intuitions, a series of studies of moral intuitions were presented in Chapter 7 of that book, and it was argued that there is a set of universal moral intuitions compatible with expectations based on evolutionary theory. The most important dimensions that drove people's decisions in the resolution of hypothetical, fantasy dilemmas were Species (people favor members of the human species over any others); Inclusive Fitness (people favor kin or members of their community over others); and membership in an abhorrent political movement—Nazis (disfavored over all other humans). Two other dimensions were of moderate importance: Numbers (a tendency to favor a number of people over a single individual); and Social Contract (a tendency to favor innocent persons over those there because it was their job).

These patterns of morality could have developed in much the same way as several other aspects of complex behavioral dispositions. Physiological, structural, and behavioral traits are adaptations that developed to enhance the reproductive success of the individual organisms involved. Not only are simple and relatively invariant response tendencies affected through the processes of differential reproduction and natural selection, but complex processes involved in perceiving the world are influenced. There is evidence that humans solve problems that involve social contracts very efficiently. There is also evidence that mating strategies and patterns of reproduction have evolved in such a way that they enhance the lifetime reproductive success of the individuals, and behavioral tendencies that support cooperation and communication are enhanced.

The development of speech and language was considered at length because the processes and mechanisms involved in their development are similar to those involved in the development of morality. This sculpting of receptivity initially serves the function of attachment and emotional bonding, but comes to lay the basis for communication through speech.

Language development (whether spoken or signed) is an adapted and evolved process utilizing context-specific learning modules. This process involves innate computational mechanisms that define the grammatical categories of such things as noun, verb, and auxiliary. Words are couched into grammatical categories rather than being stored as individual words, and people have categories built into memory and context that lead them to look for phrases, and to use them as the basic elements of analysis. There are cognitive modules, shared with those in the community, that enable children to learn the variable parts of language in a manner ensuring that their grammar is synchronized with that of their community.

Pinker (1994) noted that this way of conceiving language development makes huge chunks of grammar available to the child all at once, which means that it is not necessary to acquire dozens or hundreds of rules, but to just set a few mental "switches." Selection favored those speakers in each generation that hearers could best decode, as well as those hearers who could best decode the speakers. This process of natural selection led to what has been referred to as a Universal Grammar. Just as speech used the mechanisms serving the functions of emotional bonding to piggyback its development, the language system piggybacks on the speech system.

It is argued that morality is an evolved process developing in much the same way as does language, and it piggybacks on the prosodic system, using the language system to codify the rules and regulations that constitute moral systems. All human communities have developed codes, rules, and practices to regulate the reproductive pair-bond, to legitimize offspring, and to determine who inherits goods. These codes ensure the stability of the basic reproductive unit—the family. There are proximate mechanisms that have developed to further the ultimate interests of enhanced reproductive success and to promote inclusive fitness. The rules that develop can be different given historical accidents and resource availability, but they cannot hinder the ultimate reproductive success of the community members if the community is to survive and compete successfully with others who share their biological niche. Although natural selection might have shaped basic aspects of morality, this does not mean that evolved tendencies will result in a moral society. The cognitive principles involved in the codification of social ideologies are of paramount importance, and the task is to see that these ideologies are not used to pervert the desirable goals of evolutionary adaptation. Cognitive factors must direct the development of fundamental philosophical principles concerning human dignity and freedom. An adequate system of morality will seek to eliminate the exploitation of those humans who are in a less-favored status.

PLAN OF THE BOOK

The argument will be made that evolutionary theory provides insights into issues involving the understanding and manipulation of genes, human death, and ethical issues involved in health care. A large number of statistics have been gathered from a range of sources, and these are cited in many of the chapters. These statistics constitute the database from which several aspects of the questions under consideration can be viewed: frequencies of occurrences (diseases, medical treatments, medical facilities), finances (costs of medical procedures and of research activities, national budgets, profits), demographics (death rates, indices of health status, structure of the population), and outcomes (effectiveness of treatment procedures, quality of life). These figures are used to indicate the bases for the arguments regarding the controversial issues discussed. Those arguments and the policy recommendations developed can be more easily understood if the data on which they are based are explicit at the outset. With such understanding, it might be easier to direct disagreements to the database whenever that is appropriate. If the evidence is understood and found adequate, then the logic of the arguments and the reasons for policy recommendations can be evaluated. If the structure of evidence is flawed, then it is possible for critics to develop a more adequate database and then proceed to the structure of the argument.

The process involves the presentation of data, interpretations of their meaning, the statement of a moral hypothesis, evaluation of the evidence in terms of the hypothesis, and the development of policy recommendations that follow from that evaluation. This procedure is intended to capture the strengths of the procedures used by biological and social scientists when they attempt to understand the complex realities of human existence.

In Chapter 2, techniques used to screen the genetic structure of individuals (human and nonhuman) and ethical problems arising when these techniques are used with humans are discussed. The potential benefits, dangers, and ethical issues involved in the Human Genome Project are discussed in Chapter 3, in which an evolutionary perspective is reasserted.

In Chapter 4, the criteria by which death can be said to have occurred will be discussed. If agreement is to be reached regarding when it is permissible to take organs from the newly dead for transplant into individuals who need them (a topic discussed in Chapter 5), then objective criteria about the essential characteristics of death must be established and agreed upon. The discussion of death leads, in Chapter 6, to a consideration of the circumstances under which it could be permissible to end one's own life, and when it is permissible for others to assist one in ending one's life—the problem of euthanasia. Chapter 7 considers the moral and medical issues involved in developing policies to deal with euthanasia.

When the permissibility of suicide, euthanasia, and genetic screening is considered, questions regarding medical ethics should be raised, because physicians and medical staffs are involved whenever decisions are made to actively or passively end a life, to use the results of genetic screening, or to start or stop medical treatment. It is necessary to examine criteria that can be used to decide when costly medical procedures should be employed, under what circumstances they are justified, and whether the public should be expected to fund all such procedures. These issues are discussed in Chapter 8. Such discussions now form the core of debates regarding how to establish a more equitable health-care delivery system in the United States—a topic that will be considered in Chapters 9–14.

Health-care plans have been introduced in several states (with variable success, as discussed in Chapter 10) to contain the astronomical costs that the states, the federal government, and the U.S. public have had to face. Reforms are necessary if adequate health care is to be provided for all people without forcing the economic system of the country to a virtual collapse. The debate that took place in 1993–1994 produced no resolution in the 103rd Congress and the debate will continue in some form, and in several forums, over the next few years, because the basic problems remain unresolved. That debate is reported, in Chapter 14, as a case history in the formation of public policy. The issues discussed are not peculiar to health care, but involve a series of philosophical and social-policy questions that arise whenever allocations of limited finite resources must be made. These concerns include existing inequalities in the distribution of resources, how available resources should be expended, and how this distribution can be accomplished in the face of legitimate conflicts of interest. It is important to sort out the implications of legitimate conflicts of interest from those due to the sheer greed of the different parties.

The health-care delivery systems that exist in the United States are examined in Chapter 9 and compared to those found in other countries. This examination provides an understanding of the range of possible options available to deal with the existing paradox that medical-care costs in the United States are higher than for any other country, with the almost unique situation of many millions of people in the United States under- and uninsured. A basic moral premise will be framed—that adequate health care is a minimal requirement for a satisfactory life. If health care is a fundamental necessity, then a just society must provide a minimum level of universal health care to its people as soon as possible.

Following considerations of these medical, ethical, and economic issues, two health-care plans, the Oregon Rationing Plan and the Managed Competition plan proposed by President Clinton, will be discussed in Chapter 10. In Chapter 11, problems that pose difficulties for any health-care reform plan will be discussed. The argument will be made, in Chapter 12, that a single-payer system is the most adequate way to attain the morally required goal of universal coverage at the most reasonable cost. In Chapter 13, the moral, medical, and financial issues are evaluated, and the nature of the political infighting that continues is discussed. The debate that took place in the Senate is considered in Chapter 14, followed by a brief analysis of the November 1994 election results. These discussions should help to understand the current state of affairs and to suggest what the next moves should be in the health-care reform effort.

CHAPTER 2

Genetic Screening

First, questions that have strong implications regarding issues in reproduction are examined. These questions involve the health and well-being of existing humans, the health and survival of conceived humans prior to birth, and the nature of the germ cells that could influence future generations. These developments involving human existence have resulted from attempts to understand the structure and functioning of the human genome. It is possible to screen the genetic structure of individual embryos and, in a few instances, to manipulate that structure in the event that potential problems are detected. Advances have occurred, using procedures such as *in vitro* fertilization (IVF), that make it possible to obtain information about the genetic structure of embryos that are being cultured prior to implantation in the uterus. At the present, genetic screening is used primarily to detect genetic defects and to determine the gender of an embryo. Genetic defects can be detected that will have direct expression in the developing organism, and it is possible to identify asymptomatic carriers of defective genes. If such carriers are detected, then a couple could be counseled regarding the risks they run as carriers of genetic diseases that can be transmitted to potential offspring. All of these developments make it possible to influence the structure of the genes of individuals and to influence the genome that will be passed on to succeeding generations. As will be discussed in Chapter 8, a series of fascinating questions regarding medical ethics have resulted.

SCOPE OF THE PROBLEMS

Most procedures used in genetic screening and genetic manipulation are in early stages of development, and too few cases have been reported for there to be any large-scale implementation of many of the procedures. Initial reports of the genetic basis for medical disorders are being reported at an ever-accelerating rate. There are several stages in the development of genetic screening programs. The first one requires an adequate understanding of the locus of the genes that are involved in well-understood single-gene defects transmitted in simple Mendelian ratios (there will be a one-in-three chance that a recessive defect will be expressed). Such single-gene disorders have been identified for more than three thousand human disorders (up to 8% of hospital pediatric admissions), according to Lee (1993) in his book *Gene Future*.

The genetic locus of defects have been identified for Down's syndrome (which occurs in about 1 of every 700 births, depending on the age of the mother); Turner's syndrome (1 in 3,000 female births); cystic fibrosis (1 in 22 white Americans are carriers; the chance that two white parents from the general population are both carriers is 1 in 625; it appears in 1 of 1,800 white births, and in 1 of 17,000 births for African-Americans, with 30,000 cases in the United States); phenylketonuria (PKU; 1 in 16,000 births); Duchenne muscular dystrophy (1 in 3,500 male births); sickle-cell disease (one in 400 African-American births with a carrier frequency of 1 in 8); Huntington's chorea (1 in 2,500 births); hemophilia A (1 in 10,000 male births); hemophilia B (1 in 70,000 births); fragile X syndrome (1 in 1,000 male and 1 in 2,500 female births); Tay-Sachs disease (1 in 3,600 births among Eastern-European Jewish births, with 1 in 30 being carriers—see Handyside, Lesko, Tarin, Winston, & Hughes, 1992; Lee, 1993); and Gaucher's disease (1 in 600 Ashkenazi Jews are carriers and an estimated 20,000 Americans have the disease). There is also evidence implicating a genetic defect linked to a form of Alzheimer's disease (Corder et al., 1993; Schellenberg et al., 1992); and amyotrophic lateral sclerosis. Lee noted that the identity of the defective protein has been identified for over 600 single-gene diseases (including sickle-cell, Tay-Sachs, PKU, and hemophilia), making it likely that methods can be developed to treat them. Although each of these single gene defects are rare, they add up to a significant number of afflicted individuals who could benefit if effective treatments are developed.

In addition to single-gene defects, a number of polygenic disorders have been identified. Among them are hypertension, coronary artery disease, congenital heart disease, cleft palate, cleft lip, and spina bifida. The number of and chromosomal location of the defective genes are not known for these disorders. Davies et al. (1994) searched the human genome for genes that predispose people to Type 1 (insulin-dependent) diabetes mellitus—which is a polygenic trait in mice. At least five genes associated with the disorder were located, making it possible to determine who are at risk. Nearly 300 families were studied in which two children had the disease but neither parent did. Blood samples were taken from all individuals in order to detect if a flaw in the genetic structure was related to the complex patterns of inheritance.

Suzuki and Knudtson (1990), in their book *Genethics*, estimated that about three thousand diseases have been traced to genetic abnormalities, with most of them exceedingly rare, such as albinism, and galastosemia (a metabolic defect that blocks the digestion of certain types of sugar molecules). Abnormalities in chromosome structures or number have been estimated to affect approximately one out of every 1,000 newborn infants. Capecchi (1994) estimated that more than five thousand human disorders have been attributed to genetic defects and argued that work should continue apace to identify the genes and mutations for the disorders, and then create the same mutations in mice through a technique known as *targeted gene replacement*. These mouse models could make it possible to trace the events leading from the malfunctioning of a gene to the manifestation of disease, and hasten the development of effective therapies. Capecchi's methods are being used to study cystic fibrosis, atherosclerosis, and hypertension.

Caplan (1992) estimated that established genetic disorders now account for almost 50% of all childhood deaths in the United States, and that as many as 25% of all hospital admissions for children involve such disorders. He believes that it is the promise of treatment applications that has been critical in securing funding for the Human Genome Project (HGP), rather than an interest in obtaining knowledge for its own sake. The ultimate test of claims in bioethics should not be at level of interests of basic science, but possible pragmatic applications in Caplan's view.

The relationship between genes and disease vectors was discussed to counter objections expressed by several critics of genetic research (e.g., Annas & Elias, 1992b; Hubbard, 1995). They argue that the incidence of gene defects is not high enough to warrant spending large sums of money for genetic research and the development of treatments, and that the funds would be better spent for social interventions to improve the quality of health. The evidence suggests that there are many reasons to believe that a large number of general cellular malfunctions involve defective genes, and that effective treatment approaches are being developed rapidly. If so, then it can be asked why there is such strong opposition to gene therapies specifically, rather than a general opposition to any advances in medical technology.

Research Strategies

Suzuki and Knudtson (1990) discussed three strategies, each one more difficult to accomplish, that can be used for human gene therapy:

1. Gene insertion, which involves the insertion of one or more copies of the normal version of a gene into the chromosomes of a diseased cell.
2. Gene modification, which entails the chemical modification of the defective DNA sequence in the living cell to recode its genetic message to match that of the normal allele.
3. Gene surgery to remove a faulty gene from a chromosome, followed by its replacement with a cloned substitute.

Most of the hereditary illnesses that are likely to benefit from gene therapy are blood and immune disorders that affect bone-marrow tissues.

A research team in France (Le Gal La Salle et al., 1993) inserted new genetic material into the brain of a rat. The procedure involved replacing a harmful gene in a virus with other selected genes and inserting the modified gene into the nerve cells of the rat. Although such studies have not been attempted with humans, it appears that these genetic-engineering techniques can be potentially useful in combating disease entities.

An article in *The Boston Globe* (April 2, 1995) contained a report that researchers at Johns Hopkins University had decoded the gene involved in a common form of kidney disease that causes cysts to form in the kidney, liver, pancreas, and spleen, eventually leading to kidney failure. It was estimated that about 500 thousand people in the United States are afflicted by this polycystic kidney disease. A disturbing aspect of the announcement is that the University has joined with a biotechnology company and filed for a patent to commercialize the information. This raises an important concern that has alarmed many observers (e.g., Keller, 1992; Lewontin, 1991) of these research programs. The issues involved in patenting scientific knowledge for commercial purposes and the impact of patents on the progress and objectivity of science and of health care have disturbing implications that will be discussed in Chapter 3.

Fetal tissue could be a valuable source of healthy genetic material, because a host organism's immune system readily accepts the tissue. Genetic material for implantation can be obtained from other species of animals and even from plants and bacteria, but the use of genetic material from alien sources has led to considerable debate regarding the ethical and moral appropriateness of introducing nonhuman genes into the human germ line, a point that will be discussed in Chapter 3.

Cancer. The potential value of genetic engineering has been suggested by the possibility that genes in tumor cells can be altered so that "killer immune cells" will attack brain cancers (Trojan et al., 1993). Genetically manipulated tumor cells appear to act as cancer vaccines when they are reinjected into rats with brain cancer. The extension of gene therapy to cancer is not trivial, because one in six (one-half million people) in the United States and Europe will die of some form of cancer each year, according to Lee (1993).

Lee (1993) documented the rapid progress being made in manipulating genes, noting that by 1993, 40 clinical experiments involving gene insertion into humans have been approved, with 18 of them for the purpose of gene therapy. A major breakthrough was announced by Nabel and his colleagues (1993), who reported the results of clinical trials with five patients who had fatal skin cancer and had DNA injected directly into their cancer cells. The injection galvanized the immune system to seek out and destroy the tumor cells. The treatment was effective for all patients, causing white blood cells to aggregate around the tumor tissue and the tumors to shrink. They hope to develop more effective techniques to inject the DNA to cause the tumors to disappear altogether. This research group was reported to have developed a procedure to more effectively introduce tumor-inhibiting genes into cells through the use of a "gene gun" that shoots pure DNA directly into the cells of the tumor (Saltus, 1995a). In this way, the genes are inserted in a matter of seconds, and the procedure seems not to damage the cells or their structures.

A major advance was reported by Kamb et al. (1994), who found a gene, MTS1, that encodes a known inhibitor of cancerous growths. The gene may be involved in the basic cycle occurring in routine cell division. A mutation of this gene has been related to a variety of major cancers, including brain, bladder, breast, blood, lung, skin, bone, ovary, and kidney. The locus of the gene is known, and its function is to stop cell division. When the gene is ineffective, cells continue dividing and begin to proliferate wildly in the typical manner of cancerous cells. The mutation was found in more than half of the tumors they examined, and they are developing a screening test, at least for melanoma skin cancer, which family studies have indicated is inherited. Those who have inherited the faulty suppressor gene and are therefore susceptible could be screened frequently to find the cancer at an early stage, when it can be most easily treated.

A team of 45 scientists reported they had identified a gene, BRCA1, that could account for about one-half of the 10% of cases of breast cancer that are thought to be familial, which could mean that as many as 600 thousand women carry a defect in this gene (Miki et al., 1994). Those who carry a mutant version of the gene have an 85% lifetime risk of breast cancer; many develop breast cancer before the age of 50, and are at an elevated risk of developing ovarian cancer. The gene is an extremely complicated one, being as much as 10 times larger than the average gene, so that it will be difficult to develop a diagnostic test to screen women who come from high-risk families. Several companies are planning human trials for compounds that have promise to block the cancer-producing gene—called an *oncogene* (Saltus, 1995b). Drugs have been used successfully to treat cancer in rodents, and human clinical trials are expected to begin within a year or two.

A cautionary statement concerning the generalization of these preliminary results was made by Cairns et al. (1994), who were not able to replicate them in regard to the rate of mutations in the various cancers. They used tissue obtained from cancerous tumors rather than using cancer cells grown in the laboratory—as did Kamb and his research group. When cells from cancerous tumors were examined, the gene mutation occurred much less often than originally reported. These differences were attributed to the fact that cells grown in the laboratory tend to change substantially over time. This failure to support the conclusions of the original investigation serves as a reminder that it is always dangerous to generalize from cell lines maintained in the laboratory to cells that have been obtained directly from human tumors. It is essential to examine tissue obtained from cancerous tumors to support generalizations regarding the actions of cells in the body. It is always dangerous to generalize widely from experimental results obtained in situations where important variables have been excluded or controlled experimentally (see Petrinovich, 1989). There is still optimism, however, that there may be a suppressor gene located on the chromosome, because deletions have often been detected in that area.

Cystic Fibrosis and Alzheimer's Disease. Two research teams won the approval of the Recombinant DNA Advisory Committee that oversees gene therapy research to use gene therapy to treat cystic fibrosis in humans (Angier, 1992). The techniques have been effective in test-tube and animal experiments to correct a defective chloride flow that produces an imbalance of salt, leading to an excessive mucous buildup in the lungs, and resulting in death. The treatment involved the insertion of enough copies of the genes the patients lack to produce a beneficial effect through actions on the pulmonary tissue. Although such studies are preliminary, they suggest that this genetic engineering procedure has promise.

The Cystic Fibrosis Genotype–Phenotype Consortium (Hamosh & Carey, 1993) studied 798 patients who were either homozygous for the primary mutation involved in cystic fibrosis or who showed one of the seven most common heterozygotic mutations (different genetic alleles at the homologous loci of a diploid chromosome set). All but one of the heterozygote combinations suffered from the same pancreatic insufficiency as the homozygotes (identical genetic alleles at the homologous loci of a chromosome). This one mutation overrode the effects of any other allele that caused pancreatic insufficiency. The association between the cystic fibrosis genotype and the pancreatic phenotype was strong, but it was not possible to use information regarding the genotype to predict the severity and course of the pulmonary disease in this group. The factors producing the most severe debilitation in cystic fibrosis are by no means simple and straightforward. It was concluded that patients carrying an allele associated with pancreatic insufficiency should be followed closely for development of the pulmonary condition, whereas those who carry an allele associated with pancreatic sufficiency would not need further evaluation.

Another development suggested a possible genetic basis of Alzheimer's disease (Travis, 1993). The genes determining apolipoprotein-E (ApoE), a protein that carries cholesterol through the bloodstream, is located in the same place on Chromosome 19 as the region suspected to be involved in Alzheimer's. This finding is important, because the form of Alzheimer's is of late onset, beginning after age 65, and including more than three-fourths of all cases.

Corder et al. (1993) examined 234 people from 42 families afflicted with late-onset Alzheimer's. It was found that by age 80, almost all individuals who had two copies of the gene for ApoE4 (one of the three major versions of the protein found in humans) had developed the disease, and their overall risk factor was more than eight times greater than for those with no copies of the gene. For families in which late Alzheimer's was present by age 75, only 20% with no copies of E4 had the disease, 45% of those with one copy were affected, and 90% of those with two copies were affected. The age of onset was related to the number of copies: no copy, 84 years; one copy, 75 years; two copies, 68 years. These epidemiological findings make it possible to investigate the specific mechanisms to identify the critical defect in the actions of the cholesterol-carrying ApoE protein.

Laboratories all over the world have been reported to be exploring ApoE4's possible role in the disease, and thousands of individuals in different populations will be tracked in the years to come to determine how E4 translates into a risk for Alzheimer's. Research is also being done to determine if the effects are due to the deleterious action of E4 or to the lack of positive actions of the other two forms. As with a great deal of science, much of the battle has been won when the investigators are able to identify a set of reasonable alternative hypotheses to be evaluated.

A replication of the findings of Corder and associates identified a sex difference in the risk factor associated with E4 (Payami et al., 1994). If there is one E4 allele, the age of onset was shifted to a younger age in women, but not in men. Given these possible sex differences, they suggested that genetic testing for Alzheimer's may be premature until the possibility of sex differences is resolved through population-based incidence studies. Rapid progress has occurred toward understanding the basic genetic mechanisms involved in disease processes, as well as the epidemiological patterns.

Some Ethical Concerns

Questions have been raised regarding the appropriateness of genetic screening when no effective treatment is available, should a defect be discovered. Some worry that providing information in such cases would raise a patient's anxiety unnecessarily, and should not be done if there is no certainty the genetic defect will be expressed, or no way to gauge the level of severity if the disease is expressed. To avoid an undesirable paternalism, the patient should be free to choose to know of potential defects, providing all of the available information and facts are made accessible that would permit an informed decision.

Necessity for Counseling. If people do not want to know whether they are at risk, they should be free to make that decision. If they want full information, they should be given full knowledge of state-of-the-art estimates of the likelihood, years until onset, and possible symptoms, so they can prepare rationally and emotionally for those eventualities. The reasonableness of providing information to those who wish to know it is supported by research done by Wiggins et al. (1992), who studied the psychological consequences of predictive testing for Huntington's disease using a sample of 135 relatively well-educated, middle-aged Canadians. Testing followed by careful counseling improved the psychological well-being of those at risk because, by reducing uncertainty, it provided them with an opportunity to plan for the future. There was no evidence of catastrophic reactions, such as suicide attempts or psychiatric hospitalization, although some people did require additional counseling. Sorenson (1992) examined the evidence concerning the psychological responses of people to genetic counseling and he, too, found that people identified as carriers who received counseling did not suffer either short- or long-term psychological harm.

A major mitigating factor is that some people might not be able to understand and evaluate the information provided, or might be so unstable emotionally that medical and psychiatric professionals decide the stress would likely overwhelm the individual. In cases in which the individual is likely to be unable to make an informed decision, a degree of protective paternalism would be justified.

Even though such research results indicate predictive testing produces no adverse psychological consequences, ethicists still express alarm. Kolata (1993) discussed an interesting case history of what she posed as either a nightmare or a dream resulting from the new genetic era. The case involved communities of Ashkenazi Jews in New York and Israel. Large families with as many as 12 children are greatly desired in these communities. Mating is done mainly within the community, and there has been an effort to discourage dating and marriage between people who are at risk of having a child with the Tay-Sachs genetic disease. Because there is a high level of inbreeding within the community, the likelihood of the expression of this double recessive trait is high.

A decade ago, the community began a genetic testing program with 45 people; the next year 250 were tested, and in 1983, 8 thousand were tested. The testing is done at five centers in the United States and one in Israel, and costs only $25 per test. After being tested, each individual is given an identification number, and when a man and woman begin dating, they are encouraged to call a hotline with their identification numbers. They are told at that time whether the match is compatible (they are not at risk of having children with a genetic defect), or they are invited to come in for genetic counseling (they each carry a recessive gene that could result in a child with a genetic disease). Of the 8 thousand people tested last year, 67 couples who were considering marriage decided against it after being advised of their risk.

One of the concerns expressed by critics is that a decade ago the test was for just the Tay-Sachs disease. Now they test also for cystic fibrosis, Gaucher's disease, soon will include testing for Canavan disease, and some want to add tests for "anything available." The concern is that some of the conditions for which people are tested are untreatable, and the fear is that people will have to live the nightmare of knowing they have a distinct probability of being afflicted with a severe disease. Another concern is that of the "slippery slope": Voluntary testing will lead to strong community pressure for people to be tested, and eventually they might even be forced to be tested, whether they want to or not.

Francis Collins, Director of the Center for Human Genome Research at NIH, said that such testing takes away the sacred principle of autonomy, and some people will have the risk of being genetic wallflowers, rejected by every suitor because of their recessive genes.

Supporters of testing agree that any risk of laboratory or human error in testing should be minimized, and the likelihood of the phenotypic expression of the genetic defect should be explained carefully. When a person from the Ashkenazi community wants to marry, there has been no question that parents should assure that the son or daughter will be marrying into a religious family with appropriate status in the community, and that the proposed mate has good economic potential. One rabbi was quoted: "That you don't leave to God, so why leave this to God? God has enough to do." Supporters agree that counseling should be just that and not involve any coercion, but that the marriage should be encouraged if the testing is accurate and the fully informed couple decide their love is strong enough to take whatever gamble they wish to take regarding their offspring.

The extreme negative position is that if we wait long enough, the ultimate bottom of the slippery slope will be reached (dooming the entire human species), and there will be so many recessive disease-causing genes identified that every single marriage will be prevented. No one, however, is talking of preventing marriage, only of increasing the likelihood that the number of marriages people enter into would be done with sufficient information to determine whether they want to marry and have offspring, knowing the potential genetic defects.

The extreme positive view was that if a test exists, use it, and permit the couple to decide before they marry whether they want to risk having a baby with a given disease. Because the available evidence indicates there are few serious psychological consequences, and that the decision to marry or not is a voluntary one made by the couple, there seems to be little in the way of a nightmare vision to fear. It is likely that knowing there is no genetic risk could enhance the dreams of the couple regarding their probable reproductive bliss, or they could proceed (or not proceed) with the knowledge that there is a finite risk, and that they are willing (or not willing) to jointly undertake it.

One problem is that some people do not comprehend what probabilities mean. It has been reported that some individuals construed a 1% likelihood to mean one in a thousand, and when one presents correlation values to laypersons (as well as to scientific experts), the meaning of different magnitudes of correlation are often grossly misinterpreted. I am reminded of an instance involving my mother. She had some diagnostic testing for cancer, and the physician reassured her that everything was fine, and pointed out that, given the test results, there was less than one chance in a million she would develop cancer (which she did not). I received a distraught call from her to the effect that she was going to get cancer. After I calmed her down (aware of her pessimistic tendencies), I asked her to tell me precisely what the physician said. When she told me of the less than one-in-a-million prediction, I asked her how she could interpret that statement to mean that she will get cancer. Her answer was, "I know I will get cancer because, given my luck, I will be the one in a million that gets it."

Jones (1993), in his book *The Language of Genes*, noted a problem that affects optimists when they are informed concerning probabilities. In his view, propaganda about smoking and lung cancer is not particularly effective because those optimists exposed to the statement that only one smoker in 10 will be expected to get cancer assume that if it is only one in 10, then that smoker will be someone else.

Some of the difficulties people have regarding the meaning of probability statements were revealed in a careful study by Murphy et al. (1994), who investigated the decisions expressed by elderly patients to have cardiopulmonary resuscitation (CPR) in the event of cardiac arrest. The investigators evaluated each patient's knowledge regarding CPR procedures, determined the patient's estimates of the probability of survival, informed each about the procedures and the real probability of survival (10–17%), and then had them decide if they would opt for CPR. The discussions were quite brief, taking only 10 or 15 minutes at the end of a routine office visit. Before they were informed, 41% opted for CPR, but after learning the probability of survival, only 22% chose it. There must be careful and thoughtful counseling and information provided to patients if they are to be able to make voluntary and informed decisions about their future.

There is a risk that people will not be able to understand or will misconstrue the data, and any feedback should be given with those possibilities in mind. However, I see no compelling argument that such information should not be made available to, and interpreted for, those who want it, whether or not ameliorative treatments have been developed.

The ethicists George Annas and Sherman Elias (1992a) edited a book, *Gene Mapping*, in which they explored the legal and ethical dimensions of the new advances in genetics. In their contribution to that book (Annas & Elias, 1992b), they identified three different conceptual levels at which the ethical issues regarding genetic information should be considered. The first involved information that could be of direct use to individuals and families: what information should be collected, to whom can it be disclosed, and the possible consequences for the individual and family. The concern at this level is to protect the autonomy of individuals and to guarantee confidentiality, especially in terms of employment and insurance. To protect these interests, individuals and families must be counseled adequately to guarantee informed consent. Most experts agree that, other things being equal, genetic screening should be available to those at risk.

The bioethicist Macklin (1992) remarked that the results of such screening should be provided, because people cannot make informed choices regarding their health and well-being without adequate information. She suggested there should be an effort to convince a patient who learns about genetic information that could effect the interests of other members of the family to disclose the information to those relatives (or let the physician do so). If a physician discovers that a patient has a genetic disorder that could increase the chances of sickness or death for relatives, then the physician should disclose the information to relatives in order that they might take steps to prevent possible harmful outcomes for themselves if there are ameliorative steps available. If there are no cures or treatments for the genetic disease, Macklin suggested that the physician might decide to honor the patient's wish to preserve confidentiality. Macklin considers it permissible for a physician to breach confidentiality if the patient was informed prior to testing that disclosure would be made, providing the disclosure is made only to those who might be directly affected.

Genetic Screening and Abortion. Robertson (1992) reminds us that legal concepts of procreative freedom include the right of persons to know whether they and their mates are at risk for genetic disease, in order to decide whether to reproduce or to remain childless. If they decide to reproduce, they can choose to use donor gametes or conceive and then screen the fetus for an at-risk condition. As long as the premise of *Roe v. Wade* remains intact, a woman can obtain the results of genetic testing of the fetus and then have a legal abortion during the first two trimesters if she so desires. It is legal to abort in order to avoid having an offspring with genes that predispose it to disease, and physicians should have to honor that legality. Robertson remarked that physicians should remain free to refuse to perform abortion because they consider abortion unacceptable and "with appropriate notice" may even withhold the results of prenatal diagnostic tests to prevent abortion from occurring. I challenged such arguments in *Human Evolution, Reproduction, and Morality*; the physician has an obligation to treat a female patient within the bounds of legality, and there is no compelling reason why the physician should not respect the informed and legal choices of a patient. Physicians should not be allowed to impose their peculiar philosophical or theological beliefs on others who desire to be treated with the respect due them under the law. When a patient makes a reasoned decision to have a legal treatment, the physician should have an obligation to respect the patient's informed choice in order to maintain the quality of the physician–patient relationship that has been championed by so many physicians.

Annas and Elias (1992b) discussed the use of prenatal genetic testing to decide whether a pregnancy should be terminated. They took the paternalistic position that, as the possibility of finding a treatment for such diseases as cystic fibrosis increases, it is less and less justifiable to abort a fetus with that condition. Such a decision should be the woman's to make and not that of the physician or the ethicist—the courts have established the rule of law and all involved should have to abide by those laws, even though their personal beliefs differ. It is illegal to kill an abortionist, even though one disagrees about the morality of abortion on theological grounds, or some idiosyncratic beliefs regarding justifiable homicide, and it should be just as illegal for a licensed professional to thwart a law-abiding patient's legal interests.

Generic Consent for Genetic Screening. The type of consent that should be required to permit genetic screening was discussed by Elias and Annas (1994). Their requirements were for pretest counseling, emphasizing the right of the patient to refuse testing if the potential harm (in terms of stigma or unacceptable choices) outweighs the potential benefits. One danger is that the information overload produced by a great deal of technological information could amount to little more than misinformation, making the entire counseling process either misleading or meaningless. They suggested that the proper approach should be the same as that used when consent is obtained to perform a physical examination: A patient is told that the purpose is to locate potential problems, but not generally about all the possible abnormalities that could be detected through routine physical exams or blood testing. The questions involved in screening should be addressed directly and publicly; if the medical profession fails to take a leadership role, then the courts will set the standard, with little likelihood that the result will be better from the perspective of either the medical profession or the public.

A second ethical level identified by Annas and Elias (1992b) involved societal issues regarding population-based screening and the specter of eugenics. These concerns caused them to worry about "genomania," "genetic fixes," and biological determinism. They discussed two concerns: what percentage of the nation's research budget should be devoted to the HGP, and who will reap the benefits (assuming there are commercial involvements and considerations).

Cost Factors. The issue of the relative costs of the HGP and whether it enhances or impedes the research effectiveness of the biomedical research community is discussed in the next chapter, where it is concluded that the cost is relatively small, does not penalize small laboratories, and the project already is paying high scientific and medical dividends. The questions regarding commercial and economic factors are discussed later under the heading "Genes for Sale," in which an argument is made for public multidisciplinary and multinational cooperation, and against patenting of gene sequences. Secrecy is the enemy of legitimate science and the public communication of findings must be expedited over the commercial interests of scientists, industries, and governments.

Their third ethical level was posed in terms of whether resources might not be better used to treat other causes of disease. One concern was that if the focus is to identify and treat genetic diseases, there will be a disregard of other conditions that cause disease, such as poverty, drug and alcohol addiction, lack of housing, poor education, and lack of access to decent medical care. They want to be sure there would be an equitable distribution of the products of HGP: that the benefits should not be available only to those who can pay for them and be denied to the under- and uninsured. I question the latter position. Any advance in expensive medical technology should be available for purchase by the privileged few if it has not been produced at public expense, just as it is possible to purchase an expensive home in a safe and healthy environment or an expensive foreign car, providing the individual has the wherewithal. The concern should be that *essential* services are equally available to all members of society who can benefit, and that concern is not peculiar to genetic services and products, or to access to medical technologies. The problem of equitable distribution of health care is discussed in several later chapters.

Questions regarding the relative merits of studying the genetic aspects of disease (or of general behavioral propensities) often are a reflection of an underlying biophobia. The historian Proctor (1992) presented a list of eugenic evils that have occurred throughout history, including the forced sterilization of individuals who were considered "criminally ill," "morally dissolute," or "subnormal" in the United States, as well as the "racial hygiene" practiced in Nazi Germany. On the one hand, Proctor quotes critics who suggest that sequencing the genome is useless because there is too much "junk" DNA that will take too long to sequence, making the entire enterprise wasteful and doomed to failure. On the other hand, he suggests that the information might be so useful that it will allow people to "play God" and seek to produce "perfect babies," which some believe transcends the fundamental limits of common decency. Proctor reminds us that physicians "play God" every time they treat an illness, and the quest for perfection is the goal not only of genetics but also of all kinds of environmental or nutritional therapies used by parents to "perfect" their babies. The constructive political task should be to establish safeguards to ensure that genetic manipulation promotes, rather than limits, human liberties. I agree that we should steer a course between the Scylla of alarmed criticism and the Charybdis of technological exuberance.

Proctor concluded that the results of genetic screening by people at risk for genetic lesions should be available to those individuals who want to have this information about themselves to make a reasoned choice regarding whether and how to have children. Individuals should be free to obtain information regarding any genetic predisposition they have to develop an illness that is incurable should they choose to have that knowledge to plan their lives. Proctor's bottom line was as follows:

> From the point of view of lives saved per dollar, monies would probably be better spent preventing exposures to mutagens, rather than producing ever more precise analyses of their origins and effects. Sequencing the human genome may be a technological marvel, but it will not give us the key to life. (p. 83)

He ends by saying,

> The danger is that in a society where power is still unequally distributed ... the application of the new genetic technologies—as of any other—is as likely to reinforce as to ameliorate patterns of indignity and injustice. (p. 84)

R. C. Lewontin (1991) devoted much of his book *Biology as Ideology* to an expression of concerns similar to those raised by Proctor and defended a similar set of conclusions. The conclusions are presented with little concrete justification, and they reflect on the realities of societal inequities that are not specific to the HGP, but involved the distribution of any resources necessary to support an adequate life.

The Specter of Genetic Determinism

King (1992) worried that there is a danger that greater attention will be paid to genetic explanations than to more complex explanations for differences "to the detriment of vulnerable and disadvantaged groups" (p. 102). She, too, concluded that limited resources should be allocated to ameliorate economic, social, and environmental conditions that influence health status, because these allocations would yield "more immediate and enduring" health benefits than focusing on genetic contributions to disease prevention. This conclusion was drawn, although no justification for the judgment regarding the relative value of the different approaches was presented.

Hubbard (1995) adopted a biological and ethical position that could lead to rather peculiar recommendations if pursued to its logical conclusion. She argued against the current emphasis on genetic components in behavior, because development is "dialectical and not linear," meaning that environmental factors influence the expression of the genome throughout development. Because the environment is of unquestioned importance, and all health conditions are affected by "virtually everything that happens in our lives," she noted that it is not possible to predict with certainty in any individual case.

Because of such uncertainties, she suggested it is not sensible to have regular health "checkups"; it would be just as beneficial to have easy access to medical treatment when symptoms first appear. She conceded that it might be reasonable to have checkups for "supposedly at-risk" populations—such as Pap smears for sexually active women, regular mammograms for women over 50, and checkups for prostrate cancer for older men. She did not approve of genetic tests as predictors of potential ill health at some unpredictable time in the future because of discriminatory consequences of "geneticization" in regard to insurance and employment. She expressed the concern that people will overemphasize the importance of genetic influences on individuals and deemphasize the need for adequate social and public health policies, because it is these latter factors that contribute most to serious illness and disease the world over. It is true that there is a desperate need for enlightened social and health policies, given the millions of people in the world who die due to preventable and treatable diseases and as a result of malnutrition. But the assumption that we have a zero-sum game, in the sense that if funds used to pursue the genetic projects were eliminated or reduced, they would be directed to these other approaches, is questionable. It is valuable to do both things: to create a just world by directing aid to alleviate suffering, and at the same time, to pursue basic research on the genetic contributions to disease.

Hubbard opposed genetic screening, because it equates prediction with prevention and could lure people into believing this foreknowledge is of use, pointing out that the extent of any disability that might occur cannot be predicted for specific individuals. This same argument could be used to argue against doing those tests she suggested, such as Pap smears, as well as HIV testing, and could be used to argue against seat-belt laws, requiring cyclists to wear helmets, regulating the speed at which people can drive, or discouraging smoking. There is no certainty that any of these things will affect any particular individual, and if they do, we have no idea how serious the effects would be or when they would occur. The reason to do such tests, pass such laws, give such advice is that there is a distinct statistical probability that "bad things" can occur, and the likelihood of these bad things can be estimated. The individual should be able to evaluate the risks of genetic screening and decide whether the information is worth knowing, and if so, whether to take any action.

In *Human Evolution, Reproduction, and Morality*, the question of genetic/biological determinism was discussed at considerable length, and some of those arguments were reviewed in Chapter 1 of this book. Because biological and genetic factors are important determiners of complex organs, behaviors, and traits does not imply that these aspects of humanity are predetermined and cannot be modified through external influences. Yet, the specter of a dangerous and undesirable biological determinism is evoked over and again in terms that would anchor human fate in biology, implying that there is a gene for this and a gene for that (see Lewontin, 1991). This specter is raised despite the articulate and continued remonstrations made by evolutionary biologists and psychologists. Even a cursory examination of the books by those evolutionary psychologists who argue for the importance of evolved processes in human affairs is sufficient to give the lie to such a charge. The interested reader is referred to books by Daly and Wilson (1988); Barkow, Cosmides, and Tooby (1992); Locke (1993); Buss (1994); Pinker (1994); and Petrinovich (1995). The arguments presented in these books do not even insinuate a genetic determinism, or any tendency to revert to two-valued heredity versus environment arguments. They are systematic attempts to understand the role of evolutionary adaptations in the shaping of complex human behavioral functions.

CHAPTER 3

The Human Genome Project

GOALS AND PROGRESS

The goal of the HGP is to locate all genes of the human genome and establish the base sequences of all its DNA. Vicedo (1992) used the metaphor of an analysis of literary text to illuminate the problems in the HGP. The mapping could be considered to be the syntactic analysis of that text, aiming to identify the words used and the systematic rules and constructions that exist. A semantic analysis also is required involving an integration of biochemical data with embryology and developmental biology to interpret the genes' role in the formation of an organism. A pragmatic analysis involves analysis of particular genomes in relation to the specific environments in which they are to be expressed. Wills (1991) used a musical metaphor that nicely expresses the limitations of simply knowing the sequence of DNA and assuming that we have learned what we need to know about human beings: Even though we have looked up the sequence of notes in a Beethoven sonata, we will not have gained the capacity to play it.

The procedural aspects involved in mapping and sequencing the human genome were outlined by Caplan (1992):

1. The creation of a high-resolution genetic linkage map established by studying families to estimate the frequency with which two different traits are inherited together over generations.
2. The creation of ordered DNA clones that are genetically engineered replicas of known DNA sequences.
3. The creation of a high-resolution physical map to identify the sequence of nucleotides (the smallest unit of genetic information in a segment of DNA) from a chromosome.

Family studies have been successful in locating the gene causing Huntington's disease (Wexler, 1992). The strategy involves looking for large families, some of whose members have the disease and others who do not. If those who have the disease have one form of a genetic marker, while the unaffected relatives have another form of that marker, investigators can close in on the gene until the defect can be identified. Wexler found such an extended family in Venezuela and was able to trace Huntington's disease in that area as far back as the early 1800s to one woman who was the founder of a kindred line, now numbering 11 thousand, with 9 thousand still living, and most still under the age of 40 (see Wexler, 1992). In the pedigree, there are 371 persons with Huntington's disease and an estimated 660 asymptotic gene carriers who are too young to show symptoms, but who will be expected to die from it as the years pass. Incidentally, Dr. Wexler serves as the chair of the ethics working group for the commission on the human genome at NIH, and is at risk of having Huntington's herself.

Because of the complexity of the HGP undertaking, some have urged scientists to proceed slowly. Suzuki and Knudtson (1990) suggested that society might be better served by a slower paced, multidisciplinary approach to decipher the human genome—one that integrates DNA sequence data with studies of human family pedigrees and cell biochemistry, much as Caplan outlined and as Wexler has done.

It is estimated that there are 3–3.5 billion base pairs in the human genome (Harris, 1992) and that mapping, assuming the cost of $1 per base that has been attained in the past, would cost over $3 billion (Kevles, 1992), and could be completed within 3 to 30 years, depending on the financial support made available, and the rate of processing each base. Gilbert (1992), Nobel Prize winner in chemistry, presented figures suggesting that the cost in 10 years should drop to about 10 cents per base, reducing the cost to about $300 million, with the project taking 10 years, assuming that the best sequencing techniques available today are used. The rate of DNA sequencing has increased about 60% per year, and further increases should be achieved, reaching an ultimate level of 1 cent per base. Assuming technological advances, Gilbert argued the optimistic position that the human sequencing will be completed in the 1990s, and the genes causing heart disease, susceptibility to cancer, and high blood pressure will be found in the following decade.

Lewontin (1991, pp. 48–49) made a staggering estimate that the genome sequencing "... might take 30 years and occupy tens or even hundreds of billions of dollars." He did not indicate how he arrived at these high estimates. Elsewhere (p. 173), he cited the more usual cost of about $300 million. It is widely agreed by leading geneticists that current DNA sequencing procedures should make it possible to establish a map for between $10 million and $75 million (Dickson, 1994).

Cantor (1992), Professor of Biochemistry at the University of California, Berkeley and Chief Scientist of the Department of Energy (DOE) genome project, argued that the use of animal experiments to develop models and provide genetic material will make it possible to more effectively develop therapeutic methods. Cantor suggested that the genome project should emphasize the study of those base pairs of DNA that are biologically or medically rewarding and, when the cost per base drops significantly, the sequencing of the presumptively barren regions of the genome could be started.

There has been concern by some involved in basic health research that the costs of the HGP are too great, and that in a time of limited resources, the project should be delayed, or perhaps not continued at all, especially if there is a shortage of funds for basic medical- and health-care-related research. Some facts suggest that these concerns are not well founded. Wills (1991), Professor of Biology and a member of the Center for Molecular Genetics at the University of California, San Diego, reported that the HGP received $41 million from DOE and $60 million from NIH for 1991. These amounts represent a relatively small expenditure when compared to the $800 million granted for AIDS research in 1991—eight times the amount for the entire HGP. It was estimated that the HGP currently receives $165 million from the federal government (Fisher, 1994). NIH expenditures for the HGP accounted for only 1% of the total budget of NIH in 1991, and if the project should be funded at the level of $200 million a year that was recommended by the National Academy of Sciences, the NIH share still would be only 1.5% of the agency's budget—roughly 3% of the resources available to it for external grants (Kevles & Hood, 1992).

Lewontin (1991, p. 51) suggested a malevolent motivation when he asked the rhetorical question of why so many "powerful, famous, successful, and intelligent scientists want to sequence the human genome?" He answered that, in part, they are devoted to the ideology of single, unitary causes and do not ask themselves more complicated questions. But, he also stated that part of the answer is "a rather crass one."

> The participation in and control of a multibillion-dollar, 30-or-50 year research project that will involve the everyday work of thousands of technicians and lower-level scientists is an extraordinarily appealing project for an ambitious biologist. Great careers will be made. Nobel Prizes will be given. (p. 51)

He discussed the serious problems of scientists' commercial involvement when they become principal scientists or principal stockholders in biotechnology companies.

Yet another concern is that the HGP is the kind of "big science" project that proceeds at the expense of the many small groups of scientists scattered across the country. Kevles and Hood's analysis indicated that NIH funded 175 different genome projects at an average amount of $312,000 a year in 1991. This figure is about 1.5 times the average NIH grant for basic research and about equal to the average AIDS research grant. It appears that the HGP is not leading to a preponderance of large center grants at the expense of small laboratories. Victor McKusick (1992) characterized the HGP to be not so much big science as it is coordinated, interdisciplinary science.

Agreements have been reached to use the method of sequence-tagged sites to identify and locate genome clones, and Kevles and Hood noted that these agreements have eliminated costs of $60 million that would have been incurred over the 15-year life of the genome project. This large-scale economy, the fact that the techniques are (according to Wills, 1991), simple, cheap, and virtually foolproof, plus the fact that the sequencing rate is becoming faster and more economical each year, suggest that cost factors do not pose an insurmountable problem to pursuing the project.

Wills (1991) stated that a laboratory could be set up to sequence DNA with an investment of a few thousand dollars. Kevles and Hood (1992) contrasted this with the other technologically oriented big science projects. Giant particle accelerators and space stations cost many millions of dollars, and if they do not work, or are abandoned before completion, it is unlikely that much of scientific value will be produced. In contrast, understanding a fraction of the human genome sequence, especially if there is an initial concentration on those scientifically and medically interesting portions of the genome, could pay high scientific and medical dividends through the course of the project, as is now being realized.

As the first stage of genome mapping progresses, the second stage, a concurrent diagnosis of the abnormalities involved in genetic diseases, can begin. This diagnosis can be combined with safe, economically feasible, and medically reliable prenatal diagnoses of genetic defects.

To illustrate that this "Brave New World" scenario is not too far-fetched, consider a recent case study by Handyside et al. (1992). They used IVF to gain access to the embryos of three couples known to be carriers of the recessive genetic defect producing cystic fibrosis, and who had had at least one child with cystic fibrosis. A large number of eggs of consistent quality and a number of embryos were obtained and inseminated with the husband's sperm; the embryos were subjected to a biopsy on the third day after fertilization, and selected ones were implanted within 8 hours of the biopsy. One woman did not become pregnant. For a second woman, the only viable embryo was defective, and no implantation was attempted. For the third woman, one embryo that did not carry the recessive gene, and one that carried a single recessive, were transferred because they were the two best embryos from a morphological standpoint. This third woman became pregnant and delivered a healthy girl who, at 4 weeks of age was examined and found to have the normal genes from the homozygous normal embryo.

Although more clinical trials are necessary, these results indicate that the procedures can be beneficial therapeutically. The amniocentesis diagnosis can be done after the 13th week of pregnancy. If IVF is used without genetic screening then couples at risk to have abnormal offspring face the possibility of repeated diagnoses. If there is an abnormality, then they have to consider whether to terminate the pregnancy if they do not want a defective child. Even though some might not want to raise a defective child, they may not be willing to consider abortion after 13 weeks of gestation because of personal convictions. The use of preimplantation genetic diagnosis would preclude termination of an ongoing pregnancy.

The first report of therapeutic benefits of human gene therapy to treat a disease was reported by Angier (1994b). Scientists from the University of Pennsylvania Medical Center introduced an essential gene that was not present into the liver of a 30-year-old woman who had a potentially fatal cholesterol disorder. They removed about 15% of her liver, grew cells *in vitro* while supplying them with copies of the lacking gene. The researchers estimated that about 3–5% of her liver cells were behaving in a manner sufficient to remove cholesterol from the bloodstream—enough to lower her cholesterol levels by almost 20%—although the levels still remained high. The research community greeted these results with cautious optimism because a similar gene-therapy protocol might be useful to treat a range of other disorders, such as PKU, cystic fibrosis, immune deficiency disorder, and a number of types of cancer.

The downside is that IVF is still expensive. Simpson and Carson (1992) estimated that, in facilities such as those used by Handyside et al., the cost of each IVF procedure is approximately $5,000, with the biopsy procedures costing an additional $2,000. All procedures being developed at present must be perfected in animals and then tested using single nonembryonic human cells. These requirements impose lengthy and costly delays in understanding the processes involved and make it difficult to develop economically feasible methods.

Research using human embryos would reduce the need for research with animals to achieve desired medical advances. It was reported in the September 29, 1994 issue of *Nature* that an *ad hoc* panel of the NIH agreed that federal funds could be spent on embryo research until 14 days after fertilization—the policy in Britain and Canada. They recommended that there should be no purchase of gametes or embryos, and that aborted fetuses should not be used. However, on December 2, President Clinton ruled out the use of federal refunds to create human embryos for research purposes, but he did not specifically bar support for research that uses leftover fertilized eggs from fertilization clinics (Leary, 1994).

PERILS OF THE HUMAN GENOME PROJECT

Might Be Used to Stigmatize Individuals

One concern regarding the HGP relates to possible abuse of the knowledge obtained. Suzuki and Knudtson (1990) worried that computerized human gene banks could lead to new opportunities for wholesale genetic screening programs, many of which could be of dubious merit. Individuals could be identified who harbor genes considered to be "inferior," and this classification could lead to social injustice. It is possible that people might be required to submit to wholesale screening programs, and information regarding potential genetic defects could be used to stigmatize them.

Nelkin (1992) expressed wide-ranging concerns regarding the potential perils of genetic screening. One was that screening may be used to preserve existing social arrangements and to enhance the control of certain groups by others. There is no doubt that genetic testing can be exploited and used to violate individual personal privacy and civil liberties, and that it can lead to genetic discrimination against those who do not conform to genetic norms. The damages that could result from the misuse of medical records and of psychological test results, make it necessary to balance society's need for economic stability against the rights of the individual.

All of Nelkin's concerns should be considered carefully, and there are indications that they are being addressed in a responsible manner. It was reported that the U.S. Equal Opportunity Commission ruled that employers cannot deny a job or fire someone for genetic reasons (Saltus, 1995c). The ruling covered only employment, but could have implications for insurance coverage for any employees who receive health insurance through an employer, which many do.

Watson (1992), who won the Nobel Prize for the discovery of the structure of DNA, was the director of the NIH genome project and is considered a most effective promoter of the HGP. He achieved his goal of putting more than 3% of the genome project money into an ethics program, which is under the direction of Dr. Wexler. Watson argued that, should the need arise, he would recommend putting a higher percentage of the total budget into the ethics program.

Watson stated that he does not think anyone should be allowed access to anyone else's DNA fingerprints, that laws are needed to prevent genetic discrimination, but that genetic information should be gathered and be made available to all who desire to know their own genetic background. Wexler (1992) argued that social justice should prevail in regard to genetic screening, although she noted that genetic diseases do cross ethnic and class boundaries, while access to services, unfortunately, does not. One place to concentrate effort might be to counter the genetic illiteracy that prevails in the medical, political, and journalistic communities, and to introduce new genetic findings as a part of a reasonable health-care delivery system. Wexler agrees that there are personal, social, and economic hazards involved in genetic screening, but, considering the many who suffer from hereditary diseases, she asks, "How can we *not* proceed?"

Some have expressed concern that if genetic evidence suggests that an individual has an increased susceptibility to environmental agents that might be encountered in certain occupations, then this could be used to disqualify that individual from securing employment in those occupations. It is possible that information obtained as a result of wholesale screening would not be kept confidential, but might be made available to interested parties, much as are credit histories. These concerns are all legitimate and must be considered and addressed to provide adequate safeguards for everyone. Such concerns, however, apply not only to genetic information but also arise whenever data banks exist, as is the case for the results of psychological testing, HIV screening, as well as for academic records, credit reports, and medical records. Policies should (and can) be developed to maintain the confidentiality and security of such records in order to respect the liberty and freedom of individuals while, at the same time, furthering just policies regarding the use of medical information.

Caskey (1992), in a discussion of some of the medical and ethical issues involved in genetic screening, raised concerns regarding the use of information to the negative interests of screened individuals through the loss of insurance coverage and job opportunities. Because of such politically negative implications, he argued that genetic information must be private.

Henry Greely (1992) also considered problems that the HGP raised in terms of difficulties people might have in obtaining health insurance and possibilities of employment discrimination. As our insurance system is presently structured, people who are known to be at higher risk for genetic illness could be denied insurance or have exclusions denying them coverage in the event of such illness. Another concern is that employers could deny employment to those at risk due to the possible need for expensive benefits if the employee is incapacitated. Greely suggested that the most reasonable solution is to remove the incentive to discriminate, and that such problems provide a good reason to move to a national health insurance plan similar to the Canadian Plan.

Problems involved in genetic screening were considered by a panel of the National Academy of Sciences (Hilts, 1993c). Some American have already lost jobs and others have lost health insurance on the basis of information obtained through genetic screening. These realities led them to recommend that laws be passed to set standards for testing and monitoring laboratories to ensure that the results of testing are accurate and interpreted correctly. They recommended that testing be done only if extensive information is provided about the disease and a discussion of the options a person would have if they carry a defective gene. Such procedures would assist in making voluntary, informed decisions, and would provide support for the potential anxiety and emotional suffering caused by the presence of a genetic problem.

The panel estimated that over 160 thousand people a year have been prevented from obtaining health insurance because of existing medical conditions, and they were concerned that the availability of genetic information might greatly increase this number. It may be that the solution is not to deny individuals the information if they wish to have it, but to restrict the actions of insurance companies. Perhaps the best solution is to eliminate insurance companies from the health-care delivery system altogether, in favor a single-payer plan. Health-care delivery should not be treated as a market commodity. All people should be guaranteed access to an adequate level of minimal health care, and it should not be bought and sold. The panel recommended that health insurers should not be able to obtain information about genetic screening results. Their recommendations were similar to those made by Watson—genetic information must be considered confidential and be protected from employers and insurance companies.

The importance of enhancing human dignity, guaranteeing free and informed consent, and protecting the confidentiality of genetic data were addressed in an international meeting of the United Nations Educational Scientific and Cultural Organization's International Bioethics Committee (Butler, 1994). Two major themes emerged during the debate: (1) a warning that genome research should not "geneticize social policy" and thereby erode public support for disability health care; (2) that somatic gene therapy might lead to medical euthanasia, which could override human rights. At the same time, they argued that once gene therapies become commonplace, they will be no more dangerous than any other therapy and should be regulated in the same way. There were differences of opinion regarding the use of somatic gene therapy for "enhancements." Some argued that these should not be dismissed as unethical in all instances, whereas others decided the use should be banned for enhancement purposes, but not for therapy. There was general agreement that all screening should be voluntary, genetic information should remain confidential, abortion for cosmetic reasons (or to avoid normal traits such as less than average size) should be banned, as should abortion, whenever there is a predisposition to develop a treatable disease. The intent was to reach a consensus regarding international law that is flexible enough to provide a reference for national legislatures and laws that serve as a last resort on which victims can base an appeal. The serious ethical issues are receiving the careful attention they should as the genome research proceeds.

The Specter of Eugenics

Suzuki and Knudtson (1990) bolstered their cautionary arguments by reviewing accounts in which knowledge regarding heredity had been used to attempt genetic "improvement" of the human species—called *eugenics.* The eugenics movement worked for the passage of sterilization laws in 30 states in the United States, and these laws were framed to keep individuals considered "hereditary defectives" from reproducing. Individuals were categorized as feebleminded, alcoholic, epileptic, sexually deviant, or mentally ill, and it has been estimated that about 20 thousand persons were forcibly sterilized by January 1935, most of them in California. The movement encouraged the passage of the U.S. Immigration Act of 1924 to limit the influx of immigrants from southern and eastern Europe because of their purported genetic inferiority.

The most egregious offenses were committed in Nazi Germany, where many thousands were sterilized or killed in the interest of racial purification. Both Degler (1991) and Richards (1987) discussed the excesses of the eugenics movement within the contexts of broad social and biological considerations, and the books by these two authors should be consulted to understand the dangers inherent in simple-minded views regarding heredity. Suzuki and Knudston (1990, p. 23) wrote that

> History confirms that knowledge about heredity has always been vulnerable to exploitation by special-interest groups in society for short-sighted, self-serving, even blatantly cruel ends—often for what seem to be the noblest of motives.

They concluded that we must remain vigilant against future attempts to reshape human heredity through gene therapies that might alter the human genetic line.

Several potential perils specific to germ-line gene therapy were identified by Suzuki and Knudston. The history of eugenics indicates that once a human characteristic has been labeled a genetic defect, some will attempt to eliminate that trait in the name of genetic hygiene. One danger with genetic engineering is that it ignores the fact that gene expression differs given the nature of nutritional, climatic, or other environmental conditions known to affect gene expression. A gene could be deleterious given certain background factors, but might have no harmful effects given other circumstances. It might be possible to treat some hereditary disorders by identifying and changing environmental conditions rather than through potentially risky genetic interventions, as Proctor (1992) and King (1992) argued. Suzuki and Knudtson were concerned that the role of genes might be overemphasized as causative agents in health problems at the expense of causative social factors, as Hubbard (1995) also argued.

The aforementioned concerns are legitimate, but any increase in knowledge—be it cognitive, social, or biological—can be exploited by those with evil intent, who want to exploit advances in knowledge to gain arbitrary and self-serving ends. Ethical doubts can be raised about the wisdom of obtaining knowledge that might be of some demonstrable benefit, but which has enormous potential for misuse (Vicedo, 1992). At the same time, doubts can be raised regarding whether it is ethical to deny people who need and want access to the benefits of knowledge—especially if the benefits could be enormous and the risks only potential. A narrow focus on specific outcomes can blind investigators to dangers that might be inherent in altering basic biological factors. Large-scale interventions affecting the germ line must be considered carefully, and with adequate respect for the broad biological fabric involved. Such dangers, however, are not unique to genetics and should not be used to justify the status quo by default.

Concerns regarding germ-line alterations have become more than just academic. Brinster and Zimmermann (1994) developed a technique to alter genes through changes in the sperm-stem cells. Brinster and Avarbock (1994) demonstrated that these changed sperm could be passed to an animal's progeny, which means that these self-renewing cell types can be thought of as immortal. The procedures involved harvesting stem cells from donor testes, maintaining them as a cell culture, and transferring them to a recipient testis to establish normal spermatogenesis. These functional spermatozoa then fertilize the eggs of a female mouse and result in offspring. These procedures may make it possible to cure infertility and to make germ-line modifications, even to the extent of cross-species transfer of genetic material.

Although the potential benefits (to cure sterility and eliminate disease) could be large, the seriousness of the potential for misuse is difficult to estimate (to eliminate traits that society does not value, to enhance those it does, or to disturb the "balance of nature"). Kolata (1994b) quoted several ethicists who worried about possible moral implications, stating that it is time to discuss these issues before the research inevitably proceeds form mouse to human. As one expert remarked, "The genie was out of the bottle." Kolata (1994c) made the disturbing observation that the University of Pennsylvania, where the research was conducted, has applied for a patent of the process.

The record supports the contention that those involved in the HGP not only are aware of ethical problems, but also have taken actions to understand these problems to protect individual liberties against the unprincipled use of the information that might become available. Whenever it is suggested that genetic screening might be done, and that the results of such screening be used to direct therapy or to influence reproductive decisions, the specter of the excesses of the eugenics movement is raised. When this specter is invoked, those who view with alarm often are guilty of some of the oversimplifications involved in what Popper described in *The Poverty of Historicism* (1957). One such oversimplification involves an essentialism that invokes an unchanging essence in any attempts made to gather and use genetic data. It is argued that all such attempts are essentially the same as those that prevailed in the "ethnic purification" ideas of eugenicists of the early 20th century. That argument is one that can be challenged by comparing the stated beliefs and actions taken by those directing the HGP to those that typified the eugenicists.

A second oversimplification is that the scenarios do not take into account the fact that circumstances and conceptions have changed over time. The lack of understanding of the basic principles of genetics that prevailed 50 or 60 years ago led many well-meaning people to accept incorrect and harmful conclusions. Based on the public record, it is clear that many of those currently involved in the mainstream HGP appreciate the lessons of history and are not destined to repeat them. Also, as Kevles and Hood (1992) pointed out in their summary chapter to the book they edited (a book dedicated to investigating the scientific and social issues in the HGP), the specter of a Nazi-like eugenic program is not likely to develop in the contemporary United States as long as political democracy and the Bill of Rights continue in force. They concluded (p. 318), "If a Nazi-like eugenics program becomes a threatening reality, the country will have a good deal more to be worried about politically than just eugenics."

Suzuki and Knudtson also noted that genetic imperfection is an unavoidable characteristic of human hereditary processes, and that by eliminating genes that seem to be maladaptive at the present time, the genetic variability essential to evolutionary change could be reduced to an undesirable extent. Williams and Nesse (1991) argued that members of the medical profession, policy makers, and the general public should be educated in the principles of evolutionary biology so they are able to appreciate the importance of such concerns. They believe this education should be done by experts in evolutionary biology rather than by generalists.

Suzuki and Knudtson (1990) were concerned that most inherited human traits are not produced by single genes but are polygenic. There is danger that manipulation of single genes might interfere with delicately evolved coadapted gene complexes, and that piecemeal alterations could cause problems due to unanticipated deleterious effects on these coadapted processes. This argument does not apply to those therapies being developed to correct single-gene defects, and the available literature suggests that the researchers involved in the HGP are well informed and cautious in these regards. The existence of coadapted processes argues that people should be trained to understand gene expression within an evolutionary framework.

Genes for Sale

Serious questions are posed regarding the pursuit of scientific knowledge and the secrecy involved when commercial and economic factors intrude. Intense efforts have been made (and are continuing) to establish priority of discovery to obtain basic patents of specific gene alterations and the techniques used to produce them, with the intention to commercially market them. Suzuki and Knudtson stated the opinion that economically motivated judgments should never be permitted to override democratic principles of individual freedom and equality of opportunity. It must be remembered that secrecy is the enemy of basic science.

The HGP has already produced some scientist millionaires, and more than 12 companies are pursuing technologies related to HGP (Fisher, 1994). Investments in genome companies by venture capitalists, corporations, and stock-market investors are exceeding the $165 million a year provided through federal funding of the genome project. Some of the scientists involved in the HGP, who now are players in the genome business, claimed that all of the work they have done was being made public without patents or other limitations. They argued that the profit motive, rather than being a corrupting force, is the surest way to generate products that could save lives. Nevertheless, the economic potential should lead everyone to proceed with caution to ensure that scientific data remain in the public domain, available to all scientists interested in understanding basic physiological functions, and that products developed be available to the general public at an affordable cost.

One of the scientists involved in the business of gene mapping was discussed by Wade (1994). Venter has developed a high-technology laboratory to sequence DNA. His strategy is to focus on the tiny portion of DNA that harbors the genes, ignoring the vast stretches that have no known purpose. Backers have invested $85 million in the project and a pharmaceutical company paid $125 million for the right to market the findings—with Venter owning shares estimated to be worth about $12 million. His laboratory has analyzed over 100 thousand genetic fragments from human DNA sequences amplified in clones.

Conflicts have arisen between some scientists, such as Watson who resigned from the HGP, in part because of his disagreement with Healy (then director of NIH) over the issue of patenting the new genes. Healy argued that Japanese and European companies would obtain them, thereby realizing the profits available in medical and commercial applications of these discoveries. Watson opposed wholesale patenting of unknown genes, because it would inhibit the exchange of research information. Venter stated that he intends to publish all his DNA sequences in scientific journals. The company funding his research, however, has exclusive rights to study the sequences for at least 6 months, and for 12 more months whenever it wishes to develop the genes for commercial purposes.

Even more problematic are recent applications for patents that involve specific techniques, A patent application has been granted to NIH of the technique used for gene therapy, and NIH has given a biotechnical company exclusive rights to market the procedure. The patented procedure involves the fundamental approach to remove genes from the cells of patients, to modify them in a laboratory, and to reinsert them to correct genetic disorders such as cystic fibrosis, hemophilia, and cancer. Of the 100 human trials of gene therapy products under way, 70% rely on this technique (Day, 1995). A licensed patent can cost a company 3–5% in royalties, and these patents would be worth millions of dollars a year if everyone using the basic procedures has to pay.

Caplan (1992) cautioned that secrecy must be avoided in the HGP, as it should in all scientific research. He argued that those involved in the HGP have an obligation to share findings with other scientists in the United States and in other nations, especially in view of the fact that much of the basic research has been funded by the public. The patent status of knowledge and techniques used to manipulate the human genome should remain in the public domain. Suzuki and Knudtson concluded that the best protection against the misuse of scientific knowledge is the mandatory publication of all research findings, and that classified biomedical research should be forbidden in order that public vigilance can ensure that modern genetics does not become a tool that individual scientists might use to intentionally harm fellow human beings.

Wills (1991) reviewed genetic studies of the causes of cystic fibrosis and suggested that, because of the nature of the genetic factors involved, it might be possible to develop a simple test to identify all heterozygotes in the population—information that could generate enormous profits. Some investigators have freely distributed probes that could be used to identify chromosomes or chromosome fragments, whereas others, who have a more direct commercial intent, do so only if recipient investigators sign elaborate licensing agreements. Considerable acrimony resulted when it became economically critical to determine who had priority for the successful location of the chromosome for the cystic fibrosis gene—a priority that has immense financial implications (Wills, 1991, pp. 203–205).

Perhaps the human genome project should not be pursued full-tilt at present, because the technology used to sequence genes is developing so quickly that the cost for each base pair analyzed will be decreased greatly. The project could proceed more efficiently and economically if multidisciplinary and multinational cooperation is secured and computerized data banks are developed, before moving prematurely to a full-scale independent U.S. effort. Other countries are also involved in the HGP: Britain, the USSR, Italy, Japan, France, Canada, and Australia (Vicedo, 1992), and care should be taken to ensure that the results of all of these studies are expressed in a common scientific language to permit the construction of a useful dictionary.

These concerns support the belief that the HGP should continue, but that a perspective should be maintained, taking into consideration other basic needs in biological research. The process of allocating money to scientific and other social needs is complex, and there are never enough resources available to satisfy every need. It is important that the HGP be considered within the context of social needs as one of a number of pressing biological, technological, and social problems.

The cost–benefit question was considered by Wills (1991) in terms of whether the global HGP should proceed and continue spending the tens of millions of dollars to search for the genetic bases of certain diseases. He reminded us that worldwide, perhaps 30 to 50 thousand children a year die slow, agonizing deaths from cystic fibrosis. It has been estimated that a cure might be found in 10–15 years, but if the studies wait on the sequencing of the entire genome, then the cure could take as much as 20–30 years. These considerations support the argument that the research should continue and focus on those suspect areas of the genome. This search might not only lead to adequate gene therapies, but also undoubtedly will be accompanied by improved technologies to make the complete genome project economically and practically more feasible.

Screening for Sex

It is a relatively simple task to screen embryos for sex. Although the cry has been raised that having information regarding the sex of a fetus amounts to "playing God," few object to genetic screening for sex if there is a strong likelihood that the offspring may inherit a sex-linked disease, such as hemophilia, in which only men exhibit the disease and only women can be carriers. It is difficult to object when prospective parents want to have only daughters under such circumstances. The desire of these parents is not based on sexism, but on a preference to have a healthy child.

More problematic for some is the use of sex screening to control population levels. In countries such as China, women suffer a lower status than men, and there is a strong preference for sons. This preference is justified in terms of continuing the family line, having an heir, having sons to provide labor (in rural areas), and having sons to support parents in their old age. Studies indicate that, at the present time, most Chinese couples want no more than two children. Chinese couples express a strong desire for at least one son, and women will tend to continue childbearing until they have one healthy son.

Given these preferences, one could reduce population levels by determining the sex of the second conceived embryo for families that already have one daughter, desire a son, but do not want more than two children. If the second conceptus is male, then the pregnancy would be continued, and if female, it would be aborted. In countries such as China, where the government, people, and demographers all agree it is essential to bring population growth under control, a sex-screening policy might well make significant contributions toward population control and still respect the desires of the parents. Sex screening used in this manner might serve the ends desired by both the government and parents.

The Population Crisis Committee (Conly & Camp, 1992) estimated that an outmoded intrauterine device (IUD) accounted for 41% of total contraceptive use in China. Female sterilization accounted for 36%, and male sterilization for 12%, with such things as birth-control pills and condoms accounting for the remaining 11%. Voluntary genetic screening programs are preferable to involuntary sterilization, forced abortion, the reliance on dangerous and unreliable IUDs, or suffering the consequences of living in a country with a population too large for children to attain a satisfactory life within the limits of available environmental resources.

The sex ratio of newborn children in China is becoming quite skewed in favor of males (Kristoff, 1993). The worldwide birth ratio is normally about 105 boys for every 100 girls. In the 1953 and 1964 census in China, the sex ratio was between 104 and 105 boys for every 100 girls—roughly the expected level. The average sex ratio in China, for all of 1991 and for 9 months of 1992, was 116.5 boys to 100 girls. The increasing imbalance was attributed to the use of ultrasound machines that were installed to examine the livers of pregnant women, to check that an IUD was positioned properly, and to determine whether a fetus was developing normally. In the course of these examinations, it is possible to determine the sex of the fetus as well. Indications are that female fetuses are being aborted, and that a cottage industry has sprung up, whereby ultrasound examinations are done for the sole purpose of determining the sex of the fetus, usually at the end of the second trimester. The Chinese government acknowledged the problem and, as reported in *The New York Times* on November 15, 1994, announced that a family law to take effect in January, 1995 will ban sex screening of fetuses and forbid couples carrying serious genetic diseases to have children. The health minister said the list of such diseases would be published later, and that the termination of a pregnancy when the child is found to have a serious genetic disease or defect would need the couple's consent. Based on the track record of the Chinese government, this assurance should not assuage the fears of those concerned over the possibility of forced abortions and sterilizations.

The strongest resistance to genetic screening for sex has been expressed when it is purely a matter of personal preference on the part of the couple. Yet, we have all heard the lyrics of the song "Tea for Two," which expresses the ideal of "a boy for you, and a girl for me. Don't you see, how happy we could be." Evidence suggests that couples who do want two children tend to want one of each sex and would often prefer a male child first. Although such preferences might seem frivolous, they do seem to be strong and widespread. If characteristics such as the sex of a child are important to people, then Harris (1992, p. 158) asked, "Why not let people choose? ... Can it be right to leave such important matters to chance?"

Allowing a choice regarding the sex of a child is unlikely to lead to evolutionarily significant imbalances as a result of a surplus of males, because the number of individuals in the population who would be expected to use the abortion option for this purpose would be relatively small. If an imbalance does begin to appear, then it would be expected that normal social regulation would correct the imbalance, because daughters would be at a premium, due to the short supply, making them more valuable as a limited resource. The official press in China is beginning to warn that infant boys will be unable to find wives in 20 years (Kristoff, 1993). It is likely that the value of girl babies will increase if only because the shortage will lead to economic advantages such as expensive bride prices, and this might lead to an increased social prestige accorded to the scarce commodity represented by women, although not for reasons feminists would prefer.

The legitimacy of parental choice should be respected rather than subjecting parents to authoritarian dicta. Harris (1992, p. 161) concluded, "If free choice in reproduction begins to look as though it will produce harmful standardization we could of course revue the question of the desirability of controls. No question is ever finally closed." One can be accused of "playing God" when treatments are withheld, just as much as when there is intervention, because both require decisions.

There is little rational basis to prohibit genetic screening if it is requested by prospective parents. Avoiding the production of children who will have to live with a serious genetic defect is a reasonable goal, and parents should be free to make such choices. Genetic screening in these instances is justifiable, especially if the parents have produced a healthy alternative embryo that is available for implantation as a substitute for a discarded, defective one. The end of achieving rational population control, especially in countries whose government and citizenry want to achieve such control, is difficult to fault. People should be able to make informed choices regarding the sex of their offspring, with the understanding that if undesirable practices or imbalances in the sex ratios begin to appear, then policies can be reevaluated to decide whether to allow them to continue.

Many arguments that genetic screening will be used for frivolous reasons have little merit. The therapeutic techniques are too expensive, complicated, and intrusive to enjoy widespread use by those only of a whimsical nature, and those complexities, expenses, and complications probably will exist for the foreseeable future. Another question involves using either government or private insurance to finance genetic screening. There would be no problem if our health-care system was based on principles of cost–benefit analyses. Some procedures, such as screening sex to obtain a "boy for you, a girl for me," would be low on the priority of any health system. The cost-effectiveness of prenatal diagnosis or IVF testing, when there are reasons to suspect that a defective child will be born, can be determined. If the cost to society to maintain a defective child will be enormous, then it might be reasonable for the public to pay for screening. If the cost of maintenance is minimal, then the priority for public support might be lower relative to other needs to which limited governmental funds could be directed. These issues will be considered when the rationing of medical care is discussed.

The Bioethical Imperative

The extremely rapid advances made toward understanding the human genome, the processes of genetic expression, and the development of therapies for genetic malfunctions make it imperative that bioethicists push the medical and scientific community to take a proactive position concerning ethical issues. Matters are further complicated by the immense commercial interests involved. Fifteen biotechnical companies have made gene therapy their primary objective, and other firms are actively moving into the area (Beardsley, 1993). It has been predicted that the success of some of these therapies will result in the approval of new, commercially viable medical products within the next 2 or 3 years.

Not only have there been rapid advances, but also new research results suggest that the genes that switch on proteins inside the developing embryo have been identified in mice, zebra fish, and chickens (Angier, 1994a). These genes, called *morphogenes*, determine the destiny of the cells they influence in different locations within the embryonic body. As Angier expressed it, they give the cells their address, their fate, their identity, and their purpose in life. The events involved in shaping the central nervous system occur in humans sometime around Day 15 after fertilization, and the processes are largely finished by Day 28, with those involved in shaping the limbs beginning shortly after that time. With the discovery of morphogenes, it might not only be possible to understand reproductive development in exquisite detail, but also to find the processes by which the controlling protein stimulates the response of a master gene inside cells.

These genetic discoveries and the possibilities they engender represent only the tip of the iceberg in terms of the moral issues involved. A Scottish scientist transplanted ovaries from aborted mice fetuses into adult mice and found that the ovaries produced eggs that could be fertilized and developed into normal mice (Kolata, 1994a). It could be feasible to use the same procedure with humans. The ovaries from aborted fetuses could be transplanted into infertile women who cannot produce their own viable eggs. A 10-week-old female fetus already has manufactured all of her eggs (6–8 million), and if the implanted ovary is allowed to grow to adult size (which requires about 1 year), the eggs it contains could be fertilized naturally rather than using IVF procedures. Because egg donors are in such short supply, the fetal implant procedure could provide a bountiful and continual supply of eggs for infertile couples.

As might be expected, a storm of ethical controversy has arisen. George Annas characterized the idea of fetal implants as "so grotesque as to be unbelievable." He was quoted by Kolata (1994a) to have raised a series of questions: Should we be creating children whose mother is a dead fetus? What do you tell a child? That your mother had to die so you could exist? As Kolata noted, when fetal ovaries are used, the woman who donates the aborted fetus is now a grandmother, but she was never a mother, which certainly upsets the natural order of generations.

Caplan was concerned that it might be devastating to grow up knowing that your genetic mother was an abortus, and he concluded that no one should be able to create a child from anyone's eggs or sperm without consent. Because a fetus obviously cannot consent, he considered the procedure questionable because it treats reproduction as a commodity. But the ethicist Levine argued that most of the ethical questions pale when the good that can be done for infertile couples is considered. Even though the child might be troubled by its genesis, it almost certainly would rather have been born from a fetus's eggs than not to have been born at all.

These developments raise fundamental questions regarding the creation of life, suggesting that humans could soon have almost unlimited technological power to make choices that were hardly conceivable until recently. These new reproductive technologies, along with the ability to manipulate life as it develops, should give everyone pause for concern.

There will be active debate of these issues, and it is hoped that such debate will precede attempts at preventive political strikes based on a sense of intuitive revulsion before engaging in serious exploration and discussion of the underlying issues.

Many of the negative views expressed reflect a strong commitment to a natural fate position—that nature should not be altered, especially when such alteration involves heritable changes in the genetic line or affects the succession of generations. There is less concern with manipulations that prevent the development or transmission of defects, but whenever the creation of life is involved, strong, deep-seated emotions are evoked, and these emotions lead to prompt calls for legal prohibition. These concerns are voiced even more strongly when there is any possibility of commercialization of nonrenewable resources, such as selling human eggs, organs, or tissue; there is no objection, however, to selling renewable resources, such as blood or hair.

It is neither possible, nor desirable, to maintain value neutrality concerning the new knowledge that will result from mapping and sequencing the human genome. Caplan (1992) identified several moral questions concerning the responsibilities and duties that must be considered to protect privacy and confidentiality, to warn potential parents of risks to the health of any offspring they might choose to create, to decide when testing or screening is appropriate and when it is mandatory, and to determine the conditions or disorders that should be classified as defects, diseases, anomalies, or abnormalities.

He argued for value neutrality, and that those who seek clinical human genetics services should do so freely, without pressure or coercion. Counselors should protect the rights to privacy and confidentiality of those who seek these services, and the sole aim of clinical human genetics screening should be to provide comprehensive information to the individual. The overriding concerns should be those of safety, efficacy, reliability, and risk. Social resources to alleviate and treat genetic diseases and disorders should be considered part of the effort made to maximize the potential for health and to minimize risks of disability and disease. Such efforts, however, must be considered within the total context of improving the health and welfare of the world's population as much as possible, and should be considered in the light of the available resources.

Wilfond and Nolan (1993) analyzed problems that advances in genetic knowledge and techniques force upon us. They considered policies developed to regulate cystic fibrosis carrier screening, and their analysis provides a useful case history to summarize the concerns that should be addressed more generally. They developed their recommendations for the Hastings Center Project on Priorities in the Clinical Application of the Human Genome Research.

The major problems encountered when considering the feasibility of widespread genetic screening are those associated with coerced diagnosis, anxiety on the part of those at risk, loss of privacy, stigmatization, and discrimination. Wilfond and Nolan believe that any proposed plans should include well-defined and attainable goals, and should provide for patient education, informed consent, and counseling. There should also be assurance that the tests are reliable and valid, quality control should be guaranteed, costs should be acceptable, and there be adequate follow-up services available.

They suggested two basic models: the extemporaneous and the evidentiary. The extemporaneous model relies on the independent market to regulate professional practice, with legal and consumer forces controlling utilization and reimbursement. The evidentiary model relies on a rational analysis of data using explicit substantive criteria and goals that have been developed through public participation in the formulation and evaluation of the normative issues. The extemporaneous model has prevailed, traditionally, and much of the debate regarding a national health-care plan embraced that model through the use of managed competition and the health alliances in the health-care plan proposed by the Clinton administration, as well as in competing versions.

The evidentiary model is based on a framing of normative goals to develop priorities for different genetic diagnostic procedures. One goal of genetic diagnostic services should be to improve the ability of people to make informed personal and reproductive decisions in light of their genetic status. Another should be to reduce the incidence of disease. When the implications involved in the realization of these two goals have been understood, the task then becomes one of establishing priorities among possible diagnostic options. It might be decided that it is more justifiable to detect conditions for which some form of treatment and remediation are available, rather than provide routine screening for conditions that are not treatable at the present time.

There should be prior discussion and agreement regarding the relative importance of the different goals of screening, and if it becomes necessary to allocate limited research and treatment resources, these basic priorities can be used to guide decision making. With such procedures, quandaries that result from considering problems from a reactive stance can be minimized. The difficulty with always being reactive is that the unique specifics of each case often cloud general issues. By taking a proactive stance and developing the general issues and concerns at the outset, a system of precedents based on consistent general principles might be more easily developed. Because the decisions to be made involve some of the most profound normative values regarding the lives and experiences of people, the process must reach beyond the limits of science and medicine. It is crucial for representatives of the community to be involved centrally, because decisions regarding reproduction and health should be made by society at large and not solely by health professionals, commercial interests, lawyers, the scientific community, or ideologues of one stripe or another.

Wilfond and Nolan argued that the critical tasks should include setting criteria and standards for evaluating testing programs, making recommendations for clinical practice and health-care policies regarding which diseases should be tested for and the population that should be tested, establishing standards for informed consent, evaluating counseling and educational approaches, monitoring ongoing programs, and making recommendations regarding reimbursement and liability. Although this involves a large number of tasks, the profound nature of the decisions society will have to make should involve no less. The first and most important task is to establish the normative goals and ultimate moral values that society wants to implement. When that first step has been taken, the specific actions require little more than the fair utilization of the resources and technologies that society commands.

The Institute of Medicine of the National Academy of Sciences issued a report by a panel composed of geneticists, genetic counselors, pediatricians, ethicists, and lawyers (Marshall, 1993). It was recommended that the government create a standing committee to monitor the use of genetic screening, and this recommendation received tentative support from the director of the National Center for Human Gene Research. The panel was assembled to consider the implications of a Pennsylvania law requiring screening of every child born in the state for a battery of diseases, including Duchenne muscular dystrophy. The concern was that the parents of a Duchenne baby would be likely to learn of their child's fate, whether they wanted to know or not, and there is no known cure for the affliction.

The panel concluded that widespread testing for incurable diseases should not be done, because it will not benefit those being screened. Two other principles adopted were that parental permission should be required for all genetic tests, and that initial positive test results should always be followed by confirmatory tests, counseling, and treatment wherever possible. There was considerable controversy within the panel—some argued that testing should be mandatory whenever therapeutic procedures were available that could avoid neurological damage, such as for PKU and hypothyroidism. There was controversy within the panel surrounding a recommendation that information regarding recessive traits should not be disclosed to parents, because that particular child's health is not at issue. Some members maintained that such information should be disclosed to parents, because it might influence decisions to have another child. Examination of the discussions of this panel suggests that there should be increased debate and considerations that involve several levels of societal input and utilize the skills of a number of relevant disciplines.

Although the problem of AIDS screening does not involve genetic screening, some of the issues considered here are relevant. The Center for Disease Control (CDC) issued a proposal recommending that physicians should counsel every pregnant woman (about 4 million a year) about AIDS and urge each to be tested so that infected mothers can try to protect their unborn children (Neergaard, 1995). It is estimated that about 80 thousand heterosexual women are HIV positive and that about 7 thousand of them give birth each year. If the drug AZT is taken, the chance that a mother will infect the fetus is reduced by two-thirds. The CDC argued for mass testing on the grounds that it will save childrens' lives (those infected live only about 8 to 10 years) and reduce medical costs. Without considering the costs, it should be noted that the proposed testing was voluntary. Studies have indicated that more than 90% of pregnant women agree to testing after they receive HIV counseling. The CDC is negotiating with Medicaid to ensure such coverage as part of standard prenatal treatment, with infected women to be offered AZT therapy. These suggestions seem eminently sensible, given that they use voluntary testing, the screening detects a remediable condition, it appears to be cost effective, and it produces a humane outcome.

GENES: PROGRESS, PROBLEMS, AND SOLUTIONS

The progress of the current research into what Jones (1993) called the "language of genes" indicates that it has the potential to provide important insights into the nature of human development and functioning (both normal and abnormal). Serious questions involve the ethical, medical, and financial aspects of these research programs.

The most obvious advances are those that enable screening of parents for known single-gene defects to provide them with information regarding the likelihood that their progeny will be at risk. Such testing should be done whenever there are treatment alternatives that can prevent the expression of the gene defect. When there are no known therapies, the information should be available to individuals who want to know in order that they might make a voluntary, rational, and informed decision not to reproduce, or to go ahead with full knowledge of the risks involved.

The mapping and sequencing of the human genome that has been done, especially when information regarding extensive family histories is incorporated, has made it possible to understand some of the processes involved in the expression of polygenetic afflictions, and ways have been suggested that could make it possible to treat at-risk individuals. These medically oriented studies have led to the development of analytic and manipulative techniques that have the potential to make it possible to probe the basic workings of normal cellular functions and to understand some of the basic mechanisms of gene expression. Rapid progress is being made at the levels of basic physiological knowledge and applied medical treatments. It can be expected that basic scientific understanding will continue to advance as the genetic research continues.

Now that these optimistic and positive aspects have been considered, what are some of the problems that have led to such serious concerns among scientists, ethicists, and politicians? A number of perils have been discussed here and in Chapter 2. Foremost among the serious concerns is that information obtained through routine genetic screening could be used to stigmatize individuals in areas concerning employment and health insurance. These concerns are real, but they are no more serious than those produced when any data bank is established. There have been many precedents in which there is a need for confidentiality of records. Discrimination on the basis of data that identify potential risks must be prohibited. These do not constitute serious impediments to realizing the benefits that can be gained through genetic screening. Evidence supports the belief that the laws, rules, and regulations being proposed and implemented will minimize costs due to potential discrimination.

Another problem involves the use of genetic screening to determine the sex of the fetus. Knowing that sex-linked traits can be transmitted genetically makes such screening potentially valuable. The danger of genetic screening is not due to problems with the procedures, but with the values of members of society. It would be better to spend energy to influence underlying sexist attitudes than to deny information regarding the sex of embryos to parents who need or desire it.

Another question that has been raised since the HGP began, in fact, since the beginning of the recent round of discussions regarding the role of biological factors in complex human behavior, has been the specter of eugenics. Many are concerned, given the sordid history of past attempts to control human destiny through arbitrary manipulation of individuals deemed unacceptable by the authoritarian elements that often have controlled nations: most notorious being the formal eugenics movement in the United States that advocated forced sterilization and restrictive immigration laws, and the mass genocide practiced in Nazi Germany. I spoke against the applicability of these concerns within the contemporary climate of opinion and argued that the types of genetic analyses and manipulations being developed embody an informed recognition of the complexities involved in gene expression—an informed position that did not characterize those of the eugenicists and Nazis.

Concerns regarding the danger of sociobiological or evolutionary expressions of human psychologies are usually couched in a language that deplores biological or genetic determinism. Often those expressing these views attack the straw-man that those who stress genetic factors believe in the existence of genes for this or that complex behavior. These critics point to the indisputable fact that behavioral traits and characteristics develop within, and are strongly influenced by, an environmental complex. In Chapter 2, as well as in Part I of *Human Evolution, Reproduction, and Morality*, the accuracy and fairness of the characterizations were challenged. They do not represent the views of those of us who emphasize the importance of evolved mechanisms in the expression of complex human behavioral tendencies. Part of the objections to an emphasis on biological mechanisms is based on the belief of many critics that the environment contributes the critical influences that determine the course of behavioral propensities and development. These objections lead to recommendations that resources and concerns should be focused to correct social inequities and enhance the quality of environmental influences. This environmentalist viewpoint seems to be the result of adopting what has been called the "Standard Social Science Model," and is also compatible with an emphasis on social factors that has characterized the political views of those who espouse a liberal socialist position. It makes no sense, however, to discuss development in terms of biology versus experiential factors. The dialectical position should be embraced that such factors exist conceptually, but their expression involves a dialogue from the outset.

A serious problem involves the commercial exploitation of basic discoveries regarding the human genome and the mechanisms by which genes do their thing. There are enormous potential profits to be realized if certain genes and the techniques to influence their normal and abnormal manifestations are patented. Serious questions involve the ethical permissibility of owning a patent for the essential functioning of life systems themselves. Discoveries that are made partially at public expense should not be used for the economic gains of a few private entrepreneurs and their corporate entities. It will be documented in Chapter 11 that there have been great abuses in the research, development, production, and marketing of pharmaceuticals, and similar abuses could be on the horizon in the genetic marketplace. An examination of the sordid history of the pharmaceutical industry suggests that it would be wise to curtail such abuses before the genetic research program goes much further.

What is more alarming is the secrecy involved in commercialization. Despite the protestations by some of the leading researchers and their assurances that all data will be published in public, refereed scientific journals, the commercial considerations already provoke concern. Secrecy is the enemy of scientific progress, and open communication must be its vehicle.

Some of the ethical concerns that have been raised are not insoluble and can be resolved through open exchange of information among concerned parties. These include the development and certification of reliable and valid test procedures, the training of competent and informed counselors, and the development of adequate counseling procedures, the development of guidelines to ensure that consent for genetic screening is voluntary, and that results are confidential. The problem of using legal abortion based on the results of genetic screening, whenever parents decide they do not want to continue a pregnancy, is not an issue as long as the choices are reasoned and made in possession of relevant information. The choice to reproduce and to rear children should be made by parents enjoying their own view of morality expressed within the limits of the law.

Finally, the cost factors that have been raised as objections to pursuing the HGP are little more than a smokescreen to cover other basic moral or political objections. The economic data discussed in Chapter 2 support the conclusion that cost–benefit analyses come down on the side of continuing the project in some form.

CHAPTER 4

Death and Its Criteria

In *Human Evolution, Reproduction, and Morality* and in the last two chapters, the focus was on issues relating to the beginning of life; questions were explored about when life begins, when a human organism should be considered a person, when a person assumes the duties and responsibilities of a moral agent, and what manipulations of genetic potential are permissible. Issues were examined and policies were recommended regarding contraception, abortion, infanticide, genetics and manipulation, and the use of reproductive technologies. These issues were discussed within the perspectives of evolutionary theory, cognitive principles, and a rational liberalism seasoned with some utilitarian spices.

The principles developed will now be applied to issues that arise at the end of life. In this chapter, the defining characteristics and criteria for death will be considered, and will be followed in Chapter 5, by a discussion of philosophical and practical issues regarding organ transplants from cadavers and other organ sources.

When considering reproduction, some believe that life begins at the time an egg is fertilized by a sperm. It was argued in *Human Evolution, Reproduction, and Morality* that this event has neither moral relevance nor biological significance, and that moral considerations only come into play at the point of birth, because it is then that a public person is in the hands of society, and respect must be given to that person. Here it will be argued that moral considerations regarding death should hinge on death in the sense of a biographical life (having a life) rather than a biological life (being alive). This distinction has been developed at length by the moral philosopher Rachels (1986), and will applied here.

BIOLOGICAL CRITERIA

The biological criteria of life are based on an organism's *being alive.* The most conservative criterion of the death of a person in the biological sense is when the body is cold and pale, and breathing and heartbeat cease. The conservative standard to decide when death occurs is the point at which an individual has sustained either irreversible cessation of circulatory and respiratory functions, or there is an irreversible cessation of all functions of the entire brain—including the brain stem, which regulates basic physiological processes (Botkin & Post, 1992). The view that there is life as long as there is breath is a traditional one, and it served adequately until it became possible to keep totally, permanently, and irreversibly comatose people on life support for years. The medical ethicist Capron was quoted by Kolata (1992) to have stated that a more reasonable criterion would be that a body should be considered to have life as long as it is breathing on its own and has a heart beating on its own: referred to as the *heart–lung criterion.*

The legal definition of death in all 50 states and the District of Columbia is when whole brain death takes place; that point when there is no longer any detectable electrical activity in the brain (Angell, 1994). When electrical activity ceases in the brain, the state is irreversible and life has ended.

The most widely known case is that involving a 21-year-old, Karen Ann Quinlan, who suffered cardiopulmonary arrest in 1975 and died 10 years later, having never regained consciousness or any voluntary function. During the first 6 months after the trauma, she showed no signs of awareness of her environment or any cognitive functions, and was diagnosed to be in a persistent vegetative state (PVS—discussed in the next section). After 7 months of treatment, her father asked the courts to order the hospital to turn off the respirator, because her life was being wrongly extended by technological means. The court granted the request. The respirator was turned off, but she did not die for another 9 years, remaining in PVS and dying from an overwhelming infection.

An interesting aspect of the case is that her brain was preserved after her death and analyzed 3 years later (Kinney, Korein, Panigrahy, Dikkes, & Goode, 1994). There was little damage to the cerebral cortex or autonomic and arousal systems of the brain stem, but there was massive bilateral and symmetrical damage to the thalamus, as well as to the cerebellum. These findings were surprising, because the thalamus has not been considered to be of primary importance as the base for consciousness, being viewed as a way station for nerve impulses from the periphery, and the cerebellum is thought to be involved mainly in motor activity and coordination. Usually when consciousness is lost, there has been extensive bilateral damage to the cerebral cortex. The damage was to those thalamic nuclei closely connected with regions of the cortical association areas that are involved in "multiple and diverse cognitive functions, including selective attention" (p. 1473). It was suggested that this damage was the cause of the global impairment. The significant fact for the present discussion is the anomaly of having the midbrain structures damaged with the cerebral cortex remaining intact, placing in doubt one of the widely accepted criteria for brain death. There was electroencephalographic activity in the cortex (the criterion used to establish that the cortex is functional), yet Quinlan clearly had the reflex and behavioral signs that are used to infer that a patient is brain dead.

One reason that these niceties of definition are important is that there must be a legally binding point at which a patient can be declared brain dead in order to permit organ retrieval. The desire to maintain an organism's vital signs until organs can be recovered has made the issue of when death occurs more salient than before. Bioethicists Fox and Swazey (1992) conducted extensive field studies of transplant procedures, which they summarized in their book, *Spare Parts*. They noted that the criterion for death has not been driven by reasoned philosophical or technical concepts to signify the point of death. On some occasions when organs are to be recovered, the brain of the newly dead (neomort) is admitted to the operating room as a "beating-heart cadaver." Fox and Swazey reported that 89% of all donors are pronounced dead in intensive-care units, and that a neurologist usually participates in the diagnosis because of the need to pronounce that the individual is brain dead before organ retrieval is permissible.

Botkin and Post (1992) noted that the major advantage of the heart–lung standard is that it is easily determined and immune from confusion that death really has taken place. Using the heart–lung criterion has the disadvantage that the pool of organ donors is reduced drastically if donors are required to be dead by this criterion; some whole-brain standard would permit recovery of viable organs before they begin to deteriorate.

A major problem facing the medical profession was identified by Youngner, Landefeld, Coulton, Juknialis, and Leary (1989); only 35% of the 195 physicians and nurses who were likely to be involved in organ procurement in four university-affiliated hospitals correctly identified the legal and medical criteria for determining death, and more than half of the respondents did not use *any* coherent concept of death consistently.

It has been argued that death should be considered to have taken place when the electrical activity of the cerebral cortex has stopped, even though the brain stem regions (which regulate and maintain such things as blood pressure, respiration, and heartbeat) still exhibit electrical activity. This argument considers cognitive functioning to be the hallmark of life, at least to the extent that the individual can be considered to have interests—a position argued later when considering life in the biographical sense.

One problem with a strictly biological criterion for death is similar to those encountered when the onset of personhood is defined in terms of fetal viability (see *Human Evolution, Reproduction, and Morality*, Chapter 9). This problem occurs when the status of technology and the quality of personnel available at a particular medical facility would influence the point at which viability begins and ends. If the onset of viability (that point at which the fetus can survive if removed from the mother) is chosen to signal the start of personhood, then personhood would not reflect any biographical or social reality, but would be defined technologically. Crucial moral issues should not rest on any arbitrary technological criteria. Similar concerns are present when death is considered, because it is possible to maintain the biological functions of an organism at the end of its existence for extremely long periods of time through the use of life-support systems, even though the person may be in an irreversible coma and will expire when the medical apparatus is disconnected. Biological life (perhaps *technological life* is a better phrase) continues, but biographical life has ceased, and it only remains to be established that the cessation is irreversible.

Persistent Vegetative State

Technological advances make it possible to keep an extremely premature fetus, as well as an anencephalic baby (born with a complete absence of the cerebral cortex), alive for long periods. It is possible to keep mature people classified PVS alive for almost indefinite periods of time. To cope with the large number of problems that these changes in technological capability produced, a Multi-Society Task Force (1994) was formed to summarize current knowledge of the medical aspects of PVS in adults and children.

The cited estimates indicating that PVS is a significant problem in the United States: there are 10–25 thousand adults and 4–10 thousand children existing in PVS. The costs of caring for a patient for the first 3 months in PVS is estimated to be about $149,200, and the estimated cost of long-term care in a skilled nursing facility ranges from $350 a day ($126,000 a year) to $500 a day ($180,000 a year). For the United States, it costs somewhere between $1–7 billion a year to care for PVS adults and children.

PVS is only one form of permanent unconsciousness, and the task force identified the different forms, their medical indicators, and the prognosis for neurological recovery and survival. PVS was defined as a clinical condition with complete unawareness of the self and the environment, with sleep–wake cycles, and either complete or partial preservation of hypothalamic and brain-stem autonomic functions. The conditions range from a transient one from which there could be partial to full recovery to a permanent one with no recovery of function.

The considered the probability of recovery and survival from PVS as a function of type of injury (traumatic and nontraumatic) and age (especially children as compared to adults) and concluded that recovery of consciousness from a traumatic vegetative state is unlikely after 3 months and exceedingly rare after 12 months in both adults and children. Patients with degenerative or metabolic disorders, or who have congenital malformations and remain in PVS, are unlikely to recover consciousness, and if they do, their life span is substantially reduced–for most ranging from 2 to 5 years.

At the level of ethics, they recommended that surrogate decision makers (especially family members) should be given appropriate psychosocial and religious counseling to face decisions about termination of treatment and that these surrogate decision makers, as well as patients who left advance directions to terminate all forms of life-sustaining medical treatment, should be accorded the right to their wishes, including termination of hydration and nutrition. They recommended that more systematic data be collected regarding epidemiology, incidence, prevalence, and natural history of PVS in order to develop better clinical predictions regarding the likelihood of recovery of consciousness and survival.

Angell (1994, p. 1524), executive editor of the *New England Journal of Medicine,* in which the report was published, added that there should be recognition that

> For many families, the possibility of sustaining the life of a patient in a persistent vegetative state means that the tragedy of losing a loved one is compounded by the anguish of the daily physical reminder of what that person once was.

After considering the task force reports—the case of Karen Ann Quinlan, and the case of Baby K (an anencephalic infant with no possibility of recovery, but whose mother insisted, on religious grounds, that it be delivered and that life support be provided)—Angell suggested that it was reasonable to allow caregivers to stop treatment so that the patient will die whenever the medical decision has been made that the condi-

tion is irreversible. Angell maintained that the burden should be shifted from those who want to discontinue treatment to those who want to continue it, making it possible to establish guidelines to stop treatment after a specified time based on medical data. This policy is more satisfactory than those used now. The current policies place immense stress on medical staff and surrogate decision makers, and often result in the decision being made by the courts, rather than by either the loved ones or qualified medical personnel. This situation has been deplored by Annas (1994), who argued that it is physicians, rather than the courts, who should define the standards for medical practice. The medical community should accept responsibility to set standards, and they must follow them once they have been established.

BIOGRAPHICAL CRITERIA

Botkin and Post (1992) argued that the moment of death should not be signaled by any single physiological event, but as a moment defined by philosophical concepts that speak to what it means to be alive, and that this moment should be fixed by social consensus. Rachels (1986) maintained that it is life in the biographical sense that should be protected, rather than that in the biological sense. He argued that death occurs at the point at which consciousness is no longer possible. The moral philosopher Nagel (1979) argued that organismic survival is not the defining quality of life; it is good to be alive because of the "goods" involved in life. He identified these goods to be such things as the ability to perceive, think, desire, and act. Death, considered in this light, involves the frustration of projects, an inability to exercise intentions or to pursue aspirations, decisions, and human relationships. These goods are similar to those discussed by Regan (1983), who proposed a subject-of-a-life criterion, whereby life is characterized as having beliefs and desires; perception, memory, and a sense of the future; an emotional life, together with feelings of pleasure and pain; preferences; the ability to initiate action in pursuit of desires and goals; and an individual welfare in the sense that experiential life fares well or ill.

Kleinig (1991), in his book *Valuing Life*, suggested that a better phrasing of the matter would be to consider life in the autobiographical, rather than the merely biographical sense. Humans can be set apart from other animals by the fact that a continuous self-consciousness is involved at a level that is quite different from that for other species. He referred to a "greater mental complexity" that gives people an interest in a continued life and an ability to be the "agents of their own tomorrow."

The importance of the idea of a continued self-conscious life will be discussed in Chapters 6–7 when the moral permissibility of euthanasia is considered.

Dworkin (1993) spoke of life in terms that resemble the autobiographical conception. When considering the question of whether life support should be terminated, he argued that the important concern is to protect the patient's autonomy and best interests. The patient should be allowed to live a life structured by a continuous theme to its end; a person's life has had value because of what that life made it possible for the person to do and feel.

Dworkin emphasized the importance of dying with dignity; that it is important that life ends in a way that the death keeps faith with the way the person has lived. The right to be treated with dignity requires others to acknowledge that the person had moral standing, that it is intrinsically important how the life proceeds, and how it is ended. Dignity is considered a central aspect bestowing value on life and providing the intrinsic importance of human life.

Both Rachels and Nagel invoked the idea of dignity when considering the state of a comatose person. One must look beyond the categorical state of an individual at a specific time and consider the individual as a person identified by history and possibilities in Nagel's (1979) view. He discussed the case of an individual who had been a mature and intelligent member of society, but who suffered severe and irreversible brain damage. The individual is not dead, but lives at the level of a contented infant for whom happiness is a full stomach and a dry diaper. It is reasonable to doubt that the person who *was* can be said to exist any longer, and that a different creature is now in existence. The question is whether an individual being maintained in such a state is being allowed the dignity of personhood by being kept alive. The person that *was* is now dead, and the question is whether the new individual's life should be continued and at what cost for what future benefit. If the argument is accepted that all life is due some degree of respect, then the benefits and costs of preserving this new individual, who has limited future possibilities in terms of a complex life, must be weighed. If there are limited resources available, then one might consider preserving the more complex life of a chimpanzee, a dog, or a human infant in the same categorical state. The test to receive benefits, relative to the interests of others who also need the benefits, should be based on an individual test of the specific organisms involved. Only by using such an individual test of abilities can the traps of speciesism, racism, and sexism be avoided. This issue will be discussed at length in the next chapter, when the allocation of scarce organs for transplant is considered.

VALUING LIVES

A related issue was raised by Kamm (1993), in her book *Morality, Mortality: Volume I.* She discussed the importance of future regards when considering issues involved in estimating the relative value of lives. She posed two possible descriptions of world states: In the first, a person has had many good years of a life full of achievement and meaningful involvement but will die shortly; in the second, the person has just been born but will have a few years ahead with a moderate amount of goods. The question is: Should one prefer to be in the first or the second state of the world? She argued that it would be morally wrong to prefer the second state of affairs, because it denies the value of experiences or actions once they are over, whether the value is considered to be a product or as subjective experience. A major problem with this view is that it can lead to an elitist position, whereby in all situations one would favor creative and productive people over the lesser endowed, and that could lead to an undesirable capitalization on the natural lottery that dealt us our hand, and could lead to persisting inequalities in society.

Kamm explicitly considered the problem of elitism when discussing organ donation. She wondered whether it would be permissible to sacrifice 5 years at the end of our lives in exchange for completing a work in philosophy, and if so, would that mean one should give an organ to someone who will finish a philosophical work rather than to someone who will live for 5 years, but not complete any particular project? She decided that such structural or hedonistic experiential factors are important, but that they should not be given priority over the continuation of a person's life of a quality that is satisfactory to the individual who leads it.

Temkin (1993), in his book *Inequality*, discussed a related issue of how to view the relative value of lives. An individual's life can be considered from the perspective of a complete life, incrementing to reach a "total score" regarding the goodness of the life. Another way is to consider simultaneous segments, whereby one would divide history into a series of temporal stages and measure goodness (or inequality) in terms of the quality of life within the same temporal stages. Yet another way is to consider corresponding segments, dividing the life into a series of functional developmental stages, such as childhood, early adulthood, middle age, and old age, with goodness (or inequality) measured in terms of inequalities between the comparable stages of people's lives.

The issue of the relative value of lives was also discussed by Kamm (1993). First, she asked whether age is an important factor to take into consideration in order to determine relative need. If a 20-year-old will die in a year if not given an organ for transplant, is that person needier than a 50-year-old who will die tomorrow without the organ? After all, the 50-year-old has had 50 years of adequate conscious life, while the 20-year-old will only be 21 at death. She suggested that such questions can be approached by invoking the possibility that the same absolute number of years of adequate conscious life may be more valuable if they produce more good for some individuals than others. She argued that the period from 20 to 40 is structurally a more significant period than that from 50 to 70, and that 70 might be a reasonable cutoff point beyond which costly life saving would be denied public support. From an evolutionary perspective, some priority might be given to those of reproductive age, because the parenting efforts might make these individuals more valuable in terms of contributing to the social community than usually would be the case with the elderly.

Her conclusion was that the younger should be favored over the older, at least to some degree. She does not believe this view represents age discrimination, because everyone will be both young and old, and be assured of preferential treatment when younger at the risk of a less-favored position when older. She argued that this preference is quite different from one based on sex or race, because one does not sacrifice a totally different person in favor of oneself using this age scheme, but does with sex or race discrimination, a point also argued by Kilner (1990). Kamm noted that ordinarily we think it is worse to die when young rather than when older; that 10 years given to a 10-year-old may "swamp" 20 years given to a 40-year-old, and that one should assign a diminishing moral value to life with age being reckoned in 10- or 20-year intervals.

It is possible (and preferable) to combine the perspectives considered by Temkin with Kamm's suggestion that the relative value of life is different at different stages of life, and then to develop an evolutionarily meaningful view emphasizing both the ultimate and proximate factors involved in life. Temkin's idea of a life-span view is appealing, because it emphasizes the total achievement of a person throughout the entire course of the life. From the ultimate evolutionary perspective, that achievement would be reckoned in terms of the lifetime reproductive success of the individual (including both the direct and indirect components on which inclusive fitness is based). Our studies of people's moral intuitions (Petrinovich et al., 1993) indicated that such a view of morality is reasonable: Policies that honor the reproductive value of individuals were emphasized by most people.

Proximate factors should not be dependent solely on chronological age, as Temkin and Kamm use it when they suggest the possibility of comparing people within different age categories. A chronological series should be anchored in terms of biologically (evolutionary) meaningful segments related to the reproductive value of the individuals involved. The relative value of an individual could be weighted by an individual's reproductive potential. The weights should be at the level of individual rather than based on some holistic societal scheme, in the same way as all evolutionary processes should be viewed. Temkin defended the use of individualistic approaches in his discussion of inequalities, arguing that the complaints of those less well off should be expressed relative to those individuals who are better off, rather than using the average levels for the society.

Kamm argued to deemphasize formal and experiential goods when deciding who is to live in cases where some preference must be given to certain individuals over others. In her scheme, a differential positive weight would be given to different age classes in descending order. More biologically meaningful stages could be used, rather than mere chronological age, such as the following series:

1. Prenatal
2. Early development (when basic emotional and perceptual patterns are established—perhaps birth to 4 years of age)
3. Early development (age 4 to the onset of the age at which reproduction is to occur—perhaps 18 in the U.S. culture—hopefully not much sooner)
4. Reproductive period (18–40 years)
5. Mature, parenting period (40–60 years)
6. Nonreproductive period (60+)

Each of these stages could be assigned an arbitrary weight and used to allocate limited resources and frame issues pertaining to inequalities that exist in the allocation of societal resources. In Chapter 8, such a schema will be discussed in more detail when considering problems involved in allocating health-care resources.

Another factor related to the concept of inclusive fitness was introduced by Kamm (1993) when she discussed whether one should favor donating an organ to a father on whom five children depend rather than to a single, unrelated person. She entertained the objections to providing the organ to the father, and noted that one effect of the father's death would be to contribute to the misery (presumably both physical and emotional) of the five children. She concluded that the father should be seen as an end in himself and not as the means to contribute to his children. This conclusion can be argued to be mistaken if one considers the ultimate end, from an evolutionary perspective, to be the father's genetic replication. Saving the father contributes to that justified end and can be considered separately from what could be permanent damage to the children. It could be argued that the significant loss to the genetic replication process would require that at least two children be involved if direct fitness is concerned, given that each child has only one-half of the father's genetic complement. If we consider indirect fitness in terms of a child's potential genetic contribution to succeeding generations, then saving one child might justify a decision to sacrifice the father, because the child could contribute offspring and grandoffspring. This argument avoids favoring the father because of a "family lifestyle"—a position that some in society believe should be given weight.

DEATH AND THE SOCIAL CONTRACT

The aforementioned considerations raise issues regarding the impact of the death of an individual on surviving members of the social community. In *Human Evolution, Reproduction, and Morality*, it was argued that a social contract is struck when a neonate is born, and that it thereby acquires personhood. Similarly, a social contract is terminated when personhood ceases, be it through death or an irreversible loss of consciousness. From an evolutionary perspective, community relationships are an essential part of the fabric that holds society together. The dead cannot be wronged, only their survivors can. Death is not always a harm to the one who dies, because the continuation of a hopeless, painful existence that does not meet the minimal standards of a satisfactory life is difficult to justify. Feinberg (1984), and Steinbock (1992) both argued that individuals lacking these minimal standards have no further "ulterior interests" in being alive, and the cessation of the agony or boredom that existence has become could make death a blessing. Under these circumstances, it is difficult to consider death a tragedy. The sad and tragic qualities exist only for the loved ones and survivors, who mourn the death of the person who was. The only interests are those of the surviving persons, who remember the deceased and want to symbolize and memorialize that person's previous existence.

The dead have no interests and no moral standing, but they do have moral value similar to any insentient thing. The human cadaver itself is insignificant in its material form. Cadavers have symbolic significance for many people, as do flags, trees, and statues, and it is this symbolic significance that gives them a moral value that should be respected. Those of us who do not ascribe to a theory of human immortality still have reasons to respect the dead, even though they cannot be harmed. As

Callahan (1987) pointed out, we can do things that are wrong to the dead because of interests the dead had as living persons, which involve agreements with and expectations on the part of living people. We often feel an obligation to honor the will of the dead because of values that we, the survivors, hold regarding the value of objects, as well as the welfare rights of heirs. The obligations we have, however, are to the heirs and not to the dead. We often feel an obligation to protect the reputation of the dead from unfair disparagement, but that too is because of such things as the intellectual tradition that the dead individual represents, and the importance for those continuing that tradition.

The medical ethicist Emanuel (1991) noted that existence is ephemeral, and one way to overcome the limitations people experience in terms of their achievements and the finite time they have to remain alive is through union with other people within a community continuing the traditions and ideals after death. In this way a person is able to transcend mortality, contributing something that endures beyond the terms of one's life, and even those of the person's children.

The practices and procedures surrounding the death of an individual serve the interests of those who survive, and many are concerned that their own survivors should respect their wills and remains when dead. Feinberg (1984, p. 94) wrote, "We behave in certain appropriate ways toward the dead then because it is our duty to do so, and that duty is imposed by the rules that define certain practices that are highly useful to living people." When engaging in behaviors such as making wills and entering into insurance contracts, the interests of society are preserved, and the terms of social contracts should be honored. It is important that duties to respect the dead be honored in order to preserve the stability of the social community and to avoid the collective loss of respect for social obligations that could result if people felt unsure about the security of their bequeaths to their heirs.

The importance of considering personhood was argued by Kamm (1993), who noted that the loss of the goods of life must be the loss by some subject (person), and that total nonexistence would imply that there has been no person. She used this idea to argue that there is an asymmetry between prenatal nonexistence and the death of a person. In the former case, there is never a person who possessed goods, and prenatal nonexistence does not deprive anyone of future goods. Death, however, involves bad things happening to an existing person, such as destruction of the life and deprivation of future goods. She noted that, in the case of unavoidable terminal illness, death may best occur at the point where a disease would put an end to the goods of life, because death does not prevent future goods (none can be produced); it ends a good life, and prevents oncoming misery. If the undesirable aspect of death is that it prevents the attainment of more good, then existence in an irreversible coma should not be prolonged, given that no further good can be expected for the person.

Kamm proposed an analogy between the loss of an existing painting as compared to the deprivation of a painting that might have existed. It is worse if an existing painting is destroyed (even by natural causes) than if the world is deprived of a painting because someone does not paint it. The former deprives the community of existing goods, while the latter only deprives the community of a potential object, and we can conceive of an almost infinite number of potential objects.

The destruction of an existing painting can be considered to defeat what Temkin (1993) called the "subject desire fulfillment" of the artist. Death is bad when it adversely affects the intentions and reputation of an existing person. By destroying a painting, one is defeating the desires and intentions of the artist, and that is undesirable. Such destruction after the artist is dead indirectly defeats the desires of the now-dead artist, even though that artist no longer has conscious states. This analogy is reasonable when the destruction is considered in the light of the preferences and interests of the artist's survivors; it is they who suffer from the destruction of a painting or from slanderous statements regarding the no-longer-living artist. These survivors need not be kin, but can be the intellectual descendents of a scholar or artist, who can suffer real damage, and have their mental states adversely affected by attacks on the reputation of the dead person. Defamation of a dead person can, therefore, cause harm to the community of survivors and should be considered to be morally impermissible unless the new evaluation is based on reasonable and public arguments of relevant facts and interpretation of the significance of the past events.

Rachels (1986, p. 43) wrote,

> The "state" you will be in after you die is exactly comparable to the "state" you were in before you were born, and it is "unimaginable" to you for exactly the same reason, namely, that there is nothing to imagine.

A cadaver should not be pitied any more than should an article of furniture. The person who *was* can be an object of pity or reverence, but the considerations concern the surviving community members. The significance attached to the cadaver is based on its history and its significance to survivors.

Viewed this way, the cadaver has no rights because it cannot feel pain, has no desires or interests, and cannot be wronged. The cadaver is important only because of its history as structured by survivors, and the respect accorded to cadavers is based on the wishes of those survivors. Several motivations seem to drive the respect that is expressed in elaborate rituals, ceremonies, and modes for the disposal of human remains. The motivations underlying elaborate procedures that people engage in reflect the fear of death that many have. This fear often seems to be fueled by disappointment regarding what the survivors have accomplished in life (as compared to ideal aspirations), by the hope that what is now a miserable existence will be compensated for by a pleasant life in a hereafter (or upon resurrection), by a belief that any suffering and sacrifice in this life will be rewarded in paradise by a beneficent creator—or it could be motivated by just a simple fear of the unknown. Albert Schweitzer (1962, p. 183) exquisitely expressed the positive motivations that characterize the feelings of many people when he construed the mystery of life in the following way: "I only know that I cling to it. I fear its cessation—death. I dread its diminution—pain. I seek its enlargement—joy." This sentiment was echoed by Nagel (1979, p. 11), who wrote, "But if death is an evil, it is the *loss of life*, rather than the state of being dead, or non-existent, or unconscious, that is objectionable."

An important motivation that drives the rituals surrounding death might be a desire on the part of survivors that their own physical remains, the residuals of their own person, be accorded dignity, and they extend that respect to the remains of others to whom they bear a relationship. The harsh treatment of dead human bodies meets disapproval because these bodies are natural symbols of humanity. Neomorts have a distinctly human form, and this humanness invokes an emotional empathy and identification on the part of observers because they identify with this form. The rituals and formalities surrounding death provide survivors a way to reflect together on past associations with the deceased, and provide a forum within which to express grief and love. Kamm (1993) noted that it might be this personalization that is of paramount importance; that there would be stronger resistance to organ transplantation if it were possible to perform facial transplants—which would be instantaneously identified with the donor—than there is to the invisible transplant of internal organs. This distinction might run even deeper, extending to natural versus nonnatural parts. It is likely that there would be fewer concerns about giving a donor's wooden leg than giving the donor's real leg.

Kamm argued that it is not possible to harm the no-longer-existing dead person who has no future interests or liberty–needs. Although I can damage the physical remains, they would have rotted anyway. If someone has an old and unused toothbrush, most would consider it permissible for me to use it to save a life, even if the owner refused to donate it.

The most likely penalty I might suffer would be to have to replace it with a new toothbrush; it could have been kept as a backup should the good one be lost. I would be considered by many, however, to have acted rightly by appropriating the toothbrush to save a life. If a dead person's used organs are not taken, they will rot in the ground, so why should it not be permissible to take those organs and offer the family compensation, using a justification similar to that of the eminent domain of government? The answer, of course, is related to the sentiments and fears of survivors. Kamm makes the sensible suggestion that some of these attitudes on the part of survivors might be changed if the donation of organs after one's death is framed to be a duty to the living members of society who need them. If the family wishes to countermand the donor's wishes, or does not want to permit the donation, then action on their part should be required to override the obligation to society. At least this would make it more permissible for medical personnel to recover organs when the donor has so specified, as well as to proceed when no instructions have been left, because proceeding without consent makes it possible for the patient to perform a duty to society upon death.

Many of the rituals and practices used to mourn the dead often serve important functions for the survivors. Stroebe, Gergen, Gergen, and Stroebe (1992) explored the nature of the psychological theories and practices involved in bereavement. They argued that one function of these practices is to break emotional bonds so that survivors can recover from their state of intense emotion and return to normal functioning as quickly and efficiently as possible. A useful way to expedite the process of recovery is to break the bonds with the deceased and relinquish emotional ties.

They discussed some major cultural differences in the way these bonds are broken in different cultures. In some, such as Shinto and Buddhist cultures, it is believed that contact should be maintained with the deceased—they have joined the ranks of the ancestors and will, in turn, be joined by the survivors. On the other hand, among the Hopi of Arizona, the deceased are to be forgotten as quickly as possible, because contact with death brings pollution, and the dead person's spirit is a depersonalized entity with which contact must be avoided. Thus, in some cultures people honor and memorialize the dead, whom they believe they will join in the hereafter, while in others, all ties are relinquished and the dead are forgotten as soon as possible. Both practices result in a normal adjustment within the cultural context, and this adjustment permits the resumption of normal activities by the survivors.

It is not the state of death that is important, but the loss of life as experienced and expressed by the society that still exists. We regard the death of Keats, at age 24, to be a great tragedy, because he would have been expected to enjoy much more productivity had he lived. The sense of tragedy on Keats's behalf is coupled with regret for the loss of his probable contributions to our literary heritage and enjoyment. We regard the death of Tolstoy, at the age of 82, as being less tragic. He lived a long and full life and had the opportunity to make great and cherished contributions that are considered to be part of our literary heritage. Both Keats and Tolstoy, however, are equally dead and were at the time of their death. The difference in our feelings seems to be occasioned by our interpretations of the progress of their lives, and our sense of whether they realized their potential (setting aside the issue of their relative contributions to our cultural heritage).

Among the strongest motivations underlying the rituals used to express respect for the dead are those driven by the symbolic significance of the previously existing person to the survivors. Feinberg (1985) argued that it is important to respect such symbolic value, but cautioned that this respect should not lead us to succumb to the moral traps of sentimentality and squeamishness. The latter two emotions can diminish the values that the symbols are intended to epitomize.

Many arguments to ban autopsies, forbid research on cadavers, or deny the use of cadavers to obtain organs for transplantation are based on sentimentality. It would seem, as Feinberg (1985) argued, that the interests of those who can be helped should have greater moral weight and take precedence over "appeals to offended sentiment" whenever there is conflict between the two. People should be educated regarding the value and good that lifesaving technologies can provide in order to place symbolic meaning in a perspective that considers the welfare of the living as the most important factor.

There should be few serious concerns regarding the moral status of cadavers on theological grounds, as I understand them. Dowie (1988) summarized the positions of several scholars regarding the use of cadavers for organ recovery and transplantation. Islamic scholars stated that the donation and removal of organs is not precluded as long as the recipient assures the donor the body will not be cremated with the donor's organ in it, because resurrection involves the entire body. The Buddhist view is that donating organs so that other persons may live is a noble act, emphasizing the oneness of mankind and the universe. Similarly, many rabbis regard organ donation to be a *mitzvah*—a good deed.

As far as Western Christianity is concerned, if the point of ensoulment is taken to be when fertilization takes place, and death is considered to be the point when the soul leaves the body, then I would presume there would be little or no absolute objection to the use of organs from cadavers for transplantation. Perhaps the only theological objections would be offered by those who believe that the soul reinhabits the physical body at the Resurrection, and that the body should be preserved in the interests of that moment. Another theologically based objection is that one should not tamper with the workings of the natural universe, because that is the work of God and not the domain for human endeavor. The implications of this position are difficult to accept, because they lead to the position that no interventions should be made in the works of nature, and would question the use of medical treatment at any stage of natural life. It is questionable to use these theological positions as guidelines to set public policy that must apply to all members of society, given that the theological views often represent those of a minority of the population, and implementation of such policies would significantly diminish the liberty–interests of the majority.

Slippery Slope

The "slippery-slope" argument has been used to oppose taking cadaver organs for transplantation; if the donation of cadaver organs is permitted, this could lead to the creation of a brisk commercial enterprise to terminate human lives to obtain organs in good condition that can be sold for transplantation. Such arguments were used to prohibit the dissection of cadavers to train doctors not so many decades ago. Feinberg (1985) traced some of the arguments made in response to a bill introduced in 1828 to permit the use of corpses for scientific purposes (providing the death occurred in a poorhouse, hospital, or charitable institution that was maintained at public expense) as long as the body was not claimed within a specified time by next of kin. Although the bill was passed, it was denounced as being unfair to poor people. The slippery-slope argument was that this bill would lead the aged poor to avoid hospitals where their organs could be used, and as a result, they would die unattended in the streets—an argument that seems rather farfetched and to lack any empirical support.

It also has been argued that the routine salvaging of organs from the brain dead will lead to a weakening of the human sentiments that lead to respect for dead bodies. This weakening would cause the loss of these noble sentiments and represent a degradation of the human character. Therefore, these procedures should be prohibited because of the unacceptable decline in morality that might occur. When considering the reasonableness of such arguments, the value of preventing death and suffering of living people should be balanced against the preference for symbolic sentimentalism and conjectured harmful possibilities.

The logic involved in slippery-slope arguments was questioned at length in Chapter 2 of *Human Evolution, Reproduction, and Morality*. It was concluded that there are safeguards society can implement to level the playing slope so that one is not compelled to glide inevitably to the depths. As with all slippery-slope arguments, they have little intrinsic merit without detailed specification of the causal steps involved and an explicit justification of the presumed inevitability of the presumed decline. The evidence does not support concerns that, as the slopers fear, doctors who perform abortions tend to be cruel to their own children, that transplant surgeons are brutal in their leisure time, that biomedical researchers who use live animals for research mistreat their pets, or that people who hunt for sport are brutal to their own pet animals. Lacking such evidence, the slippery-slope argument loses in force, and the burden of proof, both in terms of empirical evidence and causal necessity, must be shifted to the slope arguers.

CHAPTER 5

Organ Transplants

THE PROBLEM

Fox and Swazey (1992) provided estimates regarding the number of organ transplants being done worldwide: There were more than 6,000 heart transplants done by 1988 (80% taking place between 1984 and 1988); in 1989, there were transplants of 1,673 hearts, 8,886 kidneys, 2,160 livers, 412 pancreases, 89 lungs, and 70 heart–lungs. There were more than 2,000 multiple-organ transplants in the 1980s, with the most common being pancreas–kidney–duodenum combinations, and heat–lung combinations.

It was estimated that more than 400 thousand organ transplants were performed in the United States in the 1980s (Caplan, 1986a). In 1993, there were 2,299 heart transplants in the United States and 6,269 patients on the waiting list for a heart (Saltus, 1994). A representative of the New England Organ Bank stated that nearly 40% of families refuse to donate a relative's organs, often after the relative had stated a wish to donate. The American Heart Association estimated that 15 thousand people age 55 or younger could benefit from heart transplants, with a total of 40 thousand if those to age 65 are included.

There were over 31 thousand people currently waiting for organ transplants at the time they wrote, yet only 14 thousand organs from approximately 4,500 donors had been available each year (Shafer et al., 1994). The number of cases in which organs were denied by medical examiner or coroners had risen from 219 (7.2%) to 363 (11.4%); 33% of those waiting for a heart, and 29% waiting for a liver will die before an organ becomes available. This can be expressed as a person dying every 4 hours while waiting for an organ. Shafer et al. examined the reasons for a denial of organs by medical examiners and found that the leading reason was to allow the state to investigate a crime, although an extensive review of case law found no case in which a state was unable to adequately investigate a crime or prosecute a criminal defendant because evidence had been impaired by organ donation. In 1992, 29.3% of all denied cases were probable child-abuse cases, a denial that is unfortunate because pediatric organs are in critically short supply and are essential for transplantations in young recipients for whom the small size of the organ is crucial. Prosecution of these cases is routinely done with evidence that is gathered through external physical examination and other laboratory tests, making the bases for the denial of the organ problematic.

The number of patients on waiting lists for organs has expanded steadily, while the number of cadaver organ donors has plateaued at about 4,500 persons a year (Shafer et al., 1994). At the end of 1991, over 1,500 people were on waiting lists seeking livers, and over 200 were seeking hearts (Ubel, Arnold, & Caplan, 1993). Ten percent of those awaiting livers and 17% awaiting hearts were removed from waiting lists, because they died before organs became available.

The United Network for Organ Sharing, an organization that distributes donor organs to hospitals nationwide, estimated that more than 2,800 people are now on its waiting list for hearts, and roughly 40% of them will die because no organ will be available (Caplan, 1994). An even more recent estimate (Young, 1994) was that 10–12 thousand people die each year in a manner that may make them useful as organ donors, but organs are collected from only about one-third of them. Young reported that 35 thousand people were on the national waiting list for organ transplants, an increase of 14% over the preceding year. The number of potential heart recipients has been estimated to be as high as 32–75 thousand per year, but many are not even placed on waiting lists for organs (Kilner, 1990).

An estimated 5,500 kidneys were transplanted in the United States in 1984 and waiting lists at dialysis centers included 10 thousand persons actively seeking a transplant, with many potential recipients not even placed on waiting lists (Caplan, 1986a). There is a considerable supply-and-demand problem, and Caplan was concerned that a "green screen" is being used to select who will or will not receive a transplant based on ability to pay.

Transplants are expensive medical procedures. There have been numerous stories in the media regarding patients who cannot meet the costs for a transplant—such as a $100,000 down payment for a liver transplant—leading them to make appeals to politicians and the public for funding and organs. It was estimated that the median cost of a heart–lung transplant is $240,000 for initial care, and approximately $47,000 a year for follow-up medication and care (Fox & Swazey, 1992). It costs

Medicare approximately $32,000 a year for each dialysis patient, compared to an average of $56,000 for the first year of a kidney transplant and $6,000 a year thereafter. About 150 thousand patients are receiving benefits for kidney transplants from Medicare, at a total treatment cost of about $4 billion a year. This means that the federal, and some state governments, have accepted a commitment to pay for organ-transplant procedures.

The overall 1-year success rate in the United States, defined as the survival and healthy functioning of the transplanted organ and the patient, is about 85% for kidney, heart, liver, and lung transplants (Fox & Swazey, 1992). The 5-year survival rate for heart and kidney transplants is estimated to be 50%, a rate comparable to, or better than, that achieved with many other surgical and medical treatments (Caplan, 1986a). Since 1986, survival after heart–lung transplants was 70% for 1 year, 66% for 2 years, and 60% for 3 years (Theodore & Lewiston, 1990).

MORAL CONCERNS REGARDING DONORS

Citizens are encouraged to indicate their willingness to serve as organ donors upon death. In many states, people can register as an organ donor by checking a box on the driver's license application and placing a donor sticker on the license—easier than paying taxes or registering to vote. If there has been no prior indication, then organ donation is permitted if free and informed consent is given by the surviving family after the donor's brain activity has ceased. Mahowald, Silver, and Ratcheson (1987, p. 264) argued that these donations are encouraged because of the therapeutic benefits for the recipients and the social value of diminishing the grief of the family by "honoring and 'extending the life' of the loved one through the 'gift of life' to others."

Medical staff, almost without exception, do not take an organ from a neomort who had signed a donor agreement unless the kin have provided written consent. Reticence to act in accordance with the wishes of the donor avoids even the appearance of routinely "harvesting" organs, honoring the sanctity of the family and medical tradition.

Kamm (1993) devoted five chapters of her book to a discussion of organ transplants, treating the issue as a case history to consider the problems that occur whenever scarce resources are to be acquired and distributed. She noted that an original donor's decision to give an organ can be overridden after the death by the family's decision not to give, a practice that is contrary to legal proceedings regarding the distribution of the decedent's other goods. This legalism is a reflection of the extreme emotionality attached to the occurrence of death. If the decedent has left no instructions, then the family is free to decide on the disposition of the remains, but the family cannot override the decedent's decision *not* to donate organs. Medical staff is reluctant to honor an individual's decision to donate organs without agreement by kin.

This practice of following the wishes of relatives has been justified on the grounds that the family is comforted by giving a dead relative's organs if there had been a decision to do so by the donor, or if the donor had made no decision. Following the family's wishes not to donate is done to respect their feelings, with the concern for the family prevailing over both the need for the organs and the wishes of the donor.

In his book *Wonderwoman and Superman*, Harris (1992) argued that the consent of neither the deceased nor the surviving relatives should be required, because the clear benefits from cadaver transplants are so great—these benefits should override "selfish and superstitious" objections of the survivors. It is reasonable to decide that organs should be used if there is no indication that they should not be; the responsibility should be for an individual to "opt out" rather than placing the burden on donors to "opt in"—a procedure followed in France, Israel, Greece, Norway, Italy, Switzerland, Finland, Sweden, and Denmark.

Perhaps everyone should be required to answer, while they are in a healthy and competent condition, the question of whether they prefer to donate organs. Dowie (1988) suggested that—as in other countries, such as Poland, Czechoslovakia, and Austria—organs should be considered the property of the state upon death. An interest in enhancing the health of living persons can be argued to take precedence over sensibilities regarding the memory of a deceased person.

CRITERIA FOR RECIPIENTS

Because organ transplants are expensive, and there is more demand for organs than there is supply—resulting in a true scarcity—several questions arise. One regards the criteria to qualify for a transplant, a second concerns who actually gets one, and a third is what safeguards are needed to prevent inequities. The criteria for eligibility to receive a lung transplant include the following: to have a lung disease; be sick enough to need the operation; well enough to survive the wait for an organ; and willing to deal with the complexities of postoperative care (Theodore & Lewiston, 1990).

Different groups have used various criteria to allocate organs: urgency of the transplant (both in terms of the time that will elapse before the patient would be expected to die and how poor the quality of life would be without the transplant); need (how long the patient already has lived); outcome (the increment in years of life expected with an organ over those without it); waiting time (how long someone has been waiting for an organ). The principles appropriate to direct resources for transplantation are discussed in more detail when health-care rationing and general considerations involving health care policy are discussed in Chapters 8 and 9.

The question of who receives organ transplants at present has been discussed by Fox and Swazey (1992), who cited evidence that women, minorities, and low-income patients do not receive transplants at the same rate as white men with high incomes. They suggested that this imbalance could be due, primarily, to unequal admission to waiting lists for organs. A number of papers that have appeared in professional journals paint a similar picture: Women, disabled persons, the retarded, and minorities are underrepresented among those receiving transplants (Caplan, 1992). Some physicians do not raise transplant as an option if the patient does not have insurance or cannot pay the costs directly.

A bizarre case, described by Caplan (1994), is that of a 33-year-old inmate of a federal prison, who is serving a 4-year term as a convicted drug dealer. The inmate is dying of heart failure but has a good chance to live a long life if he receives a transplant. The prison officials, however, will not allow him to be evaluated by a transplant team to determine if he is medically eligible to go on a waiting list for a heart transplant. Because it is probable that he will die before his 4-year sentence is served, the officials are, in effect, sentencing him to death—which the jury of his peers did not do. An interesting aspect of this case is that the inmate, the father of two, has health insurance that would pay for the transplant. The solution to problems such as this is not to kill the patient, but to make sure that every American, in or out of prison, has equal access to medical treatment, based on how likely it is he or she will benefit from the treatment.

Some philosophical principles should be kept in mind when problems regarding the equitable distribution of scarce resources are considered. The point of reference should be the welfare of individuals rather than the aggregate good to society as a whole. This focus leads to an emphasis on the good as it is related to the interests of the patient, not as related to those of society, when the value of the additional time alive is considered. The welfare of the worst-off in society should be the primary concern. This argument is based on Rawls's maximin principle: The outcome for the worst-off should be made as good as possible. Both Temkin and Kamm argued that outcomes should be considered in terms of individual rather than aggregated values. Temkin adds that the maximin principle should not be concerned with the worst-off group, but should be a special concern for the worst-off individuals in the world.

Temkin (1993) added that undeserved inequality should not be allowed to persist, because it is always morally objectionable. He favors a position based on egalitarianism—a view that attaches value to equality itself and considers equality to be a concern over and above the extent that it promotes other ideals. Equality is a means to help the worse-off, and it is permissible to have identical gains for the worse-off even if there is equal, or even greater, gains to the better-off.

KILNER'S CRITERIA FOR PATIENT SELECTION

Kilner (1990) discussed the criteria used to allocate organs and proposed criteria he defended as the most equitable. A basic premise was that health care throughout the world is, and always has been, rationed because there is a lack of adequate numbers of physicians in many locales, many people lack the ability to pay for medical care, and increasingly scarce and expensive treatment methods have been developed.

Given the costs of organ transplantation and the scarcity of available organs, it cannot be assumed that everyone who needs a transplant will receive one. It is reasonable to assume that selection will be required, and that objective criteria should be developed and used consistently. Shortages are so severe that the criteria must take into account more than judgments regarding medical justification. Kilner organized the available criteria into four main realms: social, socioeconomic, medical, and personal. The criteria will be considered in detail because they represent the first step toward developing objective methods when it is necessary to ration health care.

The social criteria are the following four:

1. *Social Value.* This involves the impact that selection decisions would have on society at large, including income, net worth, education, community service, and occupation, and probably underlies the observed preference to treat men rather than women, and to favor one race over another. The criterion is indefensible morally, and although it is used widely, it is seldom defended explicitly. There is a consensus among physicians and ethicists that it should not be used at all in patient selection. It does not consider a person as an entity with intrinsic worth, but as a means to improve the well-being of society, and it can be argued that this is an inadmissible application of utilitarianism leading to unjust discrimination based on group membership.

2. *Favored Group.* This considers such things as the geographical location of one's residence and the social categories to which one can be assigned, such as being a child or a veteran. Although the criterion is not widely supported, it is widely used. There is a tendency for organs to be distributed preferentially to members of the community of the donor, to pick recipients who live close to major transplant hospitals (partly for the pragmatic reason that organs are highly perishable), and to favor U.S. citizens. If disorders are service-related, then veterans receive priority, and some argue that non-service-related disorders should receive priority in order that people might be more willing to serve in the armed forces. This argument has been countered: Veteran preferences produce biases, because service in the armed forces traditionally has not been open to all social groups, especially women.

3. *Resources Required.* This criterion favors those who require relatively little of a limited resource over those who need more of it. It is considered to be the least objectionable of the social criteria, probably because it involves an understandable utilitarian calculation that would result in benefit to a greater number. An argument made in opposition is that physicians have the duty to save all lives with the resources at their disposal, and they should not dictate which patients will be allowed to live when all should be saved—a position that is fine as long as there are adequate resources. If resources are limited, then not making a decision could be little more than a refusal to practice responsible medicine.

Ubel et al. (1993) addressed the resources issue when they considered problems involved in retransplantation of scarce organs after an initial transplantation has failed. Retransplant recipients do not do as well as primary transplant recipients, but 10–20% of available hearts and livers are used for retransplants. Retransplant patients sometimes are given priority because transplant teams claim a special obligation not to abandon patients they have already treated, although this raises the question of the fairness of allowing a few individuals to get multiple transplants while some die awaiting their first. The higher survival rates of primary transplant patients suggests that more organs should be directed to primary transplant candidates, and that those needing a third or fourth transplant might even be removed from waiting lists.

4. *Special Responsibilities.* Some people have responsibilities for dependents—especially for dependent children—and some provide highly valuable services needed by society. Family responsibility has been used extensively to determine who receives organs and it is generally agreed that, in emergency situations, those who have medical training should be treated first in order that they might treat others. It has been suggested that patients who will donate money to produce more re-

sources that, in turn, would be made available to others in society should be favored, because treating them could produce resources adequate to save additional lives. This suggestion is reasonable as long as the test is done individually, the persons given priority are truly indispensable, those selected actually perform their critical role, and the cost to those harmed in the process outweighs the benefits.

The socioeconomic realm contains the following three criteria:

1. *Age.* This is used to exclude the oldest patients from treatment and is widely used in the United States and the United Kingdom. At least the criterion is not racist or sexist, because everyone alive will age. Without an individual test, however, age clearly is a discriminatory criterion that is uniquely applied to medical decisions: Murder is considered to be a crime of the same severity with a 65- or a 25-year-old victim. The use of age criteria will be discussed in Chapter 9 when considering questions of medical ethics.

2. *Psychological Ability.* This concerns whether the patient has the intellectual and emotional capacity to cope with treatment procedures and to cooperate with the requirements for posttreatment. Kilner suggested that a psychological criterion should be used as merely one of several factors contributing to the patient's ability to satisfy a medical-benefit criterion.

3. *Supportive Environment.* This is widely used, and it is often deemed essential that the patient have a supportive environment during and following treatment. Marital status has been used to select patients for heart transplants, but Kilner warned that this criterion may be a cover for an ability-to-pay test—the poor may always lack the means to guarantee certain kinds of supportive care and environments deemed necessary.

Kilner suggested two major criteria in the medical realm:

1. *Medical Benefit.* This concerns the likelihood that a patient will receive significant benefit as a result of treatment. This is one of the most important criteria, and it is widely advocated and used by those in the medical profession. Physicians argue they have to do everything they can for each of their patients, and individual physicians should not have to make case-by-case decisions that affect the use of resources for patients under their care. It has been suggested that there be a medical board with several physician members in order to protect all patients and establish whatever objective priorities are necessary. A major problem with this suggestion is that there are few adequate studies of the the outcomes of medical treatments, which makes it difficult for physicians to do much more than subjectively reflect on their own personal experiences when making decisions regarding probable medical benefits.

There are three subcriteria to be considered in order to estimate medical benefit. First is the likelihood that a projected benefit will be realized (the data to determine this usually are not available); second is the length of time the patient can be expected to enjoy the benefit—which is determined by such things as age, other complicating disease factors, and the patient's overall physical condition; third is the quality of the life that the benefit will make possible, a question to be considered in Chapter 9 when developing health-care rationing systems.

2. *Imminent Death.* This involves the length of time the benefit will exist and age, to some extent, but it is mentioned separately because it has been recognized formally in the law, and to be an objective condition that a physician can assess with reasonable accuracy. Death is considered imminent when it is expected within several days or weeks according to competent medical judgment. The state of imminent death is used to jump patients to the top of priority rankings to receive organs or to use scarce ICU facilities. The danger is that some physicians might move patients into the category more quickly than do others, which gives their patients an unfair advantage. A problem with the imminent death criterion is that patients wait until they have deteriorated before being given treatment, which might defeat several other criteria and make it necessary to withdraw treatment from patients who are presently receiving it, but are in less imminent danger than those replacing them.

Finally, Kilner identified three personal criteria:

1. *Willingness.* This refers to the patient's freedom to decide whether to accept treatment, and it is commonly recognized throughout Western societies as a prerequisite for treatment to occur. Its use is justified because it respects the dignity of persons and gives them a freedom that should be respected for every human agent in a pluralistic society. The importance of dignity and respect will be discussed in Chapters 6 and 7, when euthanasia is considered. The major problem involved with this criterion is to obtain an informed decision so that the patient can be considered to have made a rational and voluntary choice. This problem also will be discussed in Chapters 6 and 7.

2. *Ability to Pay.* This refers to the inescapable fact that those who have insufficient funds, or who lack health insurance, have always been unlikely to receive scarce and expensive lifesaving treatments. The extent of this problem will be spelled out in Chapters 9–13. It is sufficient, here, to assert that this state of affairs should not be allowed in a just society, especially when many of those unable to receive treatment have paid the taxes that provided public funds to develop many of the medical procedures they are being denied.

3. *Random Selection.* When other selection criteria are unacceptable or indecisive, Kilner considered random selection to be a possible alternative. There are two types: a traditional lottery, and a first-come, first-serve approach. The latter approach is what has been used widely, with a true lottery used less often. The problem with both methods is that those who have access to better health care are likely to be referred earlier, and it has been established that the underprivileged are not likely to be placed on waiting lists, or even referred to specialists, thereby tainting the selection of patients by influencing who is deemed medically suitable to enter the selection pool. If ways can be found to objectively classify individuals into comparable selection pools, then a lottery might be a good way to select those who will receive scarce treatment resources.

Kilner questioned the use of a lottery on the grounds that it is a visible process that serves to advertise the fact that society is not willing to provide the resources necessary to protect the lives of all of its citizens. But the fact that it is visible might be beneficial, because it alerts the public that the health-care delivery system is inadequate. Such recognition might make it easier to achieve meaningful health-care reform.

There are two basic orientations by which each criterion can be examined, in Kilner's view. One is a productivity orientation, which is a standard utilitarianism emphasizing the most good for the most people. The second is a person orientation, in which the ultimate focus is to enhance personal well-being. Although one can appreciate these quite different orientations, it is not necessary to use one or the other exclusively. It is possible to have a sequential test whereby productivity is used to establish categories of individuals on the basis of the equality of the crucial factors, and then use a lottery to choose among those who have been assigned to each category in order to determine who receives scarce resources.

Kilner (1990, p. 230) considered arguments for and against each of the criteria and made the following recommendations:

> 1. Only patients who satisfy the medical–benefit and willingness-to-accept-treatment criteria are to be considered eligible.
> 2. Available resources are to be given first to eligible patients who satisfy the imminent-death, special-responsibilities, or resources-required criteria.
> 3. If resources are still available, recipients are to be randomly selected, generally by lottery, from among the remaining eligible patients.

He defended these recommendations, noting that two criteria were included from each of the social, medical, and personal realms, that all that were both person and productivity oriented were incorporated, and that those criteria relevant only as supplementary factors (to be used when other criteria, especially medical benefit, have been applied) were identified.

SALE OF ORGANS FOR TRANSPLANT

Because there is a shortage of organs for transplant, it important to have safeguards, especially if people are permitted to sell their organs. Caplan (1985) suggested that clinical transplants should be done only by physicians and institutions willing to submit all research and human-subject protection protocols to public scrutiny, and to require peer review of the scientific basis of research on which clinical applications are based.

Federal law forbids buying and selling body parts, but permits payments for services connected with collection for transplantation (Young, 1994). A large transplant support industry has developed, involving 69 transplant agencies collecting body parts from voluntary donors and forwarding them to 277 transplant hospitals. In 1991, hospitals sold a kidney for an average of $15,683, a heart for $16,050, and a liver for $20,776, with a markup of the cost for the organ reaching as much as 200%. The cost of organs as a percentage of the total transplant charge for three procedures has been estimated by Young: kidney—$50,562, with 31% the cost of the organ; heart—$116,843, 13.7%; liver—$186,934, 11.2%. There is also a good market for other body parts. If a heart is found to be unsuitable for transplantation, then the heart valves are purchased for an amount sufficient to reimburse the sender for acquisition costs, which are about $450 for the average heart (Dowie, 1988). The valves are frozen and sold for $2,995 for each aortic valve, and $2,595 for the pulmonary valve: about $5,500 for a $450 heart.

The irony is that a donor's survivors cannot receive any remuneration, even to cover burial expenses. This policy can create a considerable hardship for cases in which a poor family altruistically donates body parts—a hardship that it is difficult to justify, given the profitable industry involved in procurement and implantation. The National Organ Transplant Act of 1984 (which allows reasonable payment for costs of removal, transportation, implantation, processing, preservation, quality control, and storage of a human organ) has been renewed by the House of Representatives without allowing for donor compensation. A similar version of the bill is awaiting action in the Senate. One issue is who owns the human body, and who should be paid for saving lives. Profits can be made by almost everyone except a compassionate donor and the survivors of that donor.

Opposition to allowing people to sell organs has been based on several reasons. One (not very compelling) argument is that the sale of organs might drive out donations and undermine the altruistic giving that is presently encouraged. This argument might be reasonable if there was evidence that altruistic giving provided anything approaching an adequate number of organs, but the long-term true scarcity indicates otherwise. Others worry that the large economic returns could produce a traffic in organs, with citizens from Third World countries selling organs to be taken from them for transplant, even while they are still living. Fox and Swazey documented one such attempt by a U.S. physician to set up a marketing scheme to buy and sell human kidneys from persons living in the Third World, a plan that was denounced by almost everyone. There is a traffic in human spare parts in some countries, among them India. It was reported in the February 25, 1994 *Boston Globe* that the United Nations Commission on Human Rights found that more people in India sell kidneys to strangers than in any other country. The Indian Parliament passed a law banning the sale of human organs and prohibiting the removal of organs from a living donor unless they are intended for a close relative (a decision that seems to respect the importance of inclusive fitness).

Paternalism should be avoided whenever decisions regarding the permissibility of the sale of organs for transplant are made. It could be that the poor might properly wish to sell an organ. This could be permissible because, Fox and Swazey (1992) argued, those people might be unable to feed, or to purchase necessary medical care, for their families, and sale of an organ might be a better option than death, especially if the organ is one of a pair and the individual can survive with only one, which usually is the case.

Kamm (1993) considered the possibility of the poor selling an organ to feed their children, assuming this would not impair their health. She does not believe such sales should be allowed, expressing the preference that the poor feed their children in some other way, for example, by insisting on improved welfare payments. (How they might accomplish this, given their inevitable lack of power in society, is difficult to imagine.) She suggested that laws against such sales might be justified if they served as an impetus to increase welfare payments, and that passing laws against such sales might be justified if they lead to welfare reform. She also worried that someone who does not want to sell body parts, because of personal or religious convictions, might be induced to do so if offered money, which would lead the persons to sell their personal principles. She concluded that it may be best for society not to offer the opportunity for such weakness of will. This reasoning smacks of paternalism, resting on the unlikely assumption that social benefits will be produced when members of society realize that people are suffering so badly. Such reasoning does not seem to prevail in U.S. society, at least when many millions of people are allowed to suffer because of inadequate medical care and nothing has been done to alleviate the ever-present suffering, as will be discussed in later chapters.

It has been suggested that, although the sale of organs to be taken from a living donor should be prohibited, it might be reasonable to establish a "futures market," whereby a person's organs are purchased while the donor is alive and well, with these organs to be recovered upon death. In this way, the poor and their families could profit to their own direct benefit while alive. Such a market seems more reasonable than only permitting the purchase of organs from the person's estate or next of kin—at least the person would benefit and be enabled to pursue life's goods. Kamm suggested that organ trading might be permissible as insurance, whereby people could give organs while they are healthy in exchange for the assurance that if they or loved ones need an organ in the future, it would be provided—a procedure similar to that in common use for blood donation.

Given that the poor have little chance to be referred to a transplant program, the problem is that the rich have better access to the pool of donated organs. Another problem is that the poor are subsidizing directly the transplant programs whenever public funds are used to finance transplants. They subsidize them indirectly whenever private insurance funds are used, because these costly procedures raise insurance costs to a level where it may not be possible for the poor to afford insurance coverage, with the risk of being frozen out of the recipient pool completely.

Kamm (1993) proposed an instructive scenario regarding the rich and poor. She presented the following case: A very rich person (R) and a very poor person (P) both need an organ or they will die tomorrow. Five other people are about to die, because there is no money to pay for the transplants they require to survive. If an organ that is suitable for either R or P is sold to R for a high price, P will die, but the funds from the sale can be used to finance organ transplants to five other poor people. The transplant that could have been done for P at government expense would be done for R at R's expense, and the funds generated would be used to help the other five people. My guess (based on our empirical research regarding the organization of moral intuitions described in Petrinovich, 1995) is that most would not consider this procedure morally permissible, which suggests that adequate funding and organ availability should be generated for rich and poor alike in a just society—there should not be an ability-to-pay criterion if there is a scarcity of organs.

The emphasis on expensive high-technology organ transplants is a misplaced one, given the overall health needs of society. Fox and Swazey suggested that more attention should be given to ameliorate the impoverished conditions under which people live so they would not feel compelled to resort to the option of selling their organs merely to survive. They also raised a concern about using transplant procedures for the elderly, suggesting that we might better accept the biological limits imposed by the aging process and recognize our ultimate mortality, rather than engage in medical heroics to overcome natural biological limitations. Given the natural degeneration of the organism, perhaps, in the face of the aging population, the adage should be, "If it's broke, don't fix it."

They argued that the intellectual energy and human financial resources used to transplant human organs might better be directed toward decreasing poverty, homelessness, and giving increased attention to universal access to basic health-care and disease-prevention programs. Fox and Swazey (1992, pp. 208–209) summarized their view as follows:

> Allowing ourselves to become too caught up in such problems as the shortage of transplantable organs while health care continues to be defined as a private consumption rather than a social good in American society, with the consequence that millions of people do not have adequate or even minimally decent care, speaks to a values framework and a vision of medical progress that we find medically and morally untenable.

THE DEAD VERSUS THE LIVING

The dead do not have interests and cannot be harmed, because they no longer exist as persons. The welfare and interests of living persons, as well as the positive utility of using organs from aborted fetuses and cadavers for transplantation and for basic medical research, should be accorded more utility and value than the symbolic value these sources might have. *The New York Times* (December 23, 1992) reported that the social ethics organization, Communitarian Network, advocated that the law should make everyone's organs available at death unless the person or a relative objects in advance. They took this position because a large number of people die each year awaiting organ transplants, and thousands more endure the painful and expensive treatment required to survive with defective organs. Thousands of people are awaiting heart and liver transplants and 30% of them will die while waiting. In addition, 19 thousand people are candidates for kidney transplants and 30 thousand people currently receiving dialysis could benefit from a new kidney.

The needs of a desperate patient for an organ transplant should be given high priority, if available resources will permit. If organs are removed from aborted fetuses and cadavers, the removal would be done by professionals, and there is no more indignity involved than what occurs with embalming or cremation by an undertaker, or when an autopsy is performed. The importance of social contracts should be recognized, and the desires of survivors should be respected whenever the deceased has not expressed an explicit request to donate organs. One should still insist that precedence be given to the welfare of the living over the symbolic value of the dead. Feinberg (1988) observed that if the decedent has left no testamentary instructions, then the next of kin should be considered to have all property rights to the cadaver. He argued that there would be no violation of any rights if the surviving kin decide to sell the organs for research or transplantation.

It is interesting that, in one respect, a cadaver is treated with more respect than is a pregnant woman in respect to organ donation. With informed consent, a vital organ can be taken from a cadaver and transplanted to prevent the death of a person in need. A pregnant woman who chooses to have a legal abortion, however, is not allowed to donate the tissue of her aborted fetus to be transplanted to prevent a person's death. Could it be that this inequality exists because all men will be cadavers but never have an abortion?

CHAPTER 6

Suicide and Euthanasia

Moral and Legal Issues

When discussing issues of suicide and euthanasia Joel Feinberg's (1986) scheme that considers suicide as an act involving one party/certain harm and euthanasia as one involving two party/certain harm will be used. Viewed from this perspective, the same principles that apply to suicide can be applied to euthanasia. If a person makes a rational request to die, if it is assured that the request is voluntary and that the person is in a stable, competent state that qualifies the individual as a moral agent, then it should be irrelevant whether the act of suicide is assisted. If the act is judged to be permissible, then it is proper to receive aid from a second party, with the major reservation that the second party should have no secondary (financial or personal) interests and the interests of relevant third parties are not harmed by the person's death.

In this chapter the problems and issues that suicide entails will be considered, followed by a discussion of those for euthanasia as they are viewed by the law, theologians, and the medical profession. In Chapter 5, moral and medical issues will be discussed, and policies will be recommended to consider rules, laws, and institutions to enable society to deal with the question of euthanasia in a better fashion than we do now. At the end of Chapter 7, two further issues involving death will be considered: AIDS and capital punishment, both of which have relevant ramifications.

SUICIDE

Brandt (1976) defined *suicide* as any action that produces one's own death as a result of the intention to end life or to bring about a series of events to produce death. According to this definition the "heroic death" of a soldier is an act of suicide. Brandt offered this neutral definition to avoid automatically making acts of suicide only those that were irrational or immoral, thereby not prejudging the question of whether any specific act of suicide is rational and morally justified.

Although many Western theologians have argued that suicide is immoral, Gillon (1986) noted that Shintoism and Buddhism accept suicide under a variety of circumstances, including extreme pain and incurable suffering. Hinduism accepts it if one suffers serious disease or great misfortune, and in some cases of old age. Judaism is permissive regarding suicide, considering it not to be a definitive sin. Roman Catholic theologians have condemned suicide since the time of St. Augustine. It was denounced as diabolically inspired in fifth-century Church Law, and Aquinas affirmed that, being contrary to natural law, it precluded repentance. The Islamic attitude is similar to the Roman Catholic: Suicides shall suffer in the fire of hell and shall be excluded from heaven forever. Protestants are less rigid, but a deep-rooted attitude against suicide prevails.

Flew (1986) discussed, and rejected, three secular arguments that are made to support the conclusion that suicide is always morally wrong. The first is that it is unnatural and in conflict with the instinct of self-preservation. He pointed out that this is an instance of the naturalistic fallacy, which states that the *is* should determine the *ought*.

Second is that the act deprives other people of the services that the individual suicide might have provided to society. He considers this argument to have merit as far as some suicides are concerned, but little in the case of voluntary euthanasia, especially when an individual has become a hopeless burden. Third is that suicide is in effect "self-murder," and murder in any form is impermissible. The reply to this argument is that if the logic is accepted, then marriage is really adultery—"own-wife adultery." He considers both arguments to be absurd.

The dimensions suggested by Feinberg (1986) can be used to frame the issues. Activities such as race-car driving, smoking, and mountain climbing are one-party acts with a variable probability of harm, depending on a number of contextual factors. An example of a two-party act with a variable probability of harm is the decision to agree with a surgeon's recommendation to undergo dangerous therapeutic surgery. There is no simple mathematical function to guide a consequentialist calculation to decide whether the risks involved in any specific situation are reasonable.

Brandt (1976) discussed circumstances when suicide is and is not morally justified. A similar approach will be used here by considering instances in which suicide is morally approved by everyone, moving to cases considered morally permissible by some and morally impermissible by others, and finally, to consider those that are disapproved of by most people.

Morally Permissible Suicide

Although many argue that suicide is almost always morally wrong, all approve of suicide under some circumstances, even though they often do not use the word *suicide* to characterize it. If the society is engaged in a war, sacrificing one's life to save comrades in battle qualifies for the status of posthumous hero with medal attached. If the pilot of a plane that is certain to crash decides to perish in the plane in order to steer it away from a populous neighborhood rather than to bail out and survive, that action is regarded as a brave and selfless one. An individual who undergoes torture rather than betray colleagues or renounce basic moral beliefs is raised to the status of a martyr. Many consider it permissible, if faced with inescapable dehumanizing slavery, to kill oneself as a rational, autonomous being.

Although the sanctity of life is considered to be a paramount value by many people, many theologians agree that it is permissible to sacrifice one's life in instances such as the aforementioned. Sometimes the permissibility is the result of the invocation of the doctrine of double effect; if the intent is to save others or to defend ideals, then the death is not seen as suicide, but as an unintended side effect that occurred as the result of honorable intention.

Cases Producing Disagreement

There are several instances in which there is disagreement regarding the moral permissibility of suicide. One case that many consider permissible is when an individual is terminally ill, is living a life that does not qualify as minimally satisfactory (because of great pain and an inability to function adequately), and continued existence would produce great emotional and financial harm to the survivors. This is a case of one-party/certain-harm suicide, and should be considered to be morally permissible if the act is voluntary, rationally decided, and harms no other innocent person. Those who insist on the sanctity of all life (with the exception of the types of situations described earlier) will not agree that this suicide is permissible. A decision that suicide is morally permissible is based on a consequentialist balancing of the benefits provided by extending an individual life and providing an adequate future for surviving family members, against the cost in pain, anguish, and misery to all involved. Evaluating the costs and benefits of suicide in such cases would be reasonable in light of the evolutionary ideas regarding inclusive fitness, as well with as a community-oriented social-contract model, especially if the death conserves resources for the family to continue as a reproductive unit.

More difficult cases involve individuals who are neither terminally ill nor in great pain, but who decide for whatever reasons that the quality of life is not satisfactory; they believe there is no likelihood that the quality of life will improve and, on this basis, decide on suicide. Such cases strike to the heart of the argument, and moral permissibility will depend on a number of special considerations. There are two major considerations: (1) Does the individual have a responsibility to the community? (2) Is the decision voluntary, rational, and made by a competent person?

Responsibilities to the community involve whether the individual has surviving family members to whom there is an obligation and who could be left destitute and would, therefore, have to be cared for by society. Another community concern is that the act must not endanger the safety or welfare of innocent persons.

The second consideration involves whether the decision is voluntary, rational, and made while the person is in possession of relevant facts regarding the circumstances likely to exist should the suicide occur. People suffering from chronic and painful illness often find it difficult to make rational decisions. Those in a chronic state of depression usually have difficulty assessing the probability (often of even entertaining the possibility) of a satisfactory change in life circumstances; depression overwhelms everything with its aura of doom and gloom. Those in an acute state of despair because of the loss of loved ones or economic insecurity, or who are no longer able to function at the high level they once could, may decide there is no purpose in continuing what they construe as a dreary and hopeless existence.

Before discussing these examples, consider several basic points. It is permissible to voluntarily and rationally waive one's interest in property by giving such property to others. This fact led Rachels (1986) to argue that it should be just as permissible for a person dying of a painful illness to waive interest in life. There is no reason why one should not be able to waive life interests under some circumstances, even if there is no terminal illness.

Hume (1784), in his essay "Of Suicide," maintained that one who commits suicide does no harm to society, but only ceases to do good. The individual whose life is miserable should not prolong it because of "some frivolous advantage" the public may receive. By dying, one might relieve society of a burden, allowing available resources to be used by persons of more value to the society. Although some have argued this position could encourage suicide, Hume believed the "natural horror of death" will be sufficient to keep people from committing suicide for trivial reasons.

Louis Pascal (1980, p. 114) noted that a person has "the right to cause as much unhappiness to himself as he wishes." The major concern of society should be to assume that suicide is not done when the agent is cognitively incapacitated, either temporarily or permanently. Brandt (1976) concurred that the concern should be to look at the matter in a way that will help individuals to decide whether suicide is the best thing from the viewpoint of their own welfare. It is important to ensure that there is full awareness of the facts regarding the current state of affairs, as well as the state of affairs that will likely follow the suicide. The major concern should be to convince the individual to refrain from making the irrevocable decision that suicide entails while depressed or in a state of despair. It is important that the individual be aware that depression can lead to a narrowing of perceived alternatives, and that other behavioral options might provide satisfactory alternatives to the current unsatisfactory state of affairs.

Brandt (1976, p. 332) concluded,

> A decision to commit suicide for reasons other than terminal illness may in certain circumstances be a rational one. But a person who wants to act rationally must take into account ... the various possible "errors" ... and make appropriate rectifications in his initial evaluations.

To establish that a person's decision is rational and voluntary, it might be appropriate to place the person in temporary restraint, confinement, or provide treatment and counseling. The reasons for any such restraints are to allow temporary imbalances to be righted and to determine that the person is in possession of the ability to reason—what has been called soft-paternalism by Feinberg (1986).

It must be determined that a person's decision to commit suicide has been arrived at through a process of rational reflection, and that the person has construed the facts of the situation objectively. It should be assured that the person is aware of the damage the suicide might do to others involved, such as the surviving family. If it is reasonable to assume that the decision is made with full understanding of these concerns, then the action should be allowed, even though an advisor might consider the decision mistaken. A moral agent's informed and voluntary decision should command respect and take precedence over the moral intuitions of others.

The principles embodied in rational liberalism, described in Chapter 1, lead to the conclusion that suicide is morally permissible in all of the cases described under the heading of "Cases Producing Disagreement." There should be increasing degrees of assurance regarding the presence of rationality for the different cases. The first case involved a terminally ill person in pain, whose continued existence would have a negative effect on the surviving family's welfare. It should be morally permissible for that person to commit suicide, no matter what the views of the survivors. The only safeguard necessary would be a guarantee that there has been discussion with qualified advisors able to evaluate the person's acute emotional state, and that the individual understands the reality that would exist for the survivors following the suicide. To alleviate the survivors' emotional distress (and to make second-party assistance permissible), a qualified physician should certify that the person's illness is terminal. As in all of these cases, if it has been established that the person's mental state allows rational decisions, the individual's decision should prevail.

The second case does not involve terminal illness, but the person has decided that the quality of life is not adequate to warrant continuation. Counseling and discussions about the impact on survivors and the community should be more intense, and every attempt should be made to understand the individual's reasoning, to emphasize the irrevocability of the act of suicide, and to explore the likelihood that circumstances might change for the better. It should be determined that the person is rational; otherwise the decision cannot be a voluntary one. If guarantees regarding rationality are met and the harm to survivors has been explained and considered, then the decision of that individual moral agent should prevail.

Morally Impermissible Suicide

There are cases in which few would consider suicide to be permissible. These involve instances when a person is not able to make a rational decision because of drunkenness, the influence of hallucinatory drugs, an immense personal loss causing deep despair, extended treatment with sedatives, or some recent physical trauma. In such instances, the acute condition of the person might lead to a decision that would change if the immediate debilitating events were not present.

Feinberg (1986) listed a series of moral and legal incapacities that diminish the ability to make voluntary and rational decisions: coma, motor paralysis, severe retardation, derangement, psychosis, recurrent seizures, depressions, manias, rages, addiction, infancy and immaturity, intoxication and other nonaddictive drugged states, fever, nausea, pain, extreme debility or fatigue, persisting moods, and distracting emotions.

If there are such conditions, a person should be placed in temporary custody with counseling, guidance, and therapy provided until the ability to function as a moral agent is regained. If there is no reversion to a state of rationality, because of permanent neurological damage, for example, the individual should be given the protection due a moral patient. Welfare and interests should be protected, just as we protect those of immature or incompetent people, and animals to which we have an obligation.

EUTHANASIA

Euthanasia has become a major issue in contemporary society. A pressing need for attention to this issue was signaled by the fact that Derek Humphry's book *Final Exit* (1991), which is dedicated to informing individuals how to commit self-delivered or assisted suicide, has sold, according to Humphry (1992), one-half million copies since publication. Voter initiatives have been introduced (and failed) in Washington, Oregon, and California, and the Dutch Parliament has approved a law permitting euthanasia. A revised initiative has been approved by Oregon voters in the 1994 elections, and a law has been introduced in Massachusetts.

It is questionable morally to deny assistance to a person who desires to die but is physically unable to perform the actions required to end life, or who, because of a lack of medical knowledge, might perform an action that would only further impair the level of functioning and well-being without achieving the desired fatal outcome. The critical issue is that the proper safeguards discussed when considering suicide are assured, with additional safeguards to protect against immoral acts by unscrupulous second parties.

The permissibility of euthanasia has produced some of the most active political discussions regarding death. Humphry (1992), one of the founders of the Hemlock Society, observed that few of us will have an abortion, making it easier for us to theorize about the moral rights and wrongs of abortion from objective heights, especially for men, who will never undergo the procedure. All of us accept the fact, however, that we are going to die one day, making the euthanasia issue of potential personal concern for all. Much of the discussion regarding euthanasia implicitly contrasts eternal life with eternal death, but perhaps the proper contrast is death now or death later—its inevitability cannot reasonably be denied (Barrington, 1986).

The permissibility of euthanasia should not be posed in terms of why people should be given a new legal right, but should be in terms of why people should be restrained by law from doing what they are interested in doing (Flew, 1986). Current laws prevent sufferers from achieving a quick death and often result in degrading, cruel, and painful events for the patient (who could be suffering pain that threatens all sense of dignity) and force people who care for them to watch the pointless pain hopelessly. It is permissible, and has been considered praiseworthy, to put an animal out of its misery, but not to shorten the suffering of a human—which does not constitute a respect for people, but treats people worse than "brute animals."

The Roman Catholic argument was affirmed by Pope John Paul II, who issued an *Evangelicum Vitae* (Gospel of Life) in 1995 (Bohlen, 1995). In that Gospel, the highest form of papal message to the church hierarchy, he defined *euthanasia* as an action or omission that itself and by intention causes death, with the purpose of eliminating all suffering. He stated that euthanasia should be condemned because it represents a false mercy and is a disturbing perversion of the true compassion of sharing another's pain.

The surgeon Christian Barnard (1986, p. 174) rejected the argument made by theologians that suffering can be ennobling. He remarked that, as a physician, he had never seen anyone so ennobled:

> I have never seen any nobility in a patient's thrashing around all night in a sweat-soaked bed, trying to escape from the pain that torments him day and night. I have never seen what nobility there is supposed to be in either a pain-crazed face or in the drug-saturated sedation of a patient who, while feeling no more pain, can no longer make contact with his surroundings or other people.

Barrington (1986, p. 240), a lawyer and leader of the Euthanasia Society in Britain, echoed Bernard's view:

> One can only hope that the pathetic human wrecks who lie vomiting and gasping their lives out are as sanguine and cheerful about their lamentable condition as the smiling doctor who on their behalf assures us that no one ... really wants euthanasia.

The medical status that could make it morally and legally permissible to terminate the life of an individual is of major concern. Considerations of medical status lead to a concern regarding whether procedures must be restricted to passive actions, such as disconnecting life-support systems or not introducing a treatment to prolong life, as contrasted with taking positive steps, such as administering a lethal injection to terminate life. The medical community construes its major responsibility as healing the sick, and questions are raised whether a physician should ever assist in the death of a dying individual. If physicians are viewed primarily as healers, then is physician-assisted suicide ever permissible? Some argue that the U.S. Constitution demands that states prohibit killing, but Brandt (1987) wondered in what language the Constitution demanded that all possible medical techniques must be used to save someone's life.

The Trials of Dr. Kevorkian

By November 26, 1994, Dr. Jack Kevorkian had, since June 1990, assisted 21 persons in Michigan (13 since the Michigan Legislature's action, discussed later) to commit suicide using devices that either deliver a lethal injection or allow the person to inhale carbon monoxide. *The New York Times* (December 4, 1992) reported that both houses of the Michigan Legislature passed a bill making assisted suicide a felony punishable by 4 years in prison and/or a $2000 fine. The Governor signed it, putting it into effect February 25, 1993, which makes Michigan the 29th state to have such a law. The law was to be in effect while a state commission studied the matter in order to make recommendations for final legislation. The commission represented health-care providers, the elderly, and groups such as the Hemlock Society. The legislature failed to act on the commission's recommendations, allowing the provisions of the law to lapse on November 25, 1994 (Lessenberry, 1994). It was reported in *The New York Times* (November 25, 1992) that nearly two-thirds of Michigan residents expressed support for laws permitting doctors to help people commit suicide when the patient is terminally ill and requests euthanasia. Although the prosecutor for Oakland County in Michigan has filed murder charges against Dr. Kevorkian on three different occasions, the charges have been thrown out in all instances. The legislature seems to be at odds with the public, the judiciary, and at least some members of the medical profession regarding the permissibility of physician-assisted suicide.

Following the 17th assisted suicide of a patient suffering from Lou Gehrig's disease, Dr. Kevorkian challenged the state directly. He asked to be prosecuted in order to test the state's law prohibiting assisted suicide as quickly as possible (Terry, 1993), and was acquitted of all charges on

May 3, 1994. The 22-member Michigan study commission had issued its report on March 4 of that year, recommending, by a split vote of 9 yes, 6 no, and 7 abstain, that physician-assisted suicide be legalized under certain circumstances. *The New York Times* (March 6, 1994) reported that Dr. Kevorkian was devoting his efforts to collecting signatures for a statewide referendum on a constitutional amendment to guarantee assisted suicide as a right. The U.S. Supreme Court rejected Kevorkian's argument that there is a constitutional right to assisted suicide, rejecting his appeal of a Michigan Supreme Court ruling that the Constitution creates no such right (Asseo, 1995). This action will allow the Michigan trial judge to reexamine whether two murder charges against Dr. Kevorkian should have been dismissed.

The last patient assisted by Kevorkian was a 72-year-old housewife who had severe rheumatoid arthritis, advanced osteoporosis, and other disorders that had forced the amputation of both legs and removal of one eye. She made a videotape 8 months earlier in which she said (Lessenberry, 1994), "I would like a way out." A physician had prescribed morphine patches, which gave her some relief, but over time, even they proved ineffective. She would not have qualified for assisted suicide under the terms of the law that was passed by Oregon voters, because she may have lived longer than 6 months. Kevorkian criticized the Oregon law, stating that no one can determine how much time any patient has left to live, and that quality of life, not time, should be the main factor.

Views of the Medical Profession

There have been sharp exchanges in the medical community regarding the permissibility of any physician-assisted suicide, active or inactive. Some (e.g., Gaylin, Kass, Pelegrino, & Siegler, 1988) cited the physician's obligations under the "hallowed canons" of the Oath of Hippocrates, which insist on a hard and fast distinction between ceasing useless treatments (allowing to die) and active, willful taking of life (killing). They objected to a case of euthanasia published anonymously in the *Journal of the American Medical Association*, in which a patient in intense pain had been injected with a lethal dosage of sedative. One of the surprising aspects of the argument made by Gaylin et al. (1988, p. 2139) was that "Decent folk do not deliberately stir discussion of outrageous practices, like slavery, incest, or killing those in our care." They stated that medicine should not be considered only as a trade, but must be viewed as a moral profession. Can an examination of morality take place without the free and unfettered discussion of what some might consider to be "outrageous practices" to determine where the line should be drawn between the morally permissible and impermissible, and which moral principles should be brought to bear to regulate those decisions?

Vaux (1988) considered the case and suggested that the doctrine of double effect should be used, whereby the primary intention of the physician should be to relieve pain, with the death as a result of the sedation a secondary effect. Truog, Berde, Mitchell, and Grier (1992), reviewing the use of barbiturates in the care of terminally ill patients, argued that justifications using the doctrine of double effect function as a "fig leaf" for euthanasia, because the intent of the treatment is the death of the patient, whether it is couched as being primary or secondary.

Although physicians are impeded by law and custom from giving a lethal drug dosage to a patient, some have indicated they would do so if their wife, father, or child were suffering an "end-of-life agony," even at the risk of prosecution. Vaux (1988) considered such assistance to loved ones to be a "moral and loving act," arguing that, with the life-prolonging techniques and medications that now exist, death is no longer an acute, natural, and noninterventional mode, but is a chronic, contrived, and manipulated phenomenon. He wrote (p. 2141), "We cannot modify nature and then plead that nature must be allowed to run its unhindered course."

Angell (1988) discussed some of the opposed and contradictory views that physicians have regarding euthanasia. On the one hand, there are arguments in opposition that invoke the slippery slope that euthanasia leads to widespread disregard for the value of human life, and that it might be used for incompetent as well as competent persons. Singer and Siegler (1990) argued that euthanasia is a perilous public policy because of the likelihood ("or even the inevitability") of involuntary euthanasia, that euthanasia may be urged on patients to spare their families financial or emotional strain, that incompetent patients might be killed on the basis of "substituted judgment," and that those who are members of "vulnerable groups" in American society (such as the elderly, physically handicapped, minority-group members, mentally impaired, alcoholics, drug addicts, and AIDS patients) might be "subtly coerced" into requesting euthanasia. Almost anything is possible, especially if the medical profession refuses to take responsible leadership in discussing the moral issues. Dreadful events, such as those listed earlier, can be prevented if care is taken to draft rules and regulations to prevent individuals from taking such actions and to punish severely any who transgress. The medical profession must take a responsible leadership role in such matters or legislators and judges are going to make these medical decisions for them.

Angell (1988) cited a poll of doctors, done by the Center for Health,

Ethics, and Policy of the University of Colorado at Denver, which was released June 2, 1988. Three-fifths of the physicians polled favored legalizing euthanasia, but nearly half of those would not perform it themselves. This inconsistency between attitudes raises the question that if euthanasia were legalized, then would doctors morally opposed to it be required to perform it? If they were not required to do so, would it be necessary to create a new medical profession that is committed to performing euthanasia and dealing with other aspects of dying? Angell finds this an unsavory prospect, but in Chapter 8, it will be suggested as a reasonable solution to the dilemma that many members of the medical profession face.

A study of the attitudes of physicians in Washington state regarding assisted suicide and euthanasia was reported by Cohen, Fihn, Boyko, Jonsen, and Wood (1994). Many physicians thought euthanasia never to be ethically justified (58%), while many thought it was (42%). Only 33% of the total sample stated they would be willing to perform euthanasia themselves. Thirty-nine percent thought assisted suicide was never ethically justified, 50% thought it was, but only 40% stated they would be willing to assist a patient to commit suicide. These results should be interpreted in light of the fact that under most circumstances these practices are illegal, and a practitioner could be prosecuted.

Psychiatrists were the most supportive of both voluntary and involuntary euthanasia, with hematologists and oncologists (who had the most exposure to terminally ill patients) being the strongest opponents. Of those opposed, 56% stated they were influenced by religious beliefs, as compared to only 15% of those in favor. Those opposed considered the practices to be inconsistent with the physician's role in relieving pain and suffering, whereas 91% in favor believed the practices to be ethical and consistent with the physician's role. Eighty percent of those opposed cited the potential for abuse. The overwhelming belief of those in favor was that patients' rights to self-determination should be respected, with 91% of them stating that the availability of euthanasia might reduce patients' fears of losing control or of experiencing a painful death. Those who favored legalization strongly supported the requirement that the patient's request be witnessed by an independent person who would not benefit from the patient's death (90%), the physician administering a fatal overdose should have an established relationship with the patient (84%), alternatives (such as hospice care and treatment of depression) should have been fully utilized (84%), two physicians should be in accord with the decision (81%), and there should be a specified waiting period between the request for a drug overdose and the time the request is granted (78%). The law passed in Oregon in 1994 respects these expressed concerns of those Washington state physicians in favor of the legalization of assisted suicide and euthanasia. It seems unlikely that those opposed on religious grounds would approve of the actions, given any particular safeguards that might be suggested.

Although the poll cited by Angell indicated that nearly half of the physicians polled would not perform euthanasia themselves, a survey indicated that 96% of 879 intensive-care physicians either had not begun or had withdrawn life-sustaining treatment in the course of their practice, and that nearly one-third of them had done so more than five times in one year, often without the knowledge or consent of patients or their families (Knox, 1995). Thirty-four percent of these intensive-care physicians had continued life-sustaining treatment despite the wishes of patients or their surrogate decision makers. There seems to be a great deal of confusion and inconsistency on the part of physicians regarding the definition of *futile care* and what should be done in such cases. The physician who conducted the study, Dr. David Asch of the University of Pennsylvania, remarked that the study shows that physicians are incorporating some notion of medical futility into their decision making, and that they do make choices. One would hope that these choices are based on explicit and informed criteria.

The Boston Globe, on April 13, 1995, discussed a finding that physicians often misunderstood the desires of very sick patients to undergo CPR. In half the cases, the physicians were not aware of the patient's desires, or thought they wanted the opposite of what had been requested. Physicians should worry a bit less about their obligations to heal and more about the needs and justifiable desires of those who have to trust them and communicate with them.

Legislative Actions

The Dutch Parliament passed a law on February 9, 1993 that permitted euthanasia. Simons (1993a, 1993b) discussed a study backed by the Dutch government that found there had been 2,300 deaths by voluntary euthanasia, and 440 cases of voluntary suicide in the Netherlands in 1990 (about 2% of all deaths in that year). The 440 cases of voluntary suicide had risen to 1,318 by 1992. (The overall Dutch suicide rate is relatively low—12.8 per 100 thousand people—which contrasts sharply with Europe's most suicide-prone nation, Hungary—40 per 100 thousand; Chao, 1993.) Simons suggested that doctors evidently had refused most of the 9 thousand requests that had been made. About 85% of the euthanasia cases were cancer patients, and the remainder included people suffering from AIDS and multiple sclerosis, all in the final stages of the diseases.

Dutch law considers ending a patient's life or helping in suicide a criminal offense (although no doctor has gone to prison for those offenses), and the policy is defended as a matter of self-determination: People have the right to choose their own life and death. Euthanasia is permitted if several conditions are met: The request to die is voluntary and not made under pressure; the suffering must be unacceptable with no hope of recovery; the patient must request euthanasia explicitly and over a period of time; the patient must be competent and well-informed of alternative solutions that could prolong life; another physician experienced in dealing with cases of euthanasia must see the patient independently and support the decision; and a documented written report must be drawn up outlining the history of the patient's illness and declaring that the rules have been followed.

The emotional response (which borders on hysteria) by some in the medical profession is captured in an article by Reich (1993), a psychiatrist, senior scholar, and director of the Project on Health, Science, and Public Policy at the Woodrow Wilson International Center for Scholars. To convey the tone of the article, I will quote liberally rather than paraphrase in order to avoid the appearance that I am distorting his remarks. Reich is deploring the legalization of euthanasia by the Parliament of the Netherlands. He wrote,

> You don't have to be religious to mourn the new law. All you have to be is human and alive... People kill without benefit of the law every day. Soldiers kill other soldiers legally. But societies can experience such killings and remain essentially decent. It's when they legalize the killings of their own innocent members that they remove an obstacle that blocks the all-too-easy slide of civilization into moral chaos.
>
> [It is] ... an easing girded by rules that seem tight now but that will be loosened, inevitably, in practice. The spectacle of the formalized and regular killing of such patients—resulting not in 1 Dutch death out of 50, but in 5 deaths out of 50, or 10, or 20, or even more-will have a corrupting effect, not only on the value of life in the Netherlands but also in every other democratic country.
>
> But the greatest impact of this spectacle may be in undemocratic countries, where authorities less humane than Dutch legislators ... may even provide rules that permit, encourage or even demand all kinds of killing, beginning with the killing of people who ask for it and progressing to the killing of people who are said to deserve it.
>
> However, once the medical commitment to life is undermined by legal sanction—once doctors trained to preserve life are no longer afraid of initiating death—then the very nature of the medical enterprise, and the very identity of the physician, is changed. The doctor loses the mission of caring for life and takes on the role of an amoral medical technician—one whose duty could just as well be to end life as to preserve it. (p. 178)

No point would be served by refuting these arguments point by point. Why, when soldiers are allowed legally to kill soldiers of other countries, does this not lead to indecency? What is the evidence (other than the questionable analogy with Nazi Germany, which will be challenged in Chapter 7) that voluntary euthanasia inevitably leads to involuntary euthanasia and moral chaos? Why should one expect a twentyfold increase in the rate of euthanasia, and where it is etched that the physician will lose the mission of caring for life if allowed to relieve terminal suffering?

Euthanasia in the States. Voters in Oregon passed a ballot measure by a margin of 51% to 49% that makes the state the only place in the United States allowing physicians to hasten the death of the terminally ill. The measure allows a patient, with 6 months or less to live, to ask a physician to prescribe a lethal dose of drugs to end unbearable suffering. At least two physicians must agree that the patient's condition is terminal, the patient must request the drugs at least three times (the last time in writing), and the patient must self-administer the drugs. The major difference between this measure and the ones that were defeated earlier in California, Oregon, and Washington is that those defeated measures would have allowed a physician to administer the drugs.

The Oregon measure passed, even though it faced strong opposition by the American Medical Association's national board, which warned Oregon physicians they would be performing an unethical act if they participate in assisted suicide. The Oregon Medical Association did not take any position on the measure, but it was strongly opposed by the Oregon State Pharmacists Association, the Roman Catholic Church, the Oregon Hospice Association, and *The Oregonian*, Portland's newspaper. The opponents of the measure had three times the financial support than did the advocates, with Roman Catholic sources contributing more than $1 million (O'Keefe, 1994a).

The arguments in opposition were similar in form and intensity to those regarding the abortion issue. The ethicist Caplan stated that it had all the elements of the abortion debate, having everything but fetuses (O'Neill, 1994). Caplan noted that the Oregon measure is the most carefully crafted and most thoughtful of any that have been suggested throughout the country, placing the responsibility for ending life in the patient's hands. The Catholic bishops declared the move cheapened human life in our society, stating that "the vote is a cancer more lethal than any physical ailment" (O'Keefe, 1994b), and that the Catholic hospitals in the state would refuse to participate in physician-assisted suicide. Some physicians asserted that the measure destroys the relationship between patient and physician, and the executive director of the Pharmacists Association indicated that he would refuse to dispense any lethal prescription. *The Oregonian*, in an editorial on October 20, 1994, expressed the opinion that terminally ill patients suffer from clinical depression and that most physicians are unable to recognize this depression, and therefore they do not treat it, which results in too many ill-advised requests. They also argued that "almost all physical pain in dying patients can be effectively managed," and that it "cynically protects" physicians while putting patients in peril. From the tone of the editorial, it might be suggested that all physicians would be advised to take remedial training from *The Oregonian* editorial staff.

Given the strong sentiments on both sides of the issue, as well as evidence that more states are going to consider such laws, it is not surprising that legal challenges have been made to the Oregon law. It was reported in *The Oregonian* (December 8, 1994), that a federal judge issued a temporary restraining order blocking the law from taking effect, at least until December 19, and that on December 17, he blocked the law from taking effect until the courts decide if the measure is constitutional.

There are indications that initiatives permitting euthanasia are being developed in order to be placed on the ballot in several states: California, Colorado, Connecticut, Hawaii, New Hampshire, New Mexico, Washington, and Wisconsin (Bates & Lane, 1994), and in Massachusetts (Lehigh, 1995). The bill filed in the Massachusetts legislature would allow physicians to prescribe lethal medication under the same conditions and controls specified in the Oregon law (Lakshmanan, 1995), and the proposal received instant opposition by a coalition that included the Massachusetts Medical Society (MMA), the Roman Catholic Church, Massachusetts Citizens for Life, and local hospice leaders. A past-president of the MMA stated that a physician should not be put in the position of deciding how long a patient should live (Lehigh, 1995), and the general counsel of the AMA opposed it, because it violates the fundamental ethical principle that physicians will do no harm. The executive director of the state's largest antiabortion group stated that it will cheapen human life still further—there being enough of a problem with abortion on demand. It was also stated that euthanasia blurs the distinction between the healer and the killer. (Where have I heard this song before?)

Regarding the distinction between healer and killers, Brewer (1986) noted that it would seem no different for a healer to also assist in euthanasia in the same institution than for a gynecologist to work in a fertility clinic one day and perform an abortion the next day. The most reasonable approach might be to establish separate facilities (hospice and euthanasia clinics), just as we have separate facilities for reproductive problems (fertility and abortion clinics). It is doubtful that this line of argument will impress antiabortion groups, however.

Dworkin (1994a) rejected slippery-slope arguments, commenting that it would be perverse to force competent people to accept a great and known evil, such as dying in great pain, to avoid a speculative risk. A Federal District Court judge in Washington state decided that laws against assisted suicide are unconstitutional, basing the decision on rulings that have been made regarding abortion. These rulings state that matters involving such intimate and personal choices are central to the liberty protected by the 14th Amendment. In the assisted suicide decision, it was held that a competent, dying person has the freedom to hasten death, just as a pregnant woman has the right to chose abortion. Dworkin rejected the slippery-slope argument that euthanasia inevitably will be abused, noting that states have the power to guard against requests influenced by guilt, depression, poor care, or financial worries. The safeguards in place in the Netherlands are being honored and seem to be functioning adequately.

The continuing problems regarding euthanasia must be resolved by legislative action, and the medical profession, medical ethicists, and the general public should become informed regarding the moral, psychological, social, and economic realities involved. Extreme emotionalism, based on wild conjectures of the inevitable slide down the slippery slope to degradation should be ignored. The debate should center on the relevant facts and the values of society.

CHAPTER 7

Euthanasia

Moral and Medical Issues

CONDITION OF THE ORGANISM

Kleinig (1991) argued that, when considering issues in medical ethics, the condition of the organism should be viewed in a broad context, rather than from an absolute perspective. The capacities and physical condition of individuals are different depending on whether they are at a fetal, neonatal, adolescent, adult, or senescent state. Both the potential and residual qualities of organisms should be considered. Potential qualities should be evaluated in terms of the expected normal developmental trajectory, with those individuals who have a strong likelihood to realize their normal potential helped to do so, and should emphasize the expected quality of life individuals might enjoy.

Brandt (1987) developed a similar conceptual scheme, placing defective newborns into four classes: (1) serious defectives with a prospect for a short and unpleasant life in spite of any possible medical treatment; (2) those whose lives are marginally beneficial to them, but exact a high economic and psychological cost from others; (3) those who have the capacity to survive and make a reasonably satisfactory life, but whose parents might not wish to accept responsibility to care for them; (4) those who will have only marginally beneficial lives, but whose parents want to care for them. Brandt believes an early death is permissible for those in the first class; aggressive treatment is fitting for the second class, only if society is able and prepared to assume all burdens; aggressive treatment for the third class, again by society (through adoption, preferably) if the parents are not willing to take responsibility; and aggressive treatment and family care for the fourth class.

I will adopt this approach and consider a range of organismic stages that could require a decision regarding whether it is permissible to terminate a life.

1. One extreme stage is when a person is brain dead, but with continuous life support, organ systems can be kept functioning.

2. A similar stage is when the cerebral cortex has ceased functioning for a sufficient period of time to make it impossible for recovery of normal function to occur (PVS). The individual is not conscious, shows only reflex responses to painful stimulation, but the brain stem is functioning, and there is no need for life-support systems, except for a feeding tube to provide nourishment. As discussed in Chapter 4, organisms can be kept in both stages 1 and 2, for very long periods of time—months or years.

3. Another extreme stage is represented by cases of spinal tube defect in which the neonate is anencephalic or suffers from spina bifida. The anencephalics will never attain anything more than the ability to maintain a vegetative state. They may live for a few days or weeks without life-support systems if they are nourished and given antibiotics. The spina bifida infants have an opening in the base of the skull, and even with aggressive treatment they would have a 60% chance of dying within 7 years, and would probably live with gross paralysis, frequent bone fractures, incontinence of urine and feces, mental retardation, and require repeated surgery to relieve hydrocephaly (Brandt, 1987). Without surgery, they usually die of meningitis or kidney failure within the first few years of life.

4. Another stage involves infants born with Down's syndrome (commonly referred to as *mongolism*), a genetic defect estimated to occur at a rate of about 2 in one 1,000 in Western Europe and North America, with an increasing incidence related to the mother's age. These children show varying degrees of intellectual impairment, but will be able to live for many years if given proper medical and social support. A special case is the Down's syndrome infant born with an intestinal blockage that must be surgically removed to prevent death from starvation. A special question arises when the parents decide not to have the surgery performed because they do not want to raise an intellectually defective child.

5. Several stages of terminal illness are considered. The first involves persons diagnosed as terminally ill, who have an inadequate quality of life because of such things as intense pain that cannot be alleviated. One situation would be when the person requests death, but the family does not agree with the decision. A less problematic situation would be when the person requests death and the family agrees.

6. Still another stage is when the person is not terminally ill, but is living an inadequate quality of life, with no likelihood that an adequate

level of function will be regained. Assume that the person requests death and is able to rationally and forcefully argue for assistance in implementing that decision.

7. Another stage is when the person is in what Dworkin (1993) called a "happy demented" condition. The once fully competent individual now sits contentedly, mindlessly engaging in meaningless activities with no coherent sense of self and no discernable short- or long-term aims. A further complication can be introduced if the person, when competent, left formal directives that in case of mental incapacitation, none of the person's assets should be used for care, medical treatment should not be given, and, if permissible legally, the life should be terminated. Yet, at the present time, the patient lacks the mental competency that could be judged to override the earlier directive.

8. Finally, it will be useful to consider the status of individuals who are in the latter stages of terminal starvation, would accept food if it is provided, but probably would not survive because of general deterioration. Assume that if food is provided them, then the limited supplies available would not be sufficient to nourish others more likely to survive if they received the limited amount of food available—a situation involving triage.

MORAL ISSUES INVOLVED IN EUTHANASIA

Before considering the permissibility of euthanasia in each of the cases, some basic moral issues will be considered.

Dignity and Social Contracts

Dignity is not conferred by mere species membership—personhood is also required. Dignity might not be a comprehensible concept for some humans, because they lack those capacities, abilities, and achievements that establish moral agency. *Dignity* refers to the ability to control one's own life, and to make decisions regarding that life. If a person is capable of autonomous judgment and is now the victim of a severely debilitating illness, the discontinuation of medical treatment could be an act respecting the dignity due any rational human. As Humphry (1992) argued, the most important consideration for the community is to honor the desire to be in charge of one's life and the dying process. He suggested that it should be possible for physicians to assist death, at least for terminally ill, competent adults.

The concept of dignity also includes the idea of a social contract. An individual's personhood no longer exists when the organism enters a state of irreversible coma, and it could be considered an indignity to the person who once existed to allow him or her to persist in a vegetative state. The desires expressed by a person prior to the comatose state should be considered. If treatments are continued against the patient's prior decision, then it disrespects that individual to continue treatment. The overriding concern is that it is the person who should make the decision, when competent, regarding the degree of suffering that will make life not worthwhile.

In the absence of expressed desires, decisions made to preserve biological functions (given that there is no prospect for recovery of the qualities defining human dignity) may debase the person that was. Adhering to the social contract that has been struck with the person prior to the vegetative state honors the decision of the person. There are some consequentialist concerns that should be considered when establishing an overall policy.

Acute States of the Organism

There is the general problem of evaluating acute states of an organism. As discussed when considering the permissibility of allowing an act of suicide, certain acute states make an individual unable to rationally and voluntarily request death. There is the further complication that ill people sometimes have alternating moods that lead them to desire death fervently in one mood phase, and just as strongly want to stay alive at another phase. Feinberg (1986) discussed the difficulties involved in such cases. If the person has asked to be maintained on a life-support system, but in a state of greater severity requests death, the inclination is not to honor the second request. If the person has requested death if the condition worsens, and at this later stage the request is withdrawn, the second request will be honored. These decisions on the part of caregivers are reasonable because of the irrevocable outcome when a life-support system is disconnected. If we are mistaken regarding which request is that of the "true self," there is always a next time if the patient is left on life support.

Problems could arise when a cycle begins to occur over and over: A patient wants death, but when the time comes to implement the wish, wants life. For example, the patient, when out of immediate danger, calmly and in a rational and convincing manner makes a reasonable and well-argued request that no life support be used. When the crisis state is reached, the patient, clearly in a state of disorientation, cries out that life support be maintained. Perhaps the person should be made aware of this pattern, and when calm and rational, be able to make the decision that the request for life support be denied on the next occurrence, and it is that latter request that should be honored. The calm and rational periods that occur could justify the decision not to disconnect life support. Which decision prevails? The calm, rational desire for death, or the frantic, disoriented request for life? And, most important, who should make the decision?

Dworkin (1993) introduced a further complexity when he suggested a scenario in which a Jehovah's Witness has signed a formal document stipulating that no blood transfusion is ever to be given in the face of death, even if requested out of weakness. When the moment arrives at which life or death will occur, the patient, in a state of panic, pleads for the transfusion. Dworkin argued that to respect the patient's prior autonomy as a person, the plea for the transfusion should be denied, because the patient is not competent when making the transfusion request, and when competent following the transfusion, would be appalled at having had a treatment that the deeply held religious belief considers worse than dying. The prior right of autonomy should remain in force, because no new decision has been made by a person capable of autonomy to annul the prior decision.

Consequentialism

There are problems with a straightforward consequentialist approach to the question of euthanasia. The consequentialist approach requires an evaluation of life in terms of the usefulness of the life to permit the attainment of various ends. From a consequentialist perspective, it can be argued that a permanently comatose individual should not receive the full moral regard due a normal adult, because the comatose individual can experience neither happiness nor misery, and no longer possesses any social value. A strict consequentialist would consider euthanasia morally acceptable in such cases. If a patient is terminally ill, suffering great pain, and requests death, then the consequentialist would argue that euthanasia is morally correct to decrease the amount of misery in the world. Even if the person does not want to die under such circumstances, a hard-core consequentialist might even be inclined to recommend euthanasia to end the miserable life, and, thereby decrease the total amount of misery, whether it is considered at the individual level or at the holistic level of the world. Rachels (1986) argued that a person should not be killed under these circumstances, because the act would be unjustifiable murder. Although holistic, consequentialist arguments do introduce serious questions regarding morality, it might be possible to use the consequentialist calculus as one component when discussing euthanasia.

Considerations of the interests, values, and desires of an individual lead to more satisfactory moral positions than using happiness and misery as the determining entities. A contractarian argument can be used, involving the interests of all concerned. These interests involve the values of all parties in the social network, including the wishes of the patient and survivors who desire to honor, or not to honor, their social obligations. The survivors' wishes to continue or discontinue life support might prevail when the patient is irreversibly comatose—the permanently comatose individual has lost the claim to personhood, with the only morally relevant concern being the residual personhood that depends on the history of the individual who exists only in the minds of the survivors. The survivors should take into consideration not only their own sentiments and feelings, but also the dignity of the patient. As Dworkin (1993, p. 217) observed, "Making someone die in a way that others approve but he believes a horrifying contradiction of his life, is a devastating, odious form of tyranny."

Active versus Passive Euthanasia

A significant moral distinction has been suggested, especially by theologians and some members of the medical profession, between active and passive euthanasia. Some evidence that this distinction is intuitively meaningful was indicated by the results of a study of moral dilemmas (Pertrinovich & O'Neill, in press). People were asked to choose whether they would take action or do nothing in a situation in which one set of individuals would be killed by taking action in order that another group would not be killed. When the situation was worded to indicate that taking action would kill one set of individuals, the participants chose not to act as often as when worded to indicate that taking action would save the other set of individuals. This was found even though the outcome was identical in each case. This means that, at least at the semantic level, people are more likely to agree to let something happen rather than to cause it to happen.

Rachels (1986) considered the active versus passive distinction to be morally indefensible, because the outcome of decisions are what have moral relevance. Rachels (1979) argued that if a person's death is not a good thing, then no form of euthanasia is justified. Both active and passive euthanasia are acceptable, or both are unacceptable. He disagreed with the argument made by the AMA that if the means to prolong life are extraordinary, then their cessation is permissible if there is irrefutable evidence that biological death is immanent, given that the cessation of these extraordinary means is requested by the patient or the immediate family. His disagreement was based on the belief that it is difficult to specify just what is extraordinary as contrasted with ordinary, and that it is preferable to think in terms of the justifiability of the consequences, with the ordinariness of the means being irrelevant.

Rachels (1979) maintained that the use of passive euthanasia could lead to a great deal more suffering than active euthanasia. An example would be whenever a patient has to endure horrible pain for many days, finally dying when it is decided to discontinue medical treatment. In such circumstances, active euthanasia, in the form of a fatal injection, is arguably preferable on humanitarian and consequentialist grounds. If the decision to give the lethal injection is wrong because, for example, the patient's illness is curable, then the euthanasia decision is wrong regardless of whether it is active or passive. Rachels concluded that if we are justified in assuming that the patient would be no worse off dead than alive in the present condition, then the killing is not doing harm, no matter how it is done.

McMahan (1993) agreed that it is difficult to believe that the way an act is done could be more important than the nature of the outcome itself. The fundamental intuitive difference between killing and letting die is that, in cases of killing, we assign primary causal responsibility for a person's death to the agent's intervention, whereas in cases of letting die, primary responsibility for death is attributed to factors other than the intervention by the agent. The overriding concern should not be with the causal mechanism, but with the outcome.

Exception to Rachels' position was taken by Steinbock (1979), who argued for the meaningfulness of the distinction between ordinary and extraordinary methods of treatment. She argued that the purpose of allowing people to refuse medical treatment is not to give them the right to decide whether to live or die, but to protect them from unwanted interference by others, and that the reason to discontinue treatment is to avoid extraordinary treatment that causes more discomfort while providing little hope of benefiting the patient. On this basis, she rejected the idea that the intention of discontinuing treatment is to cause the patient's death.

Are procedures ordinary when there is a reasonable hope that the patient will benefit, and extraordinary when there is no reasonable hope? Or does *ordinary* also have an economic and medical component in terms of involving no excessive expense, pain, or other suffering, whereas extraordinary involves these excesses? Such problems are encountered whenever consequentialist calculations are suggested. How can the value of the length of life be compared to the costs incurred to produce that length of life to make it possible to perform a consequentialist calculation? How can the benefit of extended life be compared to the cost of pain in order to calculate the balance? How should one evaluate the value of different times of life extension: Are 2 weeks twice as good as 1; are 2 months twice as good as 1 month and one-sixth as good as one year? Rules have not been developed to make such calculations, so any practical decisions will have to be made on a case-by-case qualitative analysis until an objective system can be developed and agreed upon. The factors that could be used to make qualitative analyses include such things as independent medical testimony to the effect that the patient is dying painfully, that the patient has made a serious and reasoned request to die, that there is little likelihood of any cure that will permit the biographical life to be resumed, and that the surviving relatives have been given full information and counseling whenever their decision is involved.

Steinbock argued that the distinction between *ordinary* and *extraordinary* lies in the physician's intention. This distinction is in the spirit of the *Evangelicum Vitae* issued by Pope John Paul II (Bohlen, 1995). Although the Pope condemned euthanasia, he stated that Catholics can refuse aggressive medical procedures; they can forego extraordinary or disproportionate means if their intention is to foster an acceptance of the human condition in the face of death, because such aggressive medical treatment imposes an excessive burden on the patient and the family. According to such views, withholding extraordinary care can be justified if based on a decision not to inflict painful treatment on a patient when there is no reasonable hope of successful recovery. Withholding ordinary care would be construed as neglect, and Steinbock considers such neglect to be medically unethical. She argued that the intentional cessation of life-support treatment does not terminate life unless the doctor has the patient's death as the purpose for stopping the treatment, an evocation of the Doctrine of Double Effect, which involves a questionable logic on which to base moral permissibility.

Steinbock discussed the spina bifida example presented as Case 3. She opposed the use of active euthanasia for these infants, even though she acknowledged that waiting for them to die "might be tough" on parents, doctors, and nurses. Her belief is that it is better to make the remaining months of the child's life comfortable, pleasant, and "filled with love," because the policy is more decent and humane than killing the child. Her opposition to active euthanasia is based on the concern that a general tendency will develop to kill children rather than to let them die by withholding treatment, and such killing she considers unethical—a familiar slippery-slope argument.

Brandt (1987) adopted a position similar to the one advocated by Rachels. He argued that it would be better to give the infant a lethal injection (what he termed a "delayed abortion" or "legal miscarriage"), considering this act kinder than allowing deterioration until the inevitable death arrives, with or without treatment. Brandt considered some interesting rules of law. If a woman learns during the first 6 months of pregnancy that the infant will be born with defects, then she is free to have an abortion, and at that time can have the abortion without giving any reason at all. If a woman is a carrier of hemophilia and an amniocentesis test indicates that the fetus is male (giving it a 50% chance to be normal), then she is free to have an abortion, and some argue that she has a moral obligation to do so. Using a lethal injection to actively terminate the life of a neonate who is born defective is not permissible by law. Brandt argued that the mere timing of the termination of life is morally trivial, and that the decision should rest on the condition of the fetus or neonate, and should not involve considerations of when the knowledge about the defect was obtained.

Emanuel (1991) examined the ordinary–extraordinary care standard in detail. One objection he made to the standard is that the fundamental ethical principles on which the distinctions are based are neither articulated nor justified. There seem to be at least two implicit ethical standards. One involves judgments about what factors are appropriate to distinguish ordinary from extraordinary care. These judgments require, not factual decisions, but ethical ones that need to be justified. The second involves the substantive definition of what constitutes excessive expense or pain: how are pain, suffering, and hope to be defined and scaled in terms of relative magnitude? There are too many conflicting and ambiguous meanings employed in actual practice, and these conflicts make the ordinary–extraordinary distinction of little use to resolve the question of what medical treatments should be provided.

In the case of defective neonates, life and death decisions should be considered by the parents in consultation with the physician and a medical ethics board. A decision to actively terminate the life of a defective neonate might even be acceptable to those who want to limit the number of abortions, because it could eliminate many elective abortions. If the parents did not want to take the risk of raising a hemophiliac male child, and did not want to incur the risks of a late abortion, they might abort before knowing the sex of the fetus. If it was legally permissible to actively terminate the life of those 50% of male neonates who were born with hemophilia, then there would be no reason to abort any females. The parents would be assured that they would not have to raise a defective male, and this should eliminate any necessity to abort early, thereby sparing all female fetuses. Infanticide might be acceptable if the alternative was to legally abort all fetuses because they might be defective, if that is the only legal alternative. This argument leaves me with an uneasy intuition regarding the fact that the social contract is being voided arbitrarily. This contract is one that should be honored for all persons at the point of birth. For this reason, it can be argued that infanticide should not be allowed.

Slippery Slope

The requisite slippery-slope argument regarding euthanasia is based on the specter of Nazism. The argument is that under the Nazi regime, euthanasia and medical experiments using humans may have begun with humane intentions that were not inspired by racist motivations. Gradually, however, advocacy developed to permit voluntary euthanasia for the terminally ill, and from this voluntary euthanasia, the descent began irrevocably, until the depths of involuntary euthanasia were imposed for anyone considered useless to society (such as the mentally retarded) or enemies of the state (especially Jews and communists), and the descent to the depths of genocide was realized.

The controversial aspects of this slippery-slope argument involve, as Walton (1992) pointed out, a questionable analogy between the social context that prevailed at the time of the Nazi state and that prevailing in North America, as well as the particular historical interpretation of how the Holocaust came about in Nazi Germany. Rachels (1986) challenged the use of the term *euthanasia* ever to have been applicable to the Nazi genocide program. He documented (pp. 175–178) that the killing did not begin because Hitler and his followers were a humane bunch of guys who permitted mercy killing from a sense of compassion at the start, and then, step by step, moved from the initial mercy killing to the mass exterminations that took place in the concentration camps. From the beginning of the Nazi movement, a virulent racist ideology was firmly in place, with the explicit intention to increase the numbers of "racially pure Aryans" and to decrease the numbers of all others. There was no slide down a slippery slope from an initial benign beginning. The initial motivation was not that of euthanasia but that of genocide based on a religious and political agenda.

Muller (1991) published an intensive study of the courts of the Third Reich in his book *Hitler's Justice*. As early as 1923, there was evidence of anti-Semitism in trials taking place in the German courts. Arguments regarding the racial purity of the German people had been a central concern of the National Socialist German Worker's Party since its founding in about 1923, and in 1924 Hitler, in *Mein Kampf*, warned about the danger of "mingling higher and lower races"—with the lower including "coloreds," Slavs, and Jews in particular. He maintained that there was to be no intermarriage or "sexual interbreeding" between members of those groups and people of German descent. As early as 1930, it was argued that the law should forbid marriages between Germans and Jews, and all sexual intercourse and rape between Germans and those of the unacceptable groups should be punished. By 1935, all intermarriage was criminalized. Hitler had proposed a plan in 1927, in which he proposed that newborn infants with physical or mental defects should be killed, and he advocated mandatory sterilization in cases of genetic disorders, which were defined by law in 1936 to include feeblemindedness, schizophrenia, manic–depression, epilepsy, degenerative chorea, hereditary blindness and deafness, severe physical deformities (which included hemophilia, harelip, cleft palate, muscular dystrophy, and dwarfism), and severe alcoholism. Muller cited estimates that about 350 thousand involuntary sterilizations were performed. It seems a far stretch of historical facts to argue that these events represented an initial set of good intentions that led, step by step, down the slippery slope to the unthinkable depth of human depravity that were maximally expressed in the 1940s. The movement began in the depths of depravity.

There is no reason to believe that the kinds of voluntary euthanasia that we have been discussing have any resemblance to what prevailed in Nazi Germany, and there is no reason, given the careful and reasoned arguments that have been made by those advocating physician-assisted suicide, that it represents the first step in an inevitable movement toward involuntary euthanasia. Yet, a "falling dominoes" argument has been advanced to the effect that the dominoes are arranged so that once voluntary euthanasia is legalized, political pressures will develop inevitably for the legalization of euthanasia, public opinion will be galvanized, and politicians will move toward the legalization of involuntary euthanasia, with the ethics of the Nazi era incubating and appearing once again. The arguments offered by Reich (1993), which were quoted in Chapter 4, explicitly represent such a slippery-slope argument.

Feinberg (1986) pointed out that the reasonableness of such arguments depends on how the dominoes are in fact placed, and this raises complicated empirical considerations that are difficult to defend. The burden is on those who argue the inevitability of any slippery-slope scenario to demonstrate that the dominoes are lined up in order, with each one destined to topple the succeeding one until the ultimate unacceptable end is reached. The particular slippery-slope metaphors used are based on an oversimplification of the current social situation and on a questionable historical reconstruction of past events. I discussed the fallacies involved in slippery-slope arguments in Chapter 2 of *Human Evolution, Reproduction, and Morality*, and the interested reader might consult that extended discussion

POLICY ISSUES

Before examining the policies appropriate to each of the seven cases outlined at the start of this discussion of euthanasia, it will be helpful to consider a few further distinctions and clarifications. It is morally impermissible to kill if a person is harmed by the act; such an act is defined as *unjustifiable murder*. The crucial nature of the moral principle of freedom from harm must be recognized to protect members of society who could be placed in jeopardy of being murder victims.

Flew (1986) spelled out three critical differences between murder and voluntary euthanasia:

1. Murder victims are almost always killed against their will, whereas a patient receives voluntary euthanasia only if it has been repeatedly and strongly requested by the patient.
2. The murderer kills the victim in a way that treats the victim as an object for disposal, whereas the object of euthanasia is to save someone, at his or her own request, from needless suffering and degradation.
3. The murderer defies the rules against harming others, whereas those performing euthanasia are operating within rules that respect the interests and dignity of the person.

A distinction can be made between sustaining the life of a person and prolonging that person's process of dying. This distinction defines one of those gray areas that must be examined, because actions might be considered morally permissible, depending on how the situation is construed. A related concern is between actions that prolong only the quantity of life (preserve it for the maximum period of time possible) and those that affect the quality of the life (in terms of future prospects). This concern is present whenever a person is living a life of extreme suffering and will merely exist for a year or so. If the person cannot spend time on any activities or projects other than lying painfully in bed, the question is whether an additional year of life provides a value worth having, given that the person requests to die.

To resolve the cases of euthanasia, the wishes of the surviving family have been honored whenever there is a conflict between their wishes and those of the now-incompetent patient. This is contrary to the policy argued when organ transplants were discussed in Chapter 3, in which the wishes of the donor were favored over those of the family. This apparent contradiction is based on the position that, in the donor case, the needs and interests of the potential organ recipient should be considered, and that person's need, combined with the wishes of the donor, should override the wishes of the family. In the case of euthanasia, with no organ donation involved, decisions to honor the needs of the survivors should be favored over those of the now-incompetent patient in order to maintain the strength of the social network.

CONSIDERATION OF THE EUTHANASIA CASES

Active euthanasia should be the preferred course of action in Case 1, in which a person is brain dead and in a vegetative state, as well as in Case 2, in which the brain stem is functioning, but there is no hope of regaining biographical life. If the survivors do not want euthanasia to occur, then their wishes should prevail as long as the maintenance of support is not at public expense and does not deprive someone with a better medical prognosis of the use of a scarce facility. If the patient expressed the desire to continue support prior to the coma, that decision should not prevail if the survivors want the life-support system disconnected in Case 1; the survivors can weigh the consequences of their decision, and the person in question no longer exists. If the same circumstances prevail in Case 2, then a lethal injection should be permissible if that is the decision of the surviving relatives, given that the safeguards against abuse suggested earlier are in place. These decisions are based on the fact that the person has ceased to exist, and the important considerations are those of the living survivors.

In Case 3, which involves anencephalic and spina bifida neonates, active euthanasia would be the best procedure, and this procedure is more humane than using passive euthanasia and allowing the inevitable debilitating degeneration to run its course. The spinal tube defects occur in the first 6 weeks of pregnancy and are due to a maternal folic acid deficiency. Although the defect occurs in the first 6 weeks of pregnancy, many women do not realize they are pregnant that early, and the defect cannot be detected until the second trimester of pregnancy. It is estimated that about 2,500 babies are born in the United States with neural tube defects, and another 1,500 or more fetuses are aborted following the diagnosis.

In cases in which there is a spinal tube defect, and it is decided that euthanasia is the proper course, it would be better to terminate the life as early as possible, rather than letting the organism degenerate gradually. Whatever benefits can come from such unfortunate circumstances should be realized by using the organs and tissues for transplant or research to benefit living members of society. Unfortunately, these benefits are often not allowed. *The New York Times* (November 15, 1992) reported that the Florida Supreme Court upheld a lower court's ruling that prevented the use of an anencephalic infant's organs for transplantation. The parents wanted the infant's short life to have meaning (it did die 9 days after birth) so that other children might live, but the Court denied the request to declare the infant dead at birth so that her organs would be usable for transplant.

There are two stages in the Down's syndrome example presented as Case 4: one in which there is an intestinal blockage requiring surgery; the other in which the infant will develop normally to reach its full, although restricted, potential. In the first case, the decision to have the surgery or not should be left to the informed decision of the parents. If they do not wish to raise a retarded infant, then the surgical intervention should not be done against their wishes, and it should be morally permissible for the infant to be allowed to die using either active or passive euthanasia. In the second case, the parents should be encouraged to care for the child if possible, and if they do not wish to, or are not able to assume such responsibility, then society should provide facilities and provisions for the sustenance and care of the child, who is entitled to the respect due a moral patient.

In Case 5, the person is living an inadequate quality of life and requests death. Such a request should be honored, given the proper confirming medical opinions, regardless of the wishes of survivors. If the person is judged to be competent and the decision is considered to be rational and voluntary, then the decision of the individual in question should be allowed to prevail in all matters involving personal survival and dignity.

In Case 6, the person is suffering but not terminally ill and choose to die. As long as the medical opinion is that no cure is possible, that the suffering will likely continue or increase, and full psychological evaluation and counseling have been provided, then the person should be assisted in active voluntary euthanasia by a physician-administered lethal injection. It is desirable to have professional assistance to guarantee the procedure is done without inflicting the indignity of further damage to the person.

In Case 7, there is a "happy demented" person who had requested, when competent, not to have the estate spent on care and to be allowed to die if mentally incapacitated. To respect the person's autonomy, as well the dignity of the biographical person, the request made while competent should be honored to the extent that it is legally possible to do so. As Dworkin phrased it, the past dignity of the person when competent should take precedence when any decisions are made.

Finally, in Case 8, resources are limited, widespread starvation is taking place, and individuals are in various stages of deterioration. Qualified medical personnel should be allowed to set priorities to distribute the available resources to those individuals who will be most likely to survive the ravages of starvation and the attendant disease processes, although other individuals will be left to die of starvation. Such decisions should be permissible as long as they are based solely on criteria concerning the physical status of the individuals, with no other social or political factors playing a role. It would be desirable to relieve the medical personnel of as much of the responsibility as possible for the actual allocation of the resources with nonmedical policy makers having the major voice in the general policy decisions. Some of these issues are similar to those involved when it is necessary to consider the prospect of rationing health care on the basis of the type of disease, prognosis for recovery, and quality of life.

These cases were resolved with an intent to preserve human dignity, honor social contracts, and minimize pain and suffering. If persons decide they would be better off dead, and their judgment is considered by experts qualified in medicine, law, and morality to be based on sound evidence and logic, and to have been made voluntarily, then it should be morally permissible for those individuals to avail themselves of assisted voluntary euthanasia administered by trained medical personnel. The moral beliefs and preferences of the medical personnel involved were not considered. Their moral beliefs should not take precedence over those of the person and survivors, as long as the actions requested are not illegal. We do not allow other members of society who perform needed services to decide which of the laws of society they will choose to honor, and the medical profession should not be allowed such an exception.

SOME ADDITIONAL ISSUES INVOLVING DEATH

There are two major concerns to be considered briefly: the AIDS crisis and capital punishment. They are slighted, not because they are unimportant, but because they are so complex that each should be considered at greater length than is possible here. Because of their importance, the issues involved should be outlined.

The Problem of AIDS

Much of the discussion regarding euthanasia and the need to establish a new medical profession to deal with hospice movements and the problems of the dying also applies to the AIDS epidemic. The special issues regarding those who are dying are such things as developing increased awareness (both within and outside the medical profession) of the medical and social problems, setting research priorities, and changing some of the regulations that usually are followed before the outcomes of medical research are applied in treatment. Concerns regarding HIV testing and the confidentiality of medical records regarding the outcome of testing pose problems concerning the balance between protecting the interests of an individual and the interests of the community to cope with an epidemic. There is obviously no simple moral answer to conflicting values, and there is a paramount need to maintain confidentiality.

The rational solution to some of the problems associated with AIDS runs afoul of the theological and moral imperatives insisted upon by many elements of society. In the early stages of the epidemic, the common wisdom was that AIDS affected homosexuals and drug users exclusively. This led some religious groups to suggest that AIDS was a punishment for sin. Konner (1993) suggested that there is a need to rethink such "blame-the-victim" moralizing: If AIDS is God's punishment for male homosexuals, then lesbians must be God's chosen people, because they have lower rates of infection than heterosexuals!

AIDS is being widely transmitted through heterosexual activity, and since that reality has been brought to light, there has been a call in the heterosexual community for research to proceed apace. Konner suggested that the panic regarding AIDS is not fueled by compassion but by fear—AIDS can get to *us*. There are other major diseases that far exceed AIDS in their impact on human options in the undeveloped world. U.S. producers actively and vigorously spread tobacco addiction to the developing world, which could have a more devastating effect than AIDS; politicians have opposed family-planning programs, despite the fact that overpopulation has led to an immense number of dead due to famine; we have not attended to the health of the world's children, even though it has been estimated that, worldwide, about 15 million children die each year from diseases that are preventable with small expenditures of funds.

Old-fashioned morality still impedes many rational solutions to the problem of the AIDS epidemic. It has been demonstrated that one of the major vehicles by which AIDS is spread through the heterosexual community is through sharing contaminated needles by IV-drug abusers. These drug abusers then infect heterosexual partners who are not drug users, especially prostitutes, who have the opportunity to infect a number of subsequent heterosexual individuals. Needle-exchange programs have been opposed at both the local and national level because they might only increase the number of addicts, a view that Konner noted has been contradicted by evidence from many countries. The "just-say-no" attitude is potent in the United States whenever behaviors that the righteous consider morally unacceptable and unconscionable are judged. As Konner observed, we will have the gratification of knowing that we held firm to our high moral standards, no matter how much agony and suffering it causes the sinners.

One biological implication that is seldom considered is the impact that AIDS could have as a major factor in evolutionary change. Demographers have predicted that, in some African nations, the death rate due to AIDS could exceed the rate of reproduction in a few years. The effects of this systematic and strong selection pressure on the ability of local populations to survive, on the resultant structure of the gene pool if they do survive, and on the distribution of the phenotypes that prevail in the population are almost imponderable. Viewed from the perspective of possible evolutionary implications, the AIDS crisis extends well beyond issues in health-care management and involve possible alterations in terms of basic evolutionary adaptation in some regions of the world.

Capital Punishment

Capital punishment has been used throughout the world for most of recorded history and has been mandated for a wide variety of crimes. Cycles of public sentiment in the United States have moved between advocating that there be an almost complete abolition of the death penalty to an insistence that it should be more extensively used for a broader range of offenses. Three major arguments are often used to justify capital punishment: First is the consequentialist argument that it effectively deters crimes of the most offensive kinds; second is the deontological argument that it should be considered to be a retribution exacted by society on the perpetrators of certain crimes; and third is a naturalistic argument that emphasizes the pure primal need for vengeance.

Consequentialists point to the presumed deterrent function of capital punishment and focus on the number of lives it saves, assuming that the more lives saved the better (Kleinig, 1991). The consequentialist argument stands or falls on the assumption that execution deters more efficiently than less severe forms of punishment. Deontologists argue that certain crimes deserve the death penalty and compare the claims of the life wrongfully taken to the claims of the life of the murderer. This Kantian argument is retributive and maintains that the murderer has forfeited the claim to continued life, because the killing was done by a rational being, and it is a rational being to whom the death penalty is applied. The claim is that the death of the murderer upholds the dignity of human life and also maintains the dignity of the killer as a human being. Kleinig considered it significant and interesting that both defenders and opponents of capital punishment frequently appeal to the value of human life as a basis for their respective positions.

The naturalistic argument breaches on the shoals of the naturalistic fallacy. Even if there is an evolved tendency to wreak vengeance for harms done to one's kin or community, this does not mean that a society ought to permit the act of killing for that reason. The motivation for vengeance runs deep, especially when barbaric acts are committed. Some of the comments made by people in Oklahoma City after the Federal Building bombing reveal the strength of such motivation. People's outrage led to suggestions that the perpetrators should be publicly tortured at the site; a quick death by execution is too good for them—it should be slow and agonizing; they should have to suffer the humiliation of living a life of bondage and servitude, facing derision and suffering humiliation at the hands of the survivors. The tragedy also unleashed a strong dose of xenophobia: There was an initial eager tendency to blame Mideastern terrorists (them!) and even greater horror at the realization that the perpetrators were Americans who caused the slaughter of innocents. Such an act of unimaginable domestic horror reveals basic evolved traits that could have been useful in the EEA. It is not reasonable, however, that they should be allowed to prevail in a moral society merely because of the strength of the anger and the need for vengeance.

The moral permissibility of taking a life has been debated at length and is an especially sensitive issue to some in the medical profession. It has been argued that the participation of physicians in capital punishment is a corruption of the healing profession's role in society, a violation of the Hippocratic Oath, and the profession should not allow physicians to participate in such killing. Truog and Brennan (1993) identified several stages of participation by physicians in regard to prisoners on death row. The only stage in the execution process in which physicians should participate, in their view, is to prescribe sedatives and tranquilizers at the request of inmates who are anxious about their impending execution. Physicians should not prepare prisoners for the execution by prescribing, procuring, or preparing the medications to be used in lethal injections, nor should they insert the catheter, pronounce that death has occurred, or certify the death. They agreed with the AMA that physicians should not recover organs for transplantation after the death has been certified, even if the prisoner has requested it. They noted that medical participation in euthanasia is almost always illegal, whereas medical participation in capital punishment is generally legal. Their opinions were based on a belief that medicine is "at heart a profession of care, compassion, and healing" (p. 1348), that capital punishment does not encompass these virtues, and that participation by physicians is subversive to the core of medical ethics. Their solution is for medical societies to consider a physician's involvement in capital punishment to be grounds for revoking that physician's license.

In a free society, physicians must work within the bounds of the laws of society, even though individual physicians might not agree with the society's laws. On the other hand, participation in capital punishment might well be allowed to remain an individual physician's choice, because there will be a sufficient number to support the relatively few executions that result from court orders. A similar argument has been made regarding allowing physicians to not be involved in performing legal abortion or euthanasia whenever their individual beliefs lead them to oppose the practice. In *Human Evolution, Reproduction, and Morality*, it was argued that physicians shirk their obligations to their patients who request a legal and therapeutically justified medical procedure whenever they allow their personal ethical beliefs to deny patients theirs. Such personal betrayal can cause serious breakdown in the trust that patients have in their physician's willingness to treat their medical needs.

Still others argue that physicians who certify the death of executed prisoners are acting in an official capacity as certified members of the medical profession, and that this profession has the duty to certify a death. Although physicians have a primary duty to their patients, they also are participating members in a society that has legislated a series of laws that all citizens, including those in the medical profession, must respect. In a democracy, the legal alternative available to all, including physicians, is to work to change the law or to have it declared unconstitutional, not to violate it with impunity.

Bedau (1982) presented statistics regarding the issues involved in capital punishment. These statistics bear on such things as changes throughout the years in the specific offenses punishable by death, the effect of capital punishment on the prevalence of murder, the number of death sentences handed down by the courts, the time elapsing between sentencing and execution, and the proportion of sentences that result in execution. He included critical evaluations by a number of experts of the research on attitudes toward the death penalty and the problems involved when attempting to evaluate the deterrent effect of capital punishment.

An examination of the statistics, the quantitative models, and the arguments presented convinced me that the issues are so involved, and the statistical questions so profound that no abbreviated treatment should be attempted. For example, the effectiveness of the death penalty to deter murder is difficult to establish. Vidmar and Ellsworth (1974) asked whether the death penalty is a deterrent at all, ever, to anyone, and is it a more effective deterrent than long-term imprisonment? There is an inconsistency between the general attitudes people express and the choices they make when asked to decide specific cases. Some people say they favor the death penalty, even though it might have no deterrent effect, and many others, because they believe in the need for vengeance, or to signal a moral condemnation of crime. When questioned about the appropriateness of the death penalty in specific cases, however, many tended to be less likely to favor it. Even those who strongly favored a mandatory death penalty were reluctant to recommend a guilty verdict leading to a death sentence when presented with specific hypothetical cases. Vidmar and Ellsworth concluded that people's willingness to endorse a mandatory death penalty in the abstract might not indicate they are willing to put such a policy into practice. Gibbs (1978) concluded that a truly grotesque view of morality would be required to argue that humans be put to death for the sake of educating the public that crime does not pay, especially because there might be other means, such as life imprisonment, to achieve that end.

The whole issue of the efficacy of punishment to control behavior is complicated, and when the issue is examined in the context of the criminal justice system, the number of factors that must be included in the tangled web of interactions is staggering. There are a number of difficult issues in terms of the consequential reasonableness of capital punishment, and others concerning its moral permissibility. In the interest of developing a consistent and universal set of moral principles, these issues must be considered. However, these issues will be considered no further here.

CHAPTER 8

Medical Ethics and Hospital Review Boards

MEDICAL ETHICS

Responsibilities of Physicians

The responsibilities and risks that physicians and other medical personnel should assume regarding living and dying have received considerable discussion. The medical profession characterizes its major responsibilities to be caring, healing, and preserving life, and argued that medical personnel should not kill people actively under any circumstances. The explicit professional code of conduct adopted by the medical profession is the 2,500-year-old Hippocratic Oath, which Rachels (1986) characterized as being a historical relic rather than an actual guide. The Hippocratic Oath construes the physician to be near godliness, as the medical anthropologist, Konner (who is both an M.D. and Ph.D.), recounted in his superb book, *Medicine at the Crossroads* (1993). Konner enumerated recommendations in the Oath that enhance the exalted image of the physician, including those to maintain an imposing appearance and mode of dress, to use decisive utterances, and to speak with great brevity. The Oath also recommended that all discussion should be conducted calmly and that most things should be concealed from the patient, with nothing being revealed of the patient's future or present condition.

The Oath emphasizes that medical mysteries should be kept mysterious, and Konner observed that such procedures would not only cause raised eyebrows, but also would be on shaky legal grounds, given the current insistence on informed consent. The model many physicians believe should be accepted as a goal for physician–patient relationships is the "patient-as-colleague" model. Given the antiquated nature of the

Hippocratic Oath, the professional code of medical ethics should be reconsidered and updated in light of advances in medical technology, recognizing that these new advances have placed great strains on the entire health-care system.

Conflicts of Interest

A major problem is that many members of the health-care community have business interests in the delivery of health care that can compromise their ability to practice medicine objectively. Physicians sometimes have financial interests in the hospitals to which they send patients, the HMOs in which they practice, and the laboratories from which they order tests. These financial interests and the fear of malpractice claims lead physicians to order large numbers of tests to rule out possible medical complications, and often result in exorbitant medical costs, especially for insured patients.

One example of the complications involves the use of fetal monitoring devices to detect early labor in pregnant women who might be at risk of premature delivery (Meier, 1993a). The company selling the devices and services using them approached physicians with an offer to become shareholders in the company for $1000; in return each physician would receive 15% of the payment for services they prescribed. The fetal monitoring device is reported to cost $100 to $300 per day—about $5,000 per pregnancy. The problem is that there is no solid research evidence indicating that monitoring makes any difference beyond what can be obtained by educating the patient or offering counseling regarding possible complications in pregnancy.

The financial conflict of interest is dealt with in some instances by federal laws that bar physicians from referring Medicaire and Medicaid patients to companies in which they have financial interest, but private patients are not protected by such laws. Some private insurers and state Medicaid plans refuse to pay claims for fetal monitoring devices, whereas most do pay them. There has been a call for research by the National Institute of Child Health and Human Development, with little interest expressed by most of the companies that provide the monitoring services. There seems to be a weakness in the scientific justification, as well as a financial conflict involved in this case, and these concerns reflect badly on the medical community. Such concerns help fuel the cynicism and resentment of the public toward the medical profession, such as those reported by Hilts (1993b) regarding negative attitudes the public has toward what they perceive as exorbitant physicians' earnings.

Clinicians argue they should have total authority to deal with matters related to patient welfare. With complete authority, however, there must be guarantees that authority is being exercised responsibly, and there are reasons to believe that, in many instances, the medical profession's self-regulating medical review boards have not adequately safeguarded the interests of either patients or society.

Thompson placed the conflict of interest problem in a general perspective that included many of the aforementioned concerns. His definition of *conflict of interest* was (Thompson, 1993, p. 572):

> a set of conditions in which professional judgment concerning a primary interest (such as a patient's welfare or the validity of research) tends to be unduly influenced by a secondary interest (such as financial gain).

Primary interests are determined by the professional duties of a physician to the health of patients, a scholar to the integrity of research, and a teacher to the education of students. Secondary interests in financial gain and economic security are not illegitimate, but must be prevented from dominating, or even appearing to dominate, the primary interests.

The purpose of explicit conflict-of-interest rules is to maintain the integrity of professional judgment and public confidence in that professional judgment. There should not even be an appearance that professional judgments are influenced by secondary interests. If there is, the trust of patients in physicians generally will suffer. Thompson argued that it is not enough to rely on the good character of individual physicians to ensure that they avoid conflicts, especially given the realities of modern medicine that involve large HMOs and impersonal encounters or distant relationships, especially between specialists and patients.

Representatives of the medical profession have argued that any regulation done should be by the profession itself. Professional organizations did not even formally address conflict of interest in medical codes until the 1980s, however, and then left the problem to the discretion of the individual physician. In 1991, the AMA declared that self-referral was "presumptively inconsistent" with a physician's obligation to patients. Even if the profession exercises regulatory standards, there is still a conflict between a primary interest to maintain the integrity of the profession and a secondary interest to promote the economic welfare of colleagues.

Thompson noted that, although some physicians dismiss all of these concerns with the claim that ethics cannot be legislated, some remedies should be instituted. He suggested (in order of increasing stringency) mediation (blind trusts to insulate the physician from secondary interests), abstention (physicians or researchers should withdraw from a case if they have secondary interests), divestiture (eliminate the secondary interest), and prohibition (require them to withdraw permanently from fields in which they have substantive secondary interests). The seriousness of potential conflict of interest has been writ large, as will be evident when the issues involved in the reorganization of medical practices and health-care delivery systems are discussed in Chapters 9–13.

A concern often expressed when any changes in the health-care delivery system are suggested is that the personal physician–patient relationship might be endangered. As Belkin (1993) noted, this idealized view of medicine as a personalized, private, one-on-one relationship between a physician and a patient facing life and death together, has not been the reality of medicine for a long time. With medicine as it is practiced today, doctors oversee the deaths of relative strangers in impersonal institutional settings, courts step in to second-guess both sides, and some of the most important medical decisions are made by groups of specialists, many of whom may never have seen the patient.

The major problem with medicine is not with doctors but with the social system within which it is embedded (Konner, 1993). Physicians are not able to treat their patients as they would like, but are subject to the control of insurance company review boards that must approve treatments in advance if the insurance company is to pay for them. The scrutiny by insurance boards has led to a massive increase in the number of employees hired by physicians and hospitals to deal with insurers. The Mayo Clinic, for example, has 70 full-time employees talking to 2,400 different insurers, with the cost of this large business staff passed on to patients (Rosenthal, 1993).

Insurers argue that the reviews save money by helping to educate physicians in cost-effective care and cause physicians to think twice before ordering tests and surgery. One wonders about the adequacy of a system designed to conduct education that is staffed by employees of insurance companies, and concern if physicians must be held accountable lest they engage routinely in excessive treatment. This latter claim implies a major lack of moral responsibility by the medical community, which, if true, should be remedied by means other than the actions of insurance company oversight reviewers.

Some argue that it makes little sense to create any new bureaucracy to manage the abuses of an old one, and that medical reviews would be better done within hospitals in a manner that would require peer review before performing surgery or ordering expensive tests. Konner (1993), as well as several other experts, suggested that the best solution for the United States would be something similar to the Canadian Health Plan

(CHP), because it keeps physicians in the private sector but puts payment of doctors in the public sector. The CHP is discussed in Chapter 12, and it will be recommended as the model for an overall U.S. National Health Plan.

Some in the medical profession have spoken out forcefully and have taken leadership, especially on matters relating to dispensing birth-control information, changing abortion laws, and using physician-assisted euthanasia for the terminally ill. Some have refused to obey the law, demonstrating immense courage by making their moral objections and actions public, resulting in considerable risk to their professional reputation and for criminal prosecution. The medical community should consider ethical issues in a responsible manner, rather than forcing individual members of the profession to expose themselves to legal and economic sanctions. Based on the past record, one can raise doubts that medical associations will take a morally responsible role whenever there is a conflict between the primary and secondary interests of physicians.

The Aging Population

Problems relating to death and dying, to euthanasia, and to the care of the elderly are going to be magnified greatly in the coming years. Caplan (1989b) estimated that the number of Americans 65 and over is projected to rise from about 30 million in 1988 (12.5% of the total population), to 39.2 million in 2010 (13.8%), and 64.6 million in 2030 (21.2%). Olshansky, Carnes, and Cassel (1993) detailed some of the demographic realities that must be considered if a reasonable system of health care is to be created. Their estimates were similar to those made by Caplan: In 1900 there were 10–17 million people aged 65 or older (less than 1% of the total population); by 1992, there were 34.2 million in that age group (6.2%), and they estimated that by 2050, about 20% of the world's projected population will be 65 or older. As medical practice becomes better able to combat infectious and parasitic diseases, those degenerative diseases associated with aging, such as heart disease, stroke, and cancer will be more prevalent. The differences in the number of persons of various ages could almost disappear, with 90% of the people born in any given year living past the age of 65, and as much as two-thirds of the population surviving to an average age of 85, with about 110 years being the maximum age. These figures could be realized if nothing more happens than a substantial reduction of death due to vascular disease, because it produces most of the deaths after 65 years of age.

Olshansky et al. (1993) argued that population aging will replace population growth as the most important phenomenon from a policy standpoint. The dilemma produced by these changes is that the current construal of medical ethics has made it necessary for physicians and researchers to develop new techniques and therapeutic interventions to postpone death, which increases the number of aged in the population without necessarily increasing the quality of life of those individuals. Such changes will force governments to restructure all of their entitlement programs if they are to remain financially solvent.

Solomon et al. (1993) studied the attitudes of health professionals regarding the care of patients near the end of life. They surveyed 687 physicians and 759 nurses in five hospitals and found gaps between the views of practicing clinicians and the prevailing medical and legal guidelines concerning the treatment of patients near the end of life. Most of the respondents were aware of the guidelines regarding patient involvement in treatment decisions, but only one-third of them were satisfied with the adequacy of patient participation. The expressed dissatisfaction stemmed primarily from concerns that there was too much, rather than to little, treatment provided to terminal patients.

Eighty-seven percent of the sample agreed that all competent patients, even if they are not terminally ill, should have the right to refuse life support, although that refusal may lead to death. The same percentage agreed that there is a moral difference between allowing patients to die by foregoing or stopping treatment as compared to assisting in their suicide. The medical personnel also placed reliance on the distinction between "ordinary" and "extraordinary" treatments, despite the fact that these guidelines are difficult to understand, and most medical commissions have recommended that this distinction should be neither ethically nor legally relevant in medical decision making.

The respondents accepted the Doctrine of Double Effect, agreeing that it is acceptable to provide large quantities of narcotic analgesics to control pain and suffering, even though it does shorten the life of the patient, so long as the latter effect is not the primary intention. Solomon et al. recommended that hospital caregivers should be educated about the agreed-upon guidelines and that this education should be done in such a way that it can enhance the likelihood that they are understood and implemented. They also considered it important to attend to both the psychological and ethical aspects of moral decision making, to encourage multidisciplinary discussions of ethical issues, and to improve the level of dialogue between ethicists and those who practice at the bedside.

Dubler (1993), a lawyer and bioethicist, commented on the preceding report. She agreed that there is a chasm separating abstract principles from "the messy reality of patient care," and argued for the recognition and reduction of that gap. She considered the task of the 1990s to be to bring a patient's preferences, the concerns of family, legal rules, and ethical principles into harmony.

The Canadian Critical Care Trials Group (Cook et al., 1995) studied the attitudes of medical staffs in 37 Canadian university-affiliated hospitals to determine the importance of the factors used to consider withdrawal of life support. Hypothetical scenarios were developed to provide reliable and sensible instruments to understand treatment policies. They asked direct questions, as well as obtaining decisions in the hypothetical scenarios. A range of treatment facilities were included; they sampled a range of staff including nurses, house staff, and attending physicians.

Several factors were important in decisions to terminate care: likelihood of long-term survival, premorbid cognitive function, and age of the patient. The same treatment option was chosen by more than 50% of the respondents in only 1 of the 12 scenarios that were presented, and opposite extremes of care were chosen by more than 10% of the respondents in 8 of the 12 scenarios.

They concluded that the most striking finding was the variability in the level of care chosen for the same scenario and the fact that this variability was associated with characteristics of the individual staff member—such as age, category of staff position, number of years since graduation, the province, and the number of hospital and ICU beds in the hospital where they worked. Idiosyncratic characteristics of the health-care providers were major determinants of decisions to withdraw care, and they suggested that most patients would find it unsatisfactory that the care they receive is highly dependent on the attitudes of the health-care providers. They recommended that clinicians be made aware of their attitudinal biases that potentially influence medical decisions. Clinicians should be led to recognize the extent to which ethical, social, moral, and religious values influence their medical decision making.

The demonstrated lack of competence, combined with an unwillingness on the part of medical practitioners to develop, maintain, and implement consistent moral positions, pose major questions regarding events that occur at the end of life. Medical educators should devote more time and attention to develop and understand the legal guidelines for death and for legal rights of patients to die with dignity, while at the same time exploring the emotional and psychological implications that health-care practitioners must face. In view of the muddle physicians are in regarding issues of patient's rights and physicians' responsibilities, the exposure to ethics and moral philosophy that medical students currently receive is not accomplishing what needs to be done.

It might be necessary, as well as wise, to establish a separate set of facilities to deal with the dying and even to authorize a different medical profession, although some thoughtful commentators, such as Angell (1988), do not consider this a satisfactory solution. Facilities could be modeled on those of the hospice, where the emphasis is not on the cure and rehabilitation of life, but on the treatment, comfort, and dignity of dying. The mandate to the medical staff would be to treat the patient until it is decided on medical grounds that there is no point, and when that point has been reached, to make dying as easy and comfortable as possible. If there are safeguards against abuse, as well as trained, able, and consenting professionals, such a proposal might resolve some of the dilemmas that face medical personnel who find the emotional burdens too great to bear, or whose own moral convictions would not allow them to participate in active euthanasia, given that it was legal.

Feinberg (1986) cautioned regarding the impossible burden that could be placed on the health profession if the existing health-care system was charged with the sole responsibility to make decisions to end a life. Those doctors who specialize in treating diseases of the most dangerous and serious nature, or who mainly treat the elderly, would be faced with the necessity, again and again, to consider and perform active or passive euthanasia. These acts could lead to repeated risks of criminal prosecution under current legal regulations. Even if active euthanasia were acceptable legally and morally (which would free physicians of criminal prosecution), there were oversight review boards to accept responsibility, and a trusting and accepting attitude on the part of the public, the emotional burden could still be enormous for the medical practitioners. A better solution might be to create a hospice-type medical profession that is willing and trained to cope with these problems in a consistent and professional manner.

Two separate disciplines work well in the area of abortion, another area where there are strong moral feelings regarding providing the service. There are clinics dedicated to abortion, and most abortions take place there on an outpatient basis. Those who do not want to perform abortions are not required to perform them, and it might be satisfactory to allow those who do not want to assist in active euthanasia to have no professional obligation to do so. The hospice program that has been established to assist the large number of people dying of AIDS also resembles the proposed model. Patients who voluntarily, rationally, and morally choose a course of legal action on an informed basis should be free to exercise choice without interference from parties not involved, and physicians who desire to assist should not suffer legal or professional sanctions.

Euthanasia Revisited. The increase in age of the human population makes it necessary to consider the problem of euthanasia more carefully. Rachels (1986), when considering issues related to euthanasia, entertained an analogy between automobile mechanics and medical practitioners. The duty of automobile mechanics can be construed to be to repair cars, not destroy them. It can be argued that it is never right for a mechanic to destroy an automobile, just as it is argued that it is never right for a human mechanic (a physician) to destroy a person. Such destruction could be construed as an act contrary to the nature of both professions. There is little reason to accept this argument as it applies to mechanics (they consign spent automobiles to the wrecking yard for parts, or to the scrap heap for recycling), and no reason to accept it as appropriate in the medical settings. Assuming responsibility for the health and welfare of society requires a broader perspective than just a concern for the sustained management of the individual patient. The desires of the patient, the welfare of other people in society, the stress on available facilities, and the good of the entire community must all be considered when decisions regarding any individual are made.

The moral responsibilities of the medical profession should be reconsidered, because the circumstances facing contemporary medical practitioners are so different from those that prevailed only a few years ago, let alone those that prevailed at the time of Hippocrates. Until recently, it was not possible to keep organisms alive or to maintain the ill for extended periods of time; it was not possible to transplant organs from organism to organism (either within or between species). The increasing quality of medical care and technology has created a large population of elderly people in varying states of health. With the enormous technological changes, as well as the almost endless possibilities for the future, the medical profession would be well advised (owing it to society as well as to itself) to examine and restructure its code of ethics.

Dworkin (1993) invoked a right to beneficence when deciding on treatment procedures for an individual who is clearly no longer competent due to the loss of a coherent sense of self, no discernible short-term aims, or when choices and demands are made that systematically or randomly contradict one another. In such instances, Dworkin argued that the individual has lost the capacity of autonomy that caregivers have a duty to respect. At this point, the individual has the right to beneficence in the sense that decisions should now be made on a paternalistic basis regarding the individual's best interests, and these decisions should be reviewed by a properly constituted medical review board.

Death and Organ Transplants. There is evidence that members of the medical profession are not coping well with the demands resulting from technological advances, because they do not understand the law, which sometimes runs counter to their own personal beliefs, and they have difficulty deciding to withhold treatment from dying patients, tending to accept the position that it is permissible to let die but not to kill. In the study by Youngner et al. (1989), mentioned in Chapter 4, a survey was administered to hospital personnel involved in organ-retrieval programs. Only 63% of the 195 people surveyed knew the legal definition of death in the state in which they were located (irreversible loss of brain function), and only 35% both knew the criterion and were able to apply it correctly to identify the legal status of two hypothetical cases. Overall, 58% did not have a coherent concept of death.

These medical personnel were also asked whether there should be a law requiring hospitals to ask families of brain-dead patients about organ donation. Thirty-eight percent favored making such requests without exception; 38% were in favor, but would grant an exception at the physician's discretion; and 23% opposed such a law. The 23% opposed tended to be those physicians who most likely would be the ones to declare death to have occurred, and would have the responsibility to talk with families about organ donation.

It was concluded that if health professionals have an unclear and inconsistent understanding of death, then it is difficult for them to effectively explain the issues surrounding death to relatives. This lack of clarity regarding whether potential donors are legally dead undoubtedly contributes to the emotional discomfort of those who manage donors in the operating room. Before policies or laws are changed, it is important for health professionals to play a leadership role in clarifying issues in this important public debate. Given the lack of clarity on the part of members of the medical profession, it is difficult for them to play a meaningful role.

Social Judgment Theory

Some ideas that are applicable whenever decisions are to be made that involve facts, values, and policy should be introduced. Problems can arise in any social situation when there is a conflict of interest between different individuals. These conflicts often exist because there are specific points of agreement and disagreement regarding facts, as well as the values, and they have not been identified. The University of Colorado psychologist Hammond has developed a general approach to problems in conflict resolution. The method can be used whenever there is a genuine interest in arriving at decisions regarding social policy. Hammond and his colleagues (e.g., Hammond, Stewart, Brehmer, & Steinmann, 1977; Hammond, Mumpower, Dennis, Fitch, & Crumpacker, 1983; Hammond & Grassia, 1985) extended the methodological ideas of the profound University of California, Berkeley psychologist and philosopher of science Brunswik (1956) and formulated what Hammond calls social judgment theory (SJT). There are many situations in which decisions must be made, and agreements reached by parties who have legitimate conflicts of interest. It is difficult to arrive at reasonable decisions if there are inherent conflicts of interests, values, and opinions, and there must be agreement on policies. Even greater difficulty is encountered when the parties involved have not formulated their own positions clearly and do not understand the positions of those with whom they are negotiating.

Often what occurs during a set of negotiations is that the parties on the different sides are able to progress quite well, each secure that their own policies are consistent and convinced that they understand the policies of the others. Then, an unexpected disagreement occurs during the negotiations, and each side suspects the other of deceit, treachery, dishonesty, self-seeking motivations, and a host of other unsavory intentions. Now, instead of being a negotiation, the situation can quickly become adversarial. Hammond demonstrated that such disagreements often occur, not because there is a deceitful change in bargaining positions, but because neither side understands the relative cognitive importance of the different factors involved for themselves or those on the other side of the bargaining table.

Some concrete examples were provided in a report by Hammond et al. (1977). One case involved a study of an actual labor–management negotiation that had just been concluded. The six negotiators (three from labor and three from management) had been involved in a long and bitter strike, and agreed to reenact the negotiations that had taken place. The points of agreement and disagreement, as they stood one week prior to the settlement, were established and both sides agreed that there were four issues at that point: contract duration, wage increases, the use of certain "special workers," and the number of strikers to be recalled.

Different combinations of the magnitude each of these issues could take were established using 25 hypothetical contracts that included different degrees of emphasis on each of the four issues. Each negotiator rated each of the contracts in terms of its acceptability, estimated the weight that each had placed on each of the issues to reach the decision regarding that acceptability, and each negotiator predicted the weight that the two sides placed on each issue. Self-understanding was poor on the part of each individual negotiator, and there was poor understanding of the policies governing the values of the other side. These misunderstandings occurred despite the fact that all of the negotiators intuitively believed they understood their own value system, as well as that of the others. Furthermore, the three union negotiators were highly uniform in their policies, but the management negotiators were not. The labor negotiators, therefore, did not face a uniform management-negotiation policy, and this state of affairs was not apparent to either side.

All participants were given feedback regarding these cognitive structures and then rated the 25 contracts once again. For this second rating, several of the contracts were judged to be acceptable to both sides, and this change in agreement occurred in but a fraction of the time required for the original negotiations. Although this is a contrived situation, it demonstrates that often people's values do not conflict; rather they lack an understanding of the underlying values (their own and those of the others) regarding the concrete issues being negotiated.

An application of these procedures by Hammond and Adelman (1976) was used to resolve an actual social dispute involving issues concerning value (what ought to be), as well as questions of fact (what is). The dispute concerned the selection of handgun ammunition for the Denver, Colorado Police Department. The Department had decided that their standard ammunition did not have adequate effectiveness to stop and incapacitate a suspect, and wanted to replace the round-nosed bullet type they used with hollow-point ammunition. Some civic groups objected strongly on the grounds that such bullets created excessive injury. The police argued that the increase in injury was minimal and there was decreased threat to bystanders, because the hollow point bullets did not ricochet or pass through the initial target. While the arguments were going on, a policeman was shot and killed by a robber firing hollow-point bullets, which led to an outcry that police needed the same effective firepower that criminals were using. An active, bitter, and polarized controversy ensued between police and civil liberties groups at both the local and state level.

At this point the Hammond research group became involved. Their first step was to study the value that members of city government, police, civil liberties groups, and the general public placed on the relative importance of the three functional characteristics of bullets that all parties agreed were the important values: level of stopping effectiveness, injury to the suspect, and threat to bystanders. Because there was disagreement among the participants regarding the relative importance of these three attributes, it was agreed to weight each of the attributes equally. After having achieved this agreement regarding the characteristics the ideal bullet should have, factual information about the physical characteristics of 80 bullets was obtained from the National Bureau of Standards.

Five ballistics and medical experts, who were external to the dispute, were convened and asked to judge the potential effectiveness of the 80 different bullets on the basis of such factual data as the weight of the bullet, muzzle velocity, and amount of kinetic energy lost in a target simulating human tissue. (The danger to bystanders due to ricochets was dropped from consideration when an examination of police records indicated that there had been no recorded injuries to bystanders due to ricocheting bullets.) Using these data, the experts made independent judgments regarding each bullet's potential stopping effectiveness and the degree of injury it caused. There was high agreement among the technical experts concerning the potential effects of the bullets. Note that the assessment of the facts was completely separated from the assessment of the desired social values, and the factual assessment was made by an independent panel of experts, acceptable to all parties in the dispute.

The next step was to statistically integrate the weight of the community's desired social policy with the mean ratings by the experts on the characteristics of the bullets. One bullet was identified that had the greatest predicted social acceptability: a hollow-point bullet (not the one originally requested by the Denver Police Department) that had more stopping effectiveness and caused less injury than the standard bullet then in use. That bullet is now the standard one used by the Denver Police Department, and all parties in the dispute were satisfied that their concerns had been addressed.

The separation of value judgments regarding social policy from those regarding fact made it possible to use the factual information in a more objective fashion than usual. The negotiators agreed on the social values that should be implemented, and a separate group of experts dealt with the ballistic facts. The solution was applauded, because the members of the community did not have to engage in arguments about the specific outcome (this bullet vs. that bullet based on belief or only partially understood data), but concerned themselves with reaching agreement on the social policies that were to be implemented. These agreements made it relatively simple to pick a bullet that best implemented the agreed upon policy.

The task was not to arrive at what the "right" facts were that would lead to the proper policy, but to help the decision makers think about the issues. Hammond and his group have developed the procedures, the statistics, and the computer support required to implement decision programs such as these, making them accessible and sensible procedures to use to develop social policy, especially if the policies involve both conflicting values and facts that might be related to those policies. If such procedures are employed, then better use can be made of whatever scientific information is available when controversial issues regarding public policy are involved.

To arrive at efficient and congenial resolutions when difficult policy decisions are to be made, it helps to decompose the decision-making process into component parts. First, it has to be recognized that there might be value systems that differ in orientation and in coherence. Second, everyone might not use a consistent value system throughout the range of decisions to be made, or they do not understand the structure of the system they use. In such cases cognitive feedback, using hypothetical combinations of the dimensions that have been identified and agreed to be the important ones, can facilitate the development of mutually acceptable decisions.

Third, it should be realized that scientific information is usually complex. Different scientists often have different social values, varying levels of integrity, and varying degrees of scientific and technical competence. All of these things pose difficulties, because the nature and meaning of scientific literature must be translated by someone with scientific credentials and knowledge to make the literature intelligible to decision makers and to the public. It improves the value and credibility of scientific information if the development is by impartial experts who are outside the adversarial arena as much as possible. Because there often are multiple sources of information competing for credibility, an attempt should be made to gain as much objective information that is free from social value as possible. Such procedures are often used by governmental agencies regarding sensitive areas of social policy when they create panels composed of members of the National Academy of Sciences, whose evaluations and recommendations generally are respected and considered to be more adequate than those produced through the use of "expert witnesses" of the adversaries.

Once policymakers understand the relevant dimensions and structure of the social policies, they can be helped to use scientific information and integrate impact assessments of possible decisions, on the basis of the testimony provided by scientists and the results of studies by experts. These policy makers might then arrive at consistent policies that can be communicated clearly to those concerned with the issues, and rational accommodations and understandings would be more likely to occur than presently is the case. The presumption of goodwill and honesty, even when there are different value systems, is a better presumption than one based on the gambits of finger-pointing, name-calling, confrontation, and attribution of evil intent. The latter strategies all too often lead to a victory based on "might makes right."

To bring the discussion back to moral issues, many of the social policies influenced by the moral principles considered throughout this book involve legitimate conflicts of interest that stem from different economic, social, and moral values. There often is a body of reliable and valid information that can be brought to bear to forecast the potential outcomes of specific social-policy decisions. For example, the question of rationing health care and assigning relative value to different medical treatments has been approached much too informally. The SJT model is ideal to help resolve some of the disputes that have arisen when a system of health-care rationing is considered. A model that incorporates a rational approach similar to that used in SJT has been applied to problems in health care (Kaplan, 1994a). The Kaplan approach has been used to establish social policies—most notably in the implementation of the Oregon Health Plan. These scaling methods will be described and discussed.

It would be better to use procedures such as those involved in SJT rather than having the general public guess regarding the prognosis that can be expected with a given medical treatment, or to have the members of the medical profession decide what the values of the public seem to be. There are methods to identify the social policies the public wants to implement. In some instances, the medical profession can provide objective information regarding the efficacy and cost of various treatment methods, as well as the prognosis for different types of patients who undergo these treatments. Governmental agencies could provide data regarding available public funds. All of this information could be brought together to arrive at equitable solutions, given the existing values, constraints, and realities. Decisions arrived at could then be communicated in a rational and understandable fashion to concerned members of society. Whether all will agree with the decisions is less important than the fact that they are able to understand that the decisions were arrived at in a rational and reasonable fashion. Those who continue to disagree can then argue about the values that were incorporated or seek to have the information they think important weighted more heavily, rather than casting aspersions on the process and disparaging the motivations of those with whom they disagree, a process that quickly descends to the level of bantering empty slogans.

When questions regarding the value of various kinds of medical research are at issue, it would be useful to have the best information possible concerning the past and expected benefits of such research, the costs to the public exchequer and to the participants on whom the research would be conducted. All of these issues should be considered in light of the moral and social values involved. A more satisfactory set of agreements might be reached using this method of problem solving than if we continue to attack the values, morals, and intentions of all with whom we disagree.

POLICY FORMATION IN MEDICINE

How could the principles and procedures outlined here be used to clarify issues in medical ethics, such as those regarding organ transplants, the termination of intensive care, or the development of a national health-care delivery system? The first step would be to identify the values to use in order to develop policies to enhance the quality of life. Kamm (1993) and Temkin (1993) suggested that one criterion is the Objective List Test, which involves the development of a list of experiences and achievements considered to be valuable. Temkin suggested that a typical list would embody the following ideals: liberty, autonomy, freedom, rights, virtue, duty, equality, and justice. Once a set of ideals has been accepted, then specific objectives can be developed. Questions could be posed (other things being equal): Should the number of people benefited influence decisions? Should decisions produce the greatest overall level of benefits or should one concentrate on benefiting the worst-off in society? Is it permissible to increase everyone's benefits a standard amount, even though existing inequalities will not be reduced? Should the better-off be compelled to sacrifice some of their benefits to relieve the disadvantages of the less well-off? Should there be a minimal level below which no one is allowed to fall, with that level used to define the minimally satisfactory life to which all members of society are entitled?

A similar approach used by Kilner (1990) was described in Chapter 3. He discussed criteria that have been used to allocate organs for transplant. After describing each, he argued for a set he considered the most equitable. This procedure is good, because it brings the various factors into public view and can serve to focus discussion regarding the merits of each. Such discussions can indicate the kinds of information that should be made available to understand the implications of various choices and to place the medical and financial realities in a clearer focus.

Temkin suggested that hypothetical cases could be constructed, such as was done by Hammond et al. (1977) when they developed the series of contracts embodying the issues in the labor–management mediation. After ideals have been identified and agreed upon, they should be stated in a clear enough way that each aspect can be scaled to give each what Temkin called its "due weight." At this point, it is possible to cope with the complex issue of how to weight the different values relative to one another. Is it important to emphasize the absolute size of the gap between the better- and worse-off, or should one emphasize the relative level of the worse-off compared to all others? What is the relative value of making a big increase in the welfare of the better-off as compared to a small increase in that of the worse-off, and how big is big and small, small? Is it worse to lack certain goods than others? Some goods must be assured to all—such things as a minimally adequate level of food, clothing, housing, and health care—and other things—such as education, higher pleasure (literature, music, art, dance, cuisine), and higher conscious states (truth, knowledge, and beauty)—should be considered only after the basic goods are guaranteed to all.

After obtaining basic agreement regarding ideals, one could construct a series of hypothetical cases in which the different ideals are involved in concert with, and pitted against, one another. The recommendations that people make can be used to understand the relative weight of one ideal compared to others, to determine whether people are consistent in their application of principles, and whether the principles that have been included capture the essence of people's decision policies. Following the initial results, some principles might be eliminated from consideration because they are not used to arrive at decisions, whereas others could be added to capture the essence of the decision process. This recursive procedure can be used as a discovery technique to identify the ideals people use to arrive at decisions and to educate them regarding their own value systems compared to those of others in the society.

It is important to capture the policies of different groups of individuals with different interests, concerns, and status. Kamm (1993) suggested that many physicians exhibit strong pride regarding outcomes, favoring treatments that produce the longest possible life and the greatest differential outcome, because their major interest is to produce good individual outcomes. Those treatments producing minimal outcomes to the neediest (which is what a true maximin physician would use in the interest of maximizing the minimal level of the worst-off) might discourage the physician because of relatively poor outcomes in terms of "cures." There is also the possibility that the costs of these treatments might not be reimbursed by the government because of relatively low outcome levels, although the progress of the most miserable could be significant to them. Kamm suggested that a tolerance for poor outcomes could influence the specialties that physicians choose. Some may have chosen a specialty that tends to produce good outcomes, for example, the treatment of infectious diseases. One problem some physicians now experience is that such things as the AIDS epidemic have changed outcome rates, with a poor outcome for these patients being the norm. AIDS patients, however, might have a different set of values and interests, and an understanding of differences in outlook between medical staff and patients might serve to direct treatment efforts and lead to an understanding of the differential structure of values.

The attitudes, ideals, and weightings used by different populations of people should be considered. Lawmakers might have one set of principles, physicians another, the indigent poor still a different set from the affluent; small business people might differ from those in big business, and the middle-income taxpayer might well view everything from yet a different perspective. All of these segments of the population have valid viewpoints, and often they do not understand, let along respect, those of other players when resources are to be allocated. It might be possible to reduce much conflict and misunderstanding, to direct argumentation toward legitimate differences in values and interests, and to reach more equitable and harmonious solutions in the face of legitimate conflicts of interest.

Kamm (1993) suggested that the legitimate differences between the interests of patients and physicians might be such that basic decisions should not be made by the patient's doctor, but by a qualified board of medical experts. This suggestion in no way diminishes the role of the physician: Whenever legitimate conflicts and different desires exist, a board of experts might enhance the strength of the relationship between the physician and patient, because they become partners in pursuit of a common course that has been recommended by a body of qualified experts.

Kamm (1993, Chapter 14) suggested that a point system could be used to regulate organ transplants. Temkin (1993, Chapter 10) also grappled with practical implications in his discussions of inequality. Both Kamm and Temkin have brought considerable philosophical sophistication to bear on the issues, and both will bear careful study by those interested in developing rational and humane policies to benefit society.

Kekes (1993) examined problems that arise when complex decisions have to be made in the face of a diversity of values, and he defended what he called a *morality of pluralism*. He argued against both monistic and liberal ethical systems, because they allow some values to override others, which he considers inadmissible. He acknowledged that his own solution to conflict resolution remains a sketchy ideal, falling short of a description at a level that could be implemented.

Many other issues could be considered regarding how the different values should be combined: Should the combination be additive; should a simple weighted score be used; should it be multiplicative; should some things always override all others? When should rational choices not be made and a lottery be used? Such methodological speculations are interesting and valuable but beyond the scope of the present discussion.

Medical Ethics According to Emanuel. Some of the philosophical and policy concerns regarding medical ethics were discussed by Emanuel (1991), and he developed an ideology he called *liberal communitarianism*, which he used to resolve many dilemmas in medical ethics. He proposed that the solution of many policy problems the public faces cannot be resolved until public policy values are considered within the framework of an overriding political philosophy. Many contemporary deliberations in medical ethics fail, he argued, because a political philosophy is not developed to provide a framework to consider substantive issues.

Emanuel's underlying assumption is that medicine operates under conditions of scarcity. Questions regarding what proportion of social resources should be distributed to health care, what medical services should be guaranteed to all people, what patient selection criteria the state should permit or prohibit, all reflect the community's values and ideas. These values and ideas are seldom discussed in a manner that allows them to be recognized explicitly, and all involve political philosophy, which he characterized as "common morality idealized." This idealization is similar to the workings of scientific theories in the sense that it provides a constant and changeable framework to guide reflection and action—the idea of "wide reflective equilibrium" discussed in Petrinovich (1995).

Emanuel accepted a basic assumption that a pluralism of beliefs must be allowed, because there is no agreed-upon view of the good life to guide the enactment of laws and policies. The ideal of liberal communitarianism is that of a political community that engages in public deliberations regarding the good life and that these deliberations serve to formulate laws and policies to regulate communal life. He adopted the optimistic view that citizens will find participation in such deliberations elevating, because in the process of exchanging views and framing policies with fellow citizens, each will become bound together with others in an ongoing community that will enlarge each of their beings and even ensure that some of their being will persist after their lives are over. People will be able to understand the views of others and make decisions, not against them, but with them.

Although it is not necessary to buy into this highly optimistic view of the human condition, it is a hopeful and useful approach. The success of those SJT theorists (and of the Kaplan group to be discussed next, who also adopt a rational perspective regarding conflict resolution among people of good will) suggests that people can transcend some of their selfish interests if they are enabled to recognize them and understand the nature of the legitimate self-interests of others.

Emanuel described a typology of issues in medical ethics, the recognition of which make it possible to pursue deliberations in a productive manner. These issues were at four levels: (1) relationships between physicians and patients; (2) selection of medical interventions; (3) allocation of medical resources; and (4) application of personally transforming technology (such as genetic engineering). It is important to establish this taxonomy to focus discussions at the appropriate level, rather than implicitly moving from one level to the other without resolving issues at any one level.

Emanuel developed his position by defining the primary goods of life that are the basic and common needs of every person. He identified six different conceptions of the good life that can be found in contemporary society and argued that an individual's best interests depend on the particular personal perspective adopted, and that the kind of good life for that person is what should be respected by the community. He used this approach to consider the problems involved in terminating care to incompetent patients, to the just distribution of medical resources, and to the development of a just health-care delivery system. The specific detail at which the issues are considered makes his contribution valuable and can help to focus productive discussions and develop a sound political theory on which to base health-care policy.

Kaplan's General Health Plan Model

A methodologically sophisticated approach to determine the value of different medical procedures is that of the HIV Neurobehavioral Research Center (HNRC) in San Diego (Kaplan et al., 1994; Kaplan, 1994a). The General Health Plan Model (GHPM) considers aspects of health status in terms of four components: (1) mortality (life expectancy); (2) morbidity (level of functioning and type of symptom); (3) utility (preference for some functional states over others); and (4) prognosis (duration of the health state).

A scale, the Quality of Well-Being Scale (QWB), was developed to estimate the value of the components of GHPM. This scale categorized individuals according to level of functioning and symptom pattern. Patients were first classified according to their level of observable functioning, rated in terms of mobility, physical activity, and social activity. A list of symptoms was developed, and the one symptom that was most undesirable was identified. These classifications were done by the participants in the HNRC after extensive review of the medical and public-health literature.

After these classifications were made, individuals were rated on a scale of wellness that ranged from 0 (*Death*) to 1.0 (*Completely Well*); the observable health states were weighted by quality ratings of the desirability of these conditions. To obtain this quality value, random samples of citizens from a metropolitan community evaluated the desirability of more that 400 case descriptions. These cases were developed to make it possible to estimate the preference structure of the public in order to assign weights to each combination of an observable state and a symptom. The obtained weights were highly reliable statistically.

The next step was to consider how long the patient would be in the health state. Kaplan (1994a) provided an example of how this is done: 1 year in a state that has been assigned the weight of 0.5 is equivalent to one-half of what he calls a Quality Adjusted Life Year (QALY). A QALY is defined as the equivalent of a completely healthy year of life, or a year of life free of any symptoms, problems, or health-related functional limitations. The QALY combines the values for morbidity, mortality, utility, and prognosis into a single-scale value that can be used to compare the relative condition of different patients who have different diseases, provides a standard by which to compare the effectiveness of different medical treatments, and a standard representation of medical benefits when making cost–benefit analyses.

This general method considers trade-offs between length of life and quality of life, which is important because the Department of Health and Human Services has stated their objective to be the increase in years of healthy life for the U.S. population, and the QALY provides such an index.

This global approach avoids a single, narrowly defined, disease-specific focus. Often a treatment or medication is evaluated in terms of its effectiveness to change a single physiological indicator. Kaplan described studies intended to determine the effectiveness of two drugs to suppress heart arrhythmia. Both drugs were found to be highly effective in that regard, but neither improved life expectancy. Quite the contrary—patients taking either of the drugs had a higher chance of dying of cardiac arrest than did control patients. Kaplan commented that focusing on only the specific physiological category of arrhythmia obscured the effects on the most important behavioral outcome—life or death.

Three different methods are used to treat prostate cancer, an extremely common cancer for men age 70 or older. The three methods involve surgery, radiation therapy, or "watchful waiting." The side effects of the surgical and radiation treatments can be debilitating compared to the waiting option—but the latter is used least, because it does not treat the cancer. An analysis in terms of QALY values indicated that the quality-adjusted life expectancy was equivalent under the three options. Kaplan (1994a) suggested that this equivalency in terms of outcome indicates that the treatment option should be a matter of patient preference. If the relative expense of the three methods is considered, waiting clearly gets the nod in terms of economy—no matter who is paying for it.

It is essential to have a method to establish the value of different treatments whenever there are scarce resources to be allocated. It will be documented in Chapter 10 that health-care rationing exists worldwide, that it is a serious problem in the United States, and that the methods used to allocate the available resources are inequitable. The GHPM has been used with considerable success to develop and implement the Oregon Health Plan, which will be discussed in Chapter 10. The quality-adjusted survival analysis is an important development that can be used in many areas of clinical decision making, public policy development, and clinical trials research. The success of this program demonstrates the value of informed, problem-oriented, social science research that brings sophisticated methodologies to bear on decisions regarding social policy.

MEDICAL REVIEW BOARDS

Medical review boards were established in response to technological advances in medicine. One of the first boards was charged to make decisions regarding who among equally ill patients would be placed on long-term kidney dialysis. The decisions were not made on medical grounds, but on social and economic ones involving the number of years of life expectancy, children to support, and the financial resources available to the individual "to make the most of the gift granted by technology" (Belkin, 1993, p. 134). Although the initial ethics committee was concerned with financial considerations, later committees have attempted to avoid financial issues as much as possible, and concentrate on what Belkin considers to be the important question: how to make patients healthier, or at least to allow them to derive some pleasure from life.

There are three compelling reasons to form medical review boards to deal with problems that have arisen with the development of new medical technologies. First is that medical professionals are not educated to cope with the complexity of legal and moral questions involved in the life-and-death decisions they make every day. Not only is the medical community not educated to resolve these issues in a logical and consistent manner, but also many of them do not want to bear the responsibility of making these decisions and welcome the possibility of delegating such matters to qualified experts, leaving the medical staff free to practice its profession.

Second, there should be safeguards for the patient and for second parties when euthanasia is considered, with a clear line drawn between mercy killing and unjustifiable murder; in order for the former to be permissible, the latter must be prevented. Such safeguards protect the individual, the community, and ethical practitioners from the malfeasance of the immoral few who would exploit desperate situations.

Third, it must be assured that a decision regarding assisted or unassisted suicide has been made on a voluntary and rational basis by a competent patient. This assurance is necessary to prevent irrevocable acts whenever there is doubt that the patient making the decision is in a state that indicates the rationality required to make an irrevocable decision voluntarily.

Review boards should make wise, compassionate, and informed decisions regarding reasonable courses of action in the face of difficult circumstances. The proceedings should be deliberative rather than adversarial. These deliberative decision makers should be chosen because they can offer informed opinion, not because every shade of opinion, or even all polar positions, are represented.

Structure of Review Boards

There have been discussions of the structure of such review boards. Brandt (1987) suggested that the structure should make available the best information concerning symptoms and medical prognosis, that a family's emotional and financial circumstances should be considered, that there is state and community support for any decisions made, and that the desires of the patient and the family are involved. The relative weight to be placed on each of these factors could vary from case to case. If the case involves a defective infant, a senile adult, or a PVS patient, then the physician should give both the opinion of kin and of the medical review board the greatest weight. If the case involves a rational person, then the person's informed wishes should be weighted most heavily, other things being equal.

Brandt suggested that for problematic cases, the board should include the attending physician, a second consulting physician, a neurologist (preferably a "psychoneurologist"), at least one social worker, a specialist in ethical decisions (preferably a philosopher attached to the medical staff, but possibly a "trained" clergyman), and a "sensible" layman. Brandt does not believe the board should include judges or involve the courts, because these agencies provide an additional level of review when there are allegations that the system has been violated or is not functioning adequately.

Belkin (1993), a medical reporter for *The New York Times*, studied the problems faced by the members of the Ethics Committee in a Texas hospital. She identified the difficulties involved in making decisions and the anguish experienced by the Committee members and health providers making decisions to withdraw life support from patients under their care. The Committee in this hospital had 23 members and tended to add one more member every other month. The large size was maintained to include representatives of all major subspecialties in the hospital, a full-time ethicist, several nurses, a minister, a priest, a rabbi, a few social workers, a couple of hospital administrators, and the hospital lawyer. Belkin noted that, due to scheduling conflicts, the same combination of people rarely met twice, which raises the question of whether adequate representation was realized for many of the decisions.

It might be better to establish a smaller and more cohesive committee in the interests of equity, consistency, and efficiency. A single medical review board, having jurisdiction over all matters of human life and death involving reproduction and living and dying well, might function better. The jurisdiction should cover a geographic region (district, city, county, state) or a health-care delivery unit of sufficient size, and the workload should not be so great that it would overwhelm the board members. The board could be appointed by professional societies representing each of the appropriate substantive areas, and individuals should receive adequate compensation for the amount of time involved. The board could consist of six members who would be charged with the responsibility to collect and combine information from panels of specialists who deal with the substantive concerns involved in each case, and should be empowered to arrive at decisions regarding the permissibility of proposed medical actions. The members should include the following:

1. A general medical expert who receives reports from the attending physician and another physician qualified to render an informed opinion, appropriate hospital administrators, technicians who can provide input regarding the nature of technical support and patient maintenance facilities that can be expected, and a medical economist who can provide an evaluation of the relative costs and benefits in terms of available resources.
2. A psychiatrist, neurologist, or neuropsychologist who is competent to evaluate neurological evidence, its relation to psychological and physical functioning, and to consider the cognitive state of the person.
3. A legal expert who will receive reports from qualified legal personnel regarding legal statutes that apply and court rulings that are pertinent to each case.
4. A medical ethicist who can lead the discussion toward the appropriate moral principles that should be considered explicitly, including a discussion of standard philosophical positions and theological concerns appropriate to the patient's religious beliefs, affiliations, and background.
5. A community expert who receives input from social workers, community officials who are able to provide information regarding available community support, and family affairs experts who can evaluate the expected amount of family support. The family affairs experts should provide information regarding values and wishes of the family.
6. An expert in the psychology of emotions and psychopathology who can provide psychological evaluations and attest to the adequacy and possibility of therapy, counseling, and guidance.

A board such as this is not too large and, with the proper spirit of cooperation, an emphasis on problem solving, and attention given to the development of general principles would develop guidelines making it possible to resolve cases efficiently in light of previously considered cases, just as is possible within the judiciary. The evidence required to consider cases of different kinds will come to be recognized, and extensive time will only have to be devoted to those cases in which the conditions and circumstances are exceptional; the existence of these cases is the reason why a review board should exist in the first place.

This proposal could respect the moral conscience and beliefs of those members of the medical profession who do not want to participate in acts of abortion or active euthanasia and, more important, protect the interests of those persons who decide to take a morally justified course of action regarding their own mortal future. Konner (1993) argued that physicians should be free to resist the formidable demands on their conscience that could be caused by administering lethal injections to terminal patients. An equally important consideration is that respect is given to the conscience of those medical professionals who decide they have an obligation to provide services, such as abortion and physician-assisted euthanasia, to individuals who need and request them.

I agree with Brandt that judges or members of the court should not be included as members of a review board. Legal counsel should be provided by experts, but the court system is yet another player in the system of checks and balances that regulates society. If legal charges are to be brought, then the courts will be called upon to adjudicate. For similar reasons, members of the legislature should not be included. Legislators frame the laws and their role is to legislate, should legal regulation be required.

I disagree with Brandt's suggestion to include clergymen as a separate category. Most of those who are members of a review board will have been raised within and accept some theological tradition or another, just as are the members of the legislature. For example, *The New York Times* (December 5, 1992) presented a breakdown of the religious affiliations of the lawmakers who were members of the 1992 U.S. Congress. Of the 533 seats that had been decided at that time, there were 141 Roman Catholics, 230 Protestants of various denominations, 42 Jews, 12 Mormons, 10 unaffiliated, and 33 of other denominations. It seems safe to assume that the theological constituency is well represented in the legislature and the judiciary, and that it will undoubtedly be so represented on professional review boards of the type recommended.

There is no reason why medical review boards should be structured to have a representative cross-section of religion, ethnicity, gender, or social class. The overriding consideration is that board members bring relevant information to bear, are able to utilize the knowledge of experts, can decide on permissible actions within the law, show respect for dominant social norms and views on morality; that their positions are public, consistent, and defensible; and that they respect the moral pluralism of contemporary society.

CHAPTER 9

Health-Care Policy

Issues

PROBLEMS IN THE UNITED STATES

In this chapter, the problems that face attempts to develop a health-care delivery system in the United States will be identified, the moral issues will be discussed, statistics regarding the financing of U.S. health-care examined, and implications of the "R" word—rationing—will be considered.

Arguments regarding health-care delivery systems provide an interesting case history in applied ethics. At the outset of the 1993–1994 debate in the United States, almost everyone agreed that all citizens have a moral entitlement to a minimal and adequate level of health care—that universal coverage is a moral necessity. Although this agreement was acknowledged at the start of the health-care debate, no progress was realized. The reasons for the inability to move toward implementing an agreed upon moral necessity were manifold. One barrier to progress was the magnitude of the economics involved in the health-care system: It has been estimated (Levit et al., 1994) that $884.2 billion were spent in the United States on health care in 1993 (13.9% of the Gross Domestic Product—an estimate that the Commerce Department subsequently revised up to $939.9 billion. This much money activates a lot of involved parties whose interest is to preserve and, if possible, enhance their economic position. There are many users of health care who have a primary interest to preserve the affordability and quality of what they now have. These various constituencies include the privately insured, those insured by their employers, veterans, the disabled, and the elderly. With such complex economic and political realities, it is easy to play on people's fears, to manipulate the interests of one group against those of others, and to throw up smoke screens founded on fear and misinformation (see Fallows, 1995).

Strong political interests were involved. The members of the Congress and state governors were facing the imminent 1994 elections, and presidential hopefuls were starting to position themselves for the 1996 elections. It was acknowledged by many of the players that there was an interest on the part of Republicans, who were then the minority in Congress, to deny the Clinton administration credit for having achieved health-care reform. In this and the next five chapters, I will develop some of the issues as they emerged, discuss the nature of the debate by characterizing some of the proposals, and conjecture regarding the future course that events might take.

Considerable effort was devoted to confuse insured people by frightening them with the possibility that what they have now will be placed in jeopardy. A string of slogans were introduced, such as the old familiar menace of socialized medicine, government takeover of health care, long waiting lines, loss of personal choice of physicians, inefficient and expensive governmental bureaucracy, cattle-car care, and a general scare that "they" are going to destroy "the finest health-care system in the world."

The Republican response by Senator Dole (1994) to President Clinton's January 25, 1994, State of the Union message provides a good capsule summary of the type of rhetoric that prevailed in response to proposals to revise the health-care system:

> We know that America has the best health care system in the world ... Our country has health care problems, but not a health care crisis ... Clearly, the President is asking you to trust the Government more than you trust your doctor and yourselves with your lives and the lives of your loved ones. More cost, less choice, more taxes, less quality, more government control and less control for you and your family—that's what the President's Government-run plan is likely to give you.

The overall debate in the media and in Congress provided a sorry spectacle, led nowhere, and the possibility of any sweeping reforms to achieve universal medical coverage in the United States seem more unlikely for the immediate future. The debate was dominated by a large number of narrow economic interests that worked their ways with politically powerful individuals and groups. I will proceed by describing some of the political, financial, and medical realities that exist, discuss some of the proposals that were considered, and focus on the moral issues involved. The ball is now in the court of the new Republican-led Congress, but the issues probably will remain unresolved or only receive a partial resolution, at least until the 1996 elections are over.

The magnitude of the problems (involving over 1/7 of the total U.S. GDP and the well-being of most of its citizens), the vested interests (especially the medical, insurance, and legal communities, let alone the taxpayers), and the seldom-considered problems of morality all make the issues extremely complicated and highly interesting. Many strengths and weaknesses of the multitiered and complex U.S. democracy are exposed by the glaring lights of conflicting interests, as well as different underlying faiths and values.

The development of new and costly diagnostic and treatment technologies is leading insurers, the public, and governmental agencies to financial disaster—and the U.S. taxpayers with them. It has been argued that we should not initiate aggressive efforts to save the lives of those who are doomed to die soon, despite heroic medical efforts. Instead of using limited resources for extraordinary and expensive treatment procedures, it has been argued that funds should be used to guarantee the availability of less costly procedures to everyone, with an effort to forestall the development of serious diseases through the use of preventive medicine, such as increased efforts to vaccinate all children and to provide prenatal care for all pregnant women. It has been suggested that government support should not be given for research that is designed to create or improve life-extending technologies for the elderly.

A case reported in *The Oregonian* (November 4, 1992) highlights the dilemma we face. A 48-year-old father arrived at a hospital in time to be treated for a near-fatal heart condition. He was informed that he needed a heart transplant but, lacking the $50,000 to pay for it, was sent home and died of the heart condition that year. In this case, financial reality made it impossible to provide the necessary treatment that probably would have prolonged the life of a young father. The tragedy of such cases is revealed by the realities described in another article in *The Oregonian* (November 2, 1992). It was estimated that there are 10 thousand PVS patients who have no hope of recovery, but are hooked to respirators that will keep them alive indefinitely, at immense financial costs. The patients' vital signs are maintained because of the medical profession's commitment to avoid active euthanasia and the fact that most of these patients are insured. Even though one might agree with the moral decisions involved, the financial burden on all who pay insurance premiums is catastrophic.

In her newspaper column, Ellen Goodman (*The Oregonian*, December 11, 1992) estimated that there are at least 14 thousand PVS patients in the United States, many of whom are too immature to indicate any preference for life or death, or are comatose adults who had not left advance directive. Physicians often do not want to make the value judgments required to draw the line between life and death, so these individuals are maintained long beyond the time when any recovery is possible. Should the funds required to save the life of a young father, who would be able to support his family if he lived, prevail over the continued existence of comatose patients on life support?

The Economics of U.S. Health Care

The United States has about 5% of the world's population, but its health expenditures are estimated to be about 40% of those in the world (World Bank, 1993). One major problem is that the United States really has no coherent health-care system. What exists is a patchwork system to finance health care. Greely, a Professor of Law at Stanford University, summarized the situation as it existed in the late 1980s, although he cautioned that complete and official government records are kept only for Medicare and Medicaid patients, with the remaining statistics drawn mainly from surveys. Greely (1992, p. 265) wrote:

> The large majority of our population of 250 million people is covered through a number of very different mechanisms.... The largest group, about 150 million, have private group insurance, usually as an employee or as an employee's spouse or dependent. Another 10 to 15 million people, many of them self-employed, rely on insurance policies they have purchased themselves. About 33 million, mainly elderly, people have health coverage through Medicare and about 23 million of the poor are covered by Medicaid.

This latter figure was increased to 32 million for 1992 by the Employee Benefit Research Institute, a nonpartisan study group, as reported in *The New York Times* on January 24, 1995. The Census Bureau reported that in 1992, only 47.2% of the poor were covered by Medicaid, because it is difficult for them to meet the low financial limit to qualify for Medicaid if they have no children (Medicaid is the national and state health-insurance program for certain groups of the poor, as well as the elderly, blind, disabled, pregnant, and parents of dependent children). There was a Congressional Budget Office (CBO) estimate that spending for Medicare and Medicaid in 1992 was $242 billion, with a projected rise to $694 billion in 2005, if the program remains as structured now (Toner, 1995).

Krueger and Reinhardt (1994) specified three methods used to transfer money from private households to providers of health care:

1. Less than 10% of the nonelderly U.S. population obtain private health insurance without public subsidy (nowhere in the world is this the predominant mode of health financing).

2. Of national health spending, 42%, covering one-fourth of the population, involves government taxation of private households, with the taxes funneled to an insurance fund, as for Medicare, Medicaid, the purely socialistic health systems of the Department of Veterans Affairs, and various delivery systems of the Public Health Service.
3. About 71% of the nonelderly population in the United States obtain insurance coverage through private and public employers who procure insurance from private and public insurance carriers.

This third approach is the backbone of social insurance systems in Europe, Latin America, and Asia, where most countries mandate employer participation in health-care financing.

Robert Pear (1994d) reported that 1993 U.S. census figures indicated that the number of uninsured rose to 39.7 million (15.3% of the population)—an increase of 1.1 million from the previous year. The Employee Benefit Research Institute analyzed a Census Bureau survey of 50 thousand households and estimated that the number of uninsured in 1993 was 41 million (16.1%). They found that 43% of the 14.5 million legal and illegal noncitizens in the United States had no health insurance; these uninsured were likely to work for small businesses and have a family income of less than $30,000. The uninsured in 1993 included 9.6 million children, an increase of 900 thousand over 1992. It is estimated that more than 22 million others are underinsured because they cannot afford large copayments and deductibles. The increase in these costs prevents people from seeking medical treatment for routine minor medical problems, often delaying diagnosis and treatment until the problems become serious, foregoing preventive steps, such as routine physical examinations, that could detect developing problems at an early stage. The increased incidence of such occurrences has placed a strain on city-hospital emergency rooms, which are being utilized more and more as a substitute for primary care, leading many metropolitan hospitals to close their emergency facilities altogether because they cannot afford to provide treatment to the large number of people who are unable to pay.

There are concerns that the elite research hospitals in New York City, which provide much of the care for the poor and uninsured, will not survive the decade (Rosenthal, 1995). The California Pacific Medical Center in San Francisco, a major treatment facility, is dropping its community services and providing less treatment of uninsured patients (Eckholm, 1995b). The poor, elderly, and uninsured will experience increasingly severe difficulties in obtaining adequate medical care, especially if the intentions of the 104th Congress to cut the Medicare and Medicaid budgets are realized.

Economics and the Aging Population

Health-care costs are especially high for the elderly, particularly in the final year of life. Caplan (1989b) noted that 12.5% of the U.S. population in 1988 was elderly (65 and older), and they accounted for 33% of all health-care expenditures. Emanuel and Emanuel (1994) noted that in 1988, the mean Medicare payment for the last year of life of a beneficiary who died was $13,316 as compared with $1,924 for all Medicare beneficiaries (a ratio of 6.9:1). They remarked that it is difficult to know at the time treatments are begun whether the costs are for care at the end of life or to save a life: This can only be known retrospectively.

It has been estimated that the percentage of the population who are elderly will increase to 13.8% by 2010 and 21.2% in 2030. Many expenditures on the elderly are for long-term, noncurative care. Caplan characterized the dilemma to be that the more successfully the medical establishment prolongs life, the more opportunities it creates for expensive care. One problem is that medical providers seldom question the wisdom of continuing medical treatment when there is medical insurance available to pay for it.

Campion (1994) examined data bearing on the "oldest old"—those 85 years or older. Census figures indicated that the number of Americans in this age group increased by 232% between 1960 to 1990, whereas the number 65 years or older increased by 89%, with the general population growing by only 39%. In this oldest age group, women outnumber men by 2.6 to 1, and although the oldest constitute only 1.2% of the population (with 10% age 65 or older), 21% of deaths occur after age 84. Of those women 84 or older, 25% live in nursing homes (15% of all men), as compared to only 1.4% of those 65–74 years. The Census Bureau projects that by the year 2040, there will be 8–13 million Americans 85 years or older, with some arguing that the number could be even greater than 24 million, given continued improvements in disease prevention and reduced mortalities from cardiovascular disease and stroke, the leading killers of the old.

Battin (1987) raised the possibility of age rationing in the distribution of health care, citing statistics similar to those described by Caplan: Three out of four deaths of persons of all ages in the United States occur as a result of degenerative diseases, with the proportion much higher in old age; people over 65 use medical services at 3.5 times the rate of those below 65; in 1980, 11% of the population over 65 used 39.3% of short-stay hospital days, and Konner (1993) noted that the accounted for 29% of the total U.S. health-care expenditure of $219.4 billion in 1980. The 4.4% of the population over 75 used 20.7% of hospital days; there were about 6 million octogenarians, and the government provided an estimated $51 billion in transfers and services to them; people 80 years of age or older consumed over 77% more medical benefits than those between 75 and 79.

Konner cited statistics indicating that in 1986, those over 65 accounted for 31% of the total U.S. expenditure of $450 billion. Estimates are that the annual Medicare (the national health-insurance program for the elderly and disabled) costs will rise from $75 billion in 1986 to $114 billion by the year 2000. Medicare costs already had reached $122.8 billion in 1991 (Letsch, 1993)! Medicare, even with these very large expenditures, covers less than one-half of the total medical expenses of the elderly (De Lew, Greenberg, & Kinchen, 1992). The elderly accounted for 13% of the population and were responsible for 34% of all spending for outpatient prescription drugs, the cost of which is not generally covered by Medicare (Pear, 1993e). By the year 2040, it has been projected that if conditions remain the same, the elderly will consume 45% of all health-care expenditures. Another way to characterize the situation is that there were five workers for each Medicare beneficiary in 1960; it is estimated that there will be three workers per beneficiary in 2000, and 1.9 by the year 2040. These ratios mean that, at the present time, there is an enormous intergenerational transfer of funds, with the problem promising to become even more significant in the future.

Such statistics led Battin to ask whether there might be a time for the elderly to die and, specifically, whether there should be rationing of health care such that the elderly should be the first to be excluded from health care if there is not enough funding to include all. She argued quite sensibly that if health care had to be rationed, the criterion used should not be solely age-based, but should be based on expected time until death; the criterion should be based on each individual's medical condition considered in terms of the cost-effectiveness of treatment procedures and the expected quality of the remaining years of life. If it became necessary to establish a rationing system, then a proportion of health-care expenses now devoted to the elderly (about one-third of the total) could be reassigned to members of younger age groups, who would be expected to benefit from more years of higher quality life per unit expenditure. The purpose of discussing these economic issues regarding the aging population is to engage the argument that there is a problem in the financing of the health-care system at the present time, it will be greater in the future, and there must be some radical changes in the way the country structures its health-care policies; mere tinkering with the details is not likely to produce an adequate solution. One suggestion is explicit health-care rationing (which many consider to be moral anathema). Problems involved in establishing a rational and morally defensible rationing or priority system will be discussed in Chapter 10.

A cautionary view regarding the alarming statistics that have just been discussed is in order. Emanuel and Emanuel (1994) placed the situation in a different perspective when discussing studies comparing hospice care to the traditional medical care given at the end of life. They characterized hospice care to include a patient's refusal of life-sustaining intervention in favor of palliative care: The treatments often are done at home, and treatment protocols are used to reduce the use of high-technology interventions at the end of life. Estimates were that in the last month of life, home hospice care saved between 31% and 64% of medical care costs. During the last 6 months of life, the mean medical costs for patients receiving hospice care at home were 27% less than for conventional care, and the savings realized if hospital-based care is used were about 15%. If each of the 2.17 million Americans who died in 1988 had executed an advance directive, chosen hospice care, and refused aggressive in-hospital interventions at the end of life, about $18.1 billion would have been saved. Although this amounts to only 3.3% of all health-care spending, they pointed out that these savings should not be dismissed lightly, even though they are less than the scores of billions of dollars that could be realized from cutting administrative waste.

They estimated that in 1993, assuming that $900 billion goes to health care (actually it was closer to $940 billion), the savings would be $29.7 billion. This relatively "small" savings led them to argue that we should not use such small amounts to justify the use of less aggressive life-sustaining treatment for dying patients who desire it or who have not left advance directives. Their overwhelming concern was to respect patients' wishes, reduce pain and suffering, and provide passionate and dignified care at the end of life. I guess it depends somewhat on whether one considers a savings of $30 billion or so to represent only a small amount of money.

The Moral Issue

Health care is one of the basic benefits to which all people are entitled, and in principle, it is generally accepted that all people should be guaranteed an adequate level of health care. Brandt (1992) referred to

Articles 22–27 of the United Nations' *Universal Declaration of Human Rights*, which affirmed, among other things, that all have a right to economic security, to equal pay for equal work, to remuneration sufficient to provide an existence worthy of human dignity, including a standard of living adequate to provide food, clothing, housing, and medical care.

These articles were approved by the General Assembly on December 10, 1948, and signed by the United States. They urged the importance of respecting human dignity five times: twice in the preamble and in three separate articles. The first clause of the preamble asserts the importance of the inherent dignity of all members of the human family and affirms a faith in the dignity and worth of the human person. Article 1 asserted that all humans are born free and equal in dignity and rights, and Article 22 affirmed the right that members of society have to social security in terms of the economic and social rights that are indispensable if one is to live with dignity, with Article 23 asserting that everyone who works has the right to just and favorable remuneration that ensures for himself and his [*sic*] family an existence worthy of human dignity. Medical care is specifically mentioned as a guaranteed right in Article 25, with the status of motherhood and childhood (whether born in or out of wedlock) entitling people to special care and assistance. Article 26 specified a right to free education at the elementary levels, and indicated that elementary education should be compulsory.

The moral necessity of these articles has been argued by many philosophers. The medical profession agrees that the United States has a moral responsibility to provide health care for all who require it. The AMA Council on Ethical and Judicial Affairs (1994) presented a report on ethical issues in health-care system reform, which was adopted by their House of Delegates on December 7, 1993. The published report identifies the AMA policy to be that society should afford every citizen access to adequate health care. The report stated that society's obligation is the product of a social contract among citizens who abide by the mutually agreed upon rules of society, and that this contract requires provision of the basic goods of food, shelter, clothing, education, and adequate health care, and they suggested a set of relevant ethical principles to guide the determination of adequacy.

The principles of liberty, freedom, equality, and dignity have been applauded by politicians and endorsed by the general public. As with many abstract principles of morality, they are unanimously endorsed until it becomes necessary to face the specific problems, such as how to pay for the required care (certainly not by increasing my taxes), and what services need to be cut in order to provide basic coverage for everyone entitled to it (certainly not anything that would cost jobs or profits in my business, reduce my professional income, or impact my own out-of-pocket expenditures).

Eckholm (1993a) asked the rhetorical question: "Whose health is number 1, that of business or of workers?" The issue was whether any change in our health-care system should be made if it harms small-business owners who fear job losses and bankruptcies if they are forced to contribute more dollars toward the cost of covering their employees and the 41 million uninsured. If businesses do not contribute more, then taxpayers must assume the burden, and many find that alternative objectionable.

The dilemma facing small businesses illustrates the difficulties encountered when health-care problems are viewed from the perspective of the marketplace rather than as a basic moral issue. Morality requires a just society to provide for basic necessities with as little delay as possible. It is improper to phrase questions in terms that pit the welfare of workers and their families against the profits of business. A familiar argument is that if business is made more and more profitable, then benefits will trickle down to the worst-off after an initial period of inequality during which the better-off prosper. Such policies were woefully inadequate when adopted during the years Presidents Reagan and Bush were in office; the rich became richer, the poor, poorer and more numerous, and the middle class paid more of the costs to maintain society. Trickle-down economics failed as an economic principle to regulate the nation's economy, and it is a questionable principle to entertain when considering changes in the health-care system.

In a just society, everyone should be guaranteed the basic necessities of life, and no one should be allowed to fall below the minimum standards required to sustain a satisfactory life. Morally, it is not permissible to pursue policies that benefit the best-off in society at the expense of the worst-off, especially if minor changes will raise the life quality of the worst-off above a threshold necessary to realize an adequate life. One irony of the prevailing situation in the United States is that the very poorest in society receive health care through the Medicaid program. Sixty-three percent with means falling below the poverty line were eligible for Medicare in 1975, but the level has now decreased to only 40%, leaving the majority of the poor uninsured. Many of those who suffer are employed, but their low wages are too high for them to qualify for Medicaid. At least one person worked full-time in half of the uninsured households and in another third, a person worked part time or part of the year (Eckholm, 1994). Members of ethnic minorities are disproportionately uninsured: 14% of white Americans under 65 were uninsured in 1992, compared to 23% of blacks and 35% of Hispanic Americans. If workers have no health-care insurance through their employers, they must gamble that there will be no sickness in the family, and if they do need health care, they must quit working in order to qualify for Medicaid. A woman who has older children can become eligible for Medicaid by becoming pregnant, because poor families with younger children, but not older ones, are eligible for Medicaid.

Census statistics for 1994 identified some of the differences between those in New England who have health insurance and those who do not (Stein, 1995). Well above 90% of families with incomes over $36,000 have health insurance, compared to only 73% with incomes between $12,000 and $36,000. More than 91% of college graduates have health insurance, compared to only 75% of high school dropouts. Ninety-three percent of those who work in financial services, 91% in manufacturing, 78% in retail, and 75% in construction have health insurance. In 1990, 6.5% of Northeastern households received insurance through Medicaid, and this had increased to 10.1% in 1994—a gain that roughly corresponds to the decline in the number of families with private health insurance.

In Massachusetts, Medicaid cases increased from 544 thousand in 1990 to 666 thousand in 1994, a statistic that would loom large if the Republican majority in the 104th Congress reduced Medicaid. A decrease in Medicaid funding, with the possibility of increases in health-insurance premiums for businesses, suggests that economic problems involved in providing adequate health care to a large proportion of the U.S. population will become more serious unless there is basic restructuring of the health-care system.

Lower income working families (dubbed the "working poor") clearly are at risk under the present system. Current estimates (Krueger & Reinhardt, 1994) are that the typical health-insurance policy costs about $5,000 for a family of four. In 1990, over 15% of families earned less that $15,000, and 50% earned less than $35,000. It is unlikely that such families could afford to spend $5,000 on health insurance. In a lead editorial on January 16, 1994, *The New York Times* characterized proposed plans that would force people to face choices between providing for daily needs or insuring against potential future disaster as "merciless."

It is interesting that the Universal Declaration of Human Rights affirmed a right to education, especially for children, and in the United States, we consider it to be every child's Constitutional right to have an equal opportunity to an adequate education—a right guaranteed and financed by the public. Although we have accepted the right to universal education, the statistics to be considered in this and the next chapter indicate that we have failed to accept an obligation to provide adequate health care. It is illegal to fail to send a child to school up to the time the child reaches a certain age, but parents are allowed to fail to provide immunizations, and society fails to provide adequate funding to support a public health system that will guarantee a "sound body" in which the obligatory "sound mind" (for which it does provide funding) is housed.

Temkin (1993) considered the philosophical arguments regarding equality to reflect moral ideals. The egalitarian wants each person in society to fare as well as possible, while still maintaining that special concern must be given to the worse-off—a position he called "extended humanitarianism." This position does not insist that resources necessarily should be redistributed from the better-off to the worse-off, but it does mean that it is the gain to the worse-off that is of paramount importance whenever profound inequalities exist. If social policies were developed to improve the condition of the worse-off, then it is of no consequence that the better-off have an equal or greater gain.

There has been considerable discussion of Rawls's (1971) maximin principle: Just policies should maximize the average level of those at the minimum—the worst-off group. At the least, one should argue a weak version to the effect that if a condition arises in which some people are below an acceptable level of existence, then most (but not necessarily all) resources should be directed toward the neediest. A stronger version can be argued that directs all public resources to the worst-off who are not above the acceptable threshold level in order that they can live a minimally satisfactory life. An adequate level of health care is one of those aspects of life to which all are entitled, and given the economics of health care in the United States, it must be assured that all have access to affordable care. Those who do not have that assurance must be brought above the threshold level, even if it is at the expense of the better-off, as long as those better-off remain above the acceptable threshold level. One broad list of a minimum benefits package was discussed in the AMA Ethical Council's (1994, p. 1058) report. This list included the following items:

> hospital care, surgical care and other inpatient physician services, physician office visits, diagnostic tests, and limited mental health services..., and preventive services including prenatal care, well-child care, mammograms, Pap smears, colorectal and prostate screening, procedures and other preventive services that evidence shows are effective relative to cost.

Temkin considered undeserved inequality always to be objectionable, especially because there are natural inequalities as a result of no intentional or deliberate actions taken by anyone. The objectionable inequalities are social ones that exist because of policies that discriminate against people on the basis of arbitrary traits and characteristics. The examination of both types of inequality should be done using a complex, essentially comparative and individualistic basis—a view quite compatible with those of the evolutionary biologist who views evolutionarily significant costs and benefits at the comparative and individual levels of analysis. The focus should not be on the average level of goods in society, but on the level of goods available to the worst-off individuals in society. Temkin called this view *individualistic egalitarianism*. From any of these perspectives—extended humanitarianism, maximin principle, the overall view of justice developed by Rawls, or individualistic egalitarianism—it is mandatory that the welfare of the worst-off be improved, and that concerns regarding the better-off members of society be given secondary consideration.

While most morally responsible people agree with principles such as Temkin's individualistic egalitarianism, are there reasons to suspect that these principles do not regulate society's health care? The first question to examine is how serious the problems are with the health-care system.

Some Distressing Statistics

Although health care is a basic necessity of life to which all people should be entitled access, it has been estimated that about 41 million in the United States have no medical insurance (about one-fourth of them children), and about 22 million more people are underinsured. There have been statements in publications such as *The Wall Street Journal* that there is no serious problem of uninsured people, because 70% of them are uninsured for less than 9 months, and others have argued that the number of uninsured could be as low as 12 million. These claims were addressed by Swartz (1994), one of the researchers who made the original estimates of the distribution of uninsured periods that the commentators used to arrive at their low estimate of 12 million. Her analyses led her to conclude that the problem is even more severe than it appeared to be at first consideration. In 1992, about 21 million people were without health insurance for periods greater than 1 year (28% of all uninsured periods), and for some of them for more than 2 years (15–18%). For 21 million people, the lack of health insurance was not a temporary or transitional phase in their lives. Over the course of the year, more than 35–37 million persons experienced at least 1 month without health insurance. Estimates ranged as high as 58 million people uninsured for at least 1 month in 1992, and she calculated that an estimated 21 million were uninsured for all of 1992. Emanuel and Dubler (1995) estimated that 63 million Americans were without insurance for at least 1 month during each 28-month period during the period 1986–1988. About 3.5 million of those who had "short" uninsured periods of 6 months or less would be expected to be hospitalized during that time, which frequently means that they receive fewer services than insured patients, run a higher risk of dying when hospitalized, and would have had about $7 billion worth of hospital expenses that must somehow be paid by the individuals, the government, or absorbed by the hospital. On the basis of such statistics, Swartz reached the conclusion that even a short uninsured period does not have the benign implications that have been ascribed to it by some pundits because it is only "a short duration."

Other estimates were cited by Eckholm (1994): On any given day, about 29 million people are uninsured and remain so for 1 year or more, with about 21 million of them remaining so for at least 2 years. These uninsured do not include the elderly (who are covered by Medicare), those below the official poverty line of about $15,000 for a family of four (many of whom are covered by Medicaid), but consisted of working families with low-to-moderate incomes—half of them working full-time and another one-third working part-time or part of the year. A state-by-state analysis revealed that the range of uninsured in 1992 ran from a high of 26.6% in Nevada to a low of 8.1% in Hawaii, with the uninsured for the entire United States being 17.4%.

Comprehensive comparisons of statistics related to health care were provided by Schieber, Poullier, and Greenwald (1993, 1994). They provided information on health spending, availability, use, and outcomes for the 24 member countries of the Organization for Economic Cooperation and Development (OECD) for the years 1980–1992. The average ratio of health expenses to GDP for the OECD countries increased from 7.0% in 1985 to 8.1% in 1992. The United States had the largest absolute and the second largest relative increase in its ratio. The United States spent 13.6% of its GDP on health care, with Canada having the next largest percentage (10.3%). In 1980, the average per capita health-care spending for the OECD countries averaged $577, ranging from $64 for Turkey to $1,068 for the United States. In 1992, the average was $1,374, with Turkey spending $156 and the United States, $3,094. Under the single-payer plan, Canada spent 7.4% of its GDP on health care in 1980 and 10.3% in 1992; its per capita spending was $727 in 1980 and $1,949 in 1992. The authors concluded that the U.S. health-care system is by far the most expensive in the world, with the gap between the United States and other countries widening. Compared with the other OECD countries, the United States is facing the highest rates of increase in health-care spending relative to the GDP, and there is excess health-care inflation that produces a concomitant loss of opportunities for consumption and investment outside the health sector.

With such massive spending, one would hope that the U.S. public gets what it pays for, having greater availability and use of health services than any other country. Politicians have assured us that U.S. health care is the best in the world. Unfortunately, this does not seem so for a large proportion of the U.S. public. Schieber et al. (1993) examined several relevant indices of availability: inpatient medical-care beds per thousand population, inpatient days per capita, admission rates, average length of stay, occupancy rates, number of employees per bed, number of physicians per thousand, and physician contacts per capita. The United States had fewer inpatient medical-care beds per thousand people, fewer days of care per person, among the shortest length of stay in hospital, one of the lowest admission rates and the lowest occupancy rate. The United States had the second highest number of employees per bed (after Australia)—a rate that was 75% above the OECD average, leading the authors to conclude that more intensive care is provided in the United States during a shorter stay. Another major difference was that, although the number of physicians per thousand in the United States (2.3) was close to the OECD average (2.5) in 1991, the United States mix is oriented more toward specialists than general physicians (GPs). Only 13% of U.S. physicians are GPs as compared to 75% in 1940 (Grumbach & Bodenheimer, 1995). If general internists and general pediatricians are included, then GPs amount to about one-third of all U.S. physicians, well below the 50% or more found in Canada and many European nations.

If U.S. spending exceeds that of the other OECD countries and there is no greater availability of services, what about health outcomes? Schieber and his colleagues considered the information available for 1991 on infant mortality, life expectancy at birth, and life expectancy at age 80. The infant mortality rate for the OECD averaged 9.4 deaths (if Turkey, with 56.5, is eliminated, the average is only 7.4); the U.S. was the fifth highest (8.9), Canada had 6.8, and the U.S. rate was exceeded only by Turkey, Portugal, Luxembourg, and Greece. The OECD average male life expectancy at birth was 72.9 for males and 79.2 for females, with the U.S. value ranking 20th for males (72.0) and 18th for females (78.9). A report of statistics for 1992 (Spector, 1994) indicated that U.S. life expectancy at birth increased to 75.6 years, which raised it to the 18th rank. An interesting statistic is that for life expectancy at age 80, the United States ranked highest for males (OECD average = 5.9; United States = 7.2;

Canada = 7.1) and second for females (OECD average = 7.4; United States = 9.1; Canada = 9.3).

Fielding and Lancry (1993) compared the French health-care system to that of the United States for 1991 and found that despite the fact that the United States spent 4.2% more of its GDP on health care, life expectancy at birth in France was 81.1 years for women and 73.0 for men, whereas in the United States it was only 78.9 for women and 72.0 for men, and that infant mortality was 7.3 per live births in France compared to 8.9 in the United States. In 1991, France had 43% lower per-capita health costs than did the United States. Schieber et al. (1994) estimated that in 1992, the per-capita cost in France was 44% lower than that in the United States.

The United States had the highest percentage of low birthweight babies (OECD average = 5.4%; United States = 7.1%; Canada = 5.4%), and low birthweight has been identified as one of the major factors responsible for the high U.S. infant morality rate (although a 1995 study by Wilcox, Skjaerven, Buekens, & Kiely suggested that the excess mortality among U.S. infants is due to a greater number of preterm births). It has also been suggested that the high technologies used to treat these infants raises the level of U.S. expenditures considerably. These researchers interpreted the low ranking of the United States in terms of infant mortality to reflect "underlying social problems" as well as lack of a coordinated and comprehensive system to provide preventive and prenatal care to the entire population.

The increased life expectancy at age 80, in the view of Scheiber and his colleagues, was due to the widespread availability of technology, coupled with the aggressive treatment of elderly patients in the United States. They asked whether the benefit of this technology was worth the substantially higher costs in the United States. They wrote (1993, p. 129), "These ... data continue to reinforce the notion that by international standards the U.S. health care system is out of control," and doubted that the U.S. system unequivocally provides the best quality care in the world. They (1994, p. 111) concluded,

> The real question for American decision makers is whether rationality can override politics, as the United States approaches the twenty-first century devoting one-fifth of its economy to an expanding health sector in which inefficiency and inequity abound.

The percentages of the GDP devoted to health care are lower in all of the other countries surveyed, despite the fact that they all have universal health coverage, which the United States does not, and all have lower copayments and deductibles (Barr, 1993). In the United States, the level of both copayments and deductibles has increased without slowing the escalation of costs.

Letsch (1993), an economist in the Office of the Actuary, Health-Care Financing Administration, presented a detailed statistical report regarding national health-care spending in 1991. U.S. medical costs had risen to 13.2% ($751.8 billion) of the GDP in 1991 (in 1960 it was only 5.3%—$27.1 billion—and in 1980 it was 9.2%—$250.1 billion). This percentage will be over 14% in 1992. Another way to express the increase is that national health-care expenditures per capita were $143 in 1960, $1,068 in 1980, and $3,094 in 1992 (Pear, 1993a). The free-market approach to health care in the United States does not seem to have produced effective cost containment.

Caplan (1989b) remarked that technological advances and an increased use of medical technologies have accounted for 30–40% of health-care cost increases. Rublee (1994) examined the high-technology resources available in the United States (1992), Canada (1993), and Germany (1993). The technologies considered were those for open-heart surgery, cardiac catheterization, organ transplantation, megavoltage radiation therapy, extracorporeal shock-wave lithotripsy (centers using high-technology devices to crush calculi within the bladder), and magnetic resonance imagining (MRI). All of these technologies were more available in the United States than in the other two countries, ranging from a more than a tenfold difference (compared to MRI facilities in Canada) to a less than a twofold difference (organ transplantation facilities in Canada). France had 1.23 MRI machines per million people, compared to 3.8 per million people in the United States in 1990 (Fielding & Lancry, 1993). Rublee suggested that, although these resources are invaluable in a variety of clinical circumstances, their proliferation adds to medical-care costs and may even be associated with poor patient outcomes, because patients receive treatments to justify the equipment, even in cases where the therapeutic prognosis is poor. He concluded (p. 116), "The greater proliferation in the U.S. could mean higher quality health care or more wasteful or possibly harmful health care."

The costs of drugs are excessive in the United States as compared to other industrialized countries such as Canada, which has prompted governmental action, and both denials and justification of the cost differentials by pharmaceutical manufacturers. In 1960, $4.2 billion was spent for drugs and other nondurable medical products, of which $2.7 billion was for prescription drugs. In 1980 these figures were $21.6 billion and $12.0 billion, respectively, and in 1960, they were $60.7 billion and $36.4 billion, respectively. Pharmaceutical expenditures per capita were $182 in the United States, an amount exceeded only by France ($196) and

Germany ($258) among eight European countries (Hutton, Borowitz, Olesky, & Luce, 1994). In 1993, $75 billion were spent on retail purchases of drugs (a 5.9% increase over 1992), with $48.8 billion of that for prescription drugs (Levit et al., 1994).

Medicaid spends more than $6.7 billion a year on drugs prescribed for people outside hospitals, and it has been estimated that establishing a list of approved drugs would save the federal government at least $275 million over 5 years, with the states saving almost as much (Pear, 1993b). The drug industry has organized to fight any changes, on the grounds that drugs would be selected on the basis of cost not quality, and this would condemn poor people to inferior health care—a curious concern, given the low quality of health care received by the poor, uneducated, and uninsured because of the costs of such things as drugs.

Reinhardt (1992a), a Princeton professor of political economy, noted that the Congressional Budget Office projected that under the current system, it is possible for health spending to increase to as much as 18% of the GDP by the year 2000. All of the statistics cited earlier are particularly interesting, given that no other industrialized country spends as much as 11% of their GDP on health care at the present, and the highest estimates are that other countries will reach only 12% by the year 2000. The figures are even more striking, given the low ranking of the United States in male life expectancy, female life expectancy, and infant mortality compared to other industrialized nations.

The health-care statistics for the United States are appalling, given the size of medical expenditures. The United Nations Children's Fund (Johnson, 1993) issued a report noting that one-fifth of American children live below the poverty line—four times the rate of most industrialized countries and twice that of Britain, the next worst performer. Census Bureau data for 1992 indicated that the number of children under 6 that were below the poverty line had increased to 25%, with nearly one-half of black children below the line. It was reported in *The Boston Globe* (January 30, 1995) that the National Center for Children in Poverty estimated that 6 million children under 6 (26%) were living in poverty in 1992, an increase of 1 million since 1987 and 2.6 million since 1972. Fifty-eight percent of these children had parents who worked at least part-time and fewer than one-third of the families relied entirely on public assistance. The U.S. child mortality rate, according to the UN report, was 11 per 1,000, placing it 19th in rank among industrialized nations, with the death rate for black American children more than twice that of whites. The immunization rate for children in the United States was lower than that of dozens of developing countries.

Konner (1993) noted that if you disregard infant mortality and consider life expectancy of a preschool-age child, the life expectancy of children is worse in Harlem than in Bangladesh. More than half of the excess deaths among blacks were caused by the same diseases that claim lives in "better" neighborhoods in the United States—heart attack, stroke, and cancer. Preliminary World Bank statistics indicated that the probability of death between the ages of 15 and 60 for black males in the United States is 30%, a figure higher than that for underdeveloped nations such as Gambia, India, and El Salvador (Murray, 1990). There is little doubt that the poor in the United States suffer an inadequate health-care system.

There has been a $350 billion increment in annual health-care expenditures over the decade, but the poor and homeless did not benefit appreciably, because major cities have experienced drastic insufficiencies in the quality of medical care as indexed by longer waiting times, fewer physicians, and less choice of physicians (Ginzberg, 1994). The American public appears to be indifferent to questions of equity as long as their personal access to health-care service remains satisfactory.

Much of this apathy is due to a lack of concern by the media, a concerted effort not to know by the public, and a game of self-serving partisanship by politicians. All have shown little interest to find out and consider what is happening, making it easier to avoid issues and to fail to take corrective action. As the discussions regarding a national health-care plan proceeded, there was more obfuscation than clarification of issues, and the media made a considerable contribution to the lack of clarity regarding issues and facts. A minor example was provided by a front-page article in *The New York Times* on July 10 (Berke, 1994, p. A1). He developed the point that

> Even some democrats who still support the Clinton plan say they are concerned that universal coverage, although viewed by most Americans as a laudable goal, is not a particularly compelling message, given that most people have some coverage.

Three paragraphs later, he reported that, according to the latest Gallup poll, most Americans favor universal coverage (77%) and believe employers should pay either all or most of the insurance premium (52%). These poll results hardly support the thesis of the piece that people regard the issue of universal coverage to be not particularly compelling. The journalistic spin doctors forced issues to fit preconceived positions, especially when universal coverage, employer mandates, and the single-payer health plan were concerned. Reinhardt (1994b, p. 24), after considering the recent debate on health-care reform in Congress, raised the question of whether the politically preferred vision for American health care described by members of Congress

> faithfully reflects the independent preferences of the grass roots, or whether it is being foisted on an unsuspecting grass roots by a small, powerful policy-making elite that knows how to manipulate grass-roots "preferences" through skillfully structured information and misinformation.

De Facto *Health-Care Rationing*

The statistics reviewed here support the argument that health care is not dispensed equitably at present, and that the United States has *de facto* health-care rationing. Grumbach, Bodenheimer, Himmelstein, and Woolhandler (1991) found that in 1990, those individuals whose income was in the lowest 10% received 1.3% of total income, but paid 3.9% of total health-care costs (the comparable figures for Britain were 1.7% of income and 1.7% of costs). Those in the top 10% of U.S. income had 33.8% of total income, but paid only 21.7% of health-care costs (for Britain the figures were 24.9% of income and 25.6% of costs). Not only do the poor receive proportionally less medical care, they pay proportionally more for what they do receive.

The Oregonian (November 2, 1992) cited statistics indicating that infant mortality for blacks is twice as high as that for whites, that 80% of white women received prenatal care in the first trimester of pregnancy compared to only 61% of black women, and that 23% of 2-year-olds did not receive the recommended doses of polio vaccine in 1985. *The New York Times* (October 8, 1993) reported that, because of inadequate governmental funding to combat tuberculosis (TB), what had been a successful defeat of a killer disease has reverted to epidemic proportions. Until 1985, the incidence of TB had been steadily declining for 30 years because an effective treatment was available. In 1985, 22,000 cases were reported, and this increased every year, reaching 27,000 new cases in 1992. The CDC had proposed a plan in 1989 that could have halted the TB epidemic at a cost of $36 million per year. The program would have reduced the disease to a level low enough that TB could have been considered eradicated in the United States. Each year, however, the Bush White House eliminated funding from the budget, permitting the epidemic to occur. This epidemic is now complicated by the fact that several types of bacteria have appeared that are resistant to several antibiotics, making it even more difficult and expensive to treat. It is estimated that an effective plan to halt the epidemic would cost $484 million next year, rather than the $36 million per year that would have prevented the epidemic and avoided the enormous suffering that is occurring. The Clinton Administration recommended only $124 million for the program, and the Congressional conference committee authorized only $111 million. It should be no surprise that the TB bacterium is spread most easily in crowded conditions, with the poor and homeless suffering the most. Cases such as this support Konner's (1993) argument that we do have health-care rationing in the United States at present, at least for the uninsured, and that in the world of limited resources, poor, uninsured people are denied treatments that would benefit them, because the available resources are too often used for needless services for the insured fortunate.

Several reports have documented the connection between socioeconomic status and health. In the United States, in 1986, those with a yearly income under $9,000 had a death rate that was 3 to 7 times higher than those with a yearly income of $25,000 or more (Pappas, Queen, Hadden, & Fisher, 1993). The differences in mortality between blacks and whites were eliminated after the figures were adjusted for income differential, although the causes of death were quite different for blacks and whites. The differential death rate between the high- and low-socioeconomic groups increased between the years 1960 and 1986.

The death rates in 1989 and 1990 for men and women between the ages of 25 and 44, who did not have a high school education, was three times the rates for college graduates. Infant mortality dropped from 8.9 in 1991, to 8.5 per 1,000 live births in 1992, but the rate among black infants was double that of whites (as the UN also reported). In 1991, the infant morality rate was 17.6 per 1,000 live births for blacks and 7.3 per 1,000 for whites. [Note that these infant mortality rates, expressed per live birth, are lower than the rates for child mortality reported by the UN Children's Fund.]

Education was more strongly related to death rate due to coronary disease than was race (Keil et al., 1993), and was more strongly related to the life expectancy of older Americans than was race (Guralnik, Land, Blazer, Fillenbaum, & Branch, 1993). Angell (1993b) noted, in an editorial commentary on these studies, that similar disparities in health between socioeconomic classes is found in other Western countries, even in those in which access to preventive health care is universal. Angell was appalled by the realization that people already burdened by poverty and lack of education also have to carry a disproportionate share of illness.

Some of the forces that make the U.S. system so faulty were identified by Konner (1993, p. 57):

> the steady advance of science; the cultural traditions of medical training (*I did it, so you have to do it*); the nation's need to use trainees to care for the poor; the insurance companies' misplaced sense of responsibility—greater toward their stockholders than toward patients; the bizarre patterns of reimbursement insurers maintain; the consequent disproportionate income and influence of procedure-oriented specialities; colossal administrative waste; greed and fraud; and the enormous impact of soaring malpractice litigation on medical decision making.

If equitable funding for medical care is to be realized, it is necessary to reform the entire health-care delivery system in order to contain medical costs and provide universal health care. One possible solution was proposed by the Clinton administration, and competing plans were suggested. The AMA for the first time expressed a willingness to participate in discussions aimed at cost containment in the health-care system, perhaps because some reforms seemed to be inevitable (Hilts, 1993a).

Although everyone agreed that adequate health care should be guaranteed to all citizens, the evidence indicates that the United States is spending astronomic sums on health care, and these sums are much higher per capita than those spent by any other industrialized nation. Yet, the United States does not have a health-care system adequate to care for all of its people, and there are signs that the situation is becoming worse as more and more people are unable to afford adequate health insurance, or have none at all.

Evidence that fears regarding the worsening situation are reasonable is provided by occurrences such as the threat by Empire Blue Cross and Blue Shield (hereafter referred to as the Blues) of New York to seek increases of as much as 35% for 125 thousand customers or cancel their policies altogether (Meier, 1993b). These customers have difficulty finding insurance because they are in a pool that includes the worst insurance risks. Indications are that as many as half of the 7.6 million customers of the Blues could be at risk to lose their private insurance and be forced into managed care plans. Such moves by private insurers have two impacts on the health-care system: (1) They move many high-risk customers onto plans that must be subsidized by taxpayers; (2) Others will be forced into low-cost HMOs that do not allow patients the luxury of choosing their own physicians, a prerogative that is dear to many physicians and citizens.

Four possible solutions were suggested by President Clinton in a speech to the National Governors Association on August 16, 1993. One, which he found inadequate, was not to worry about the problems in the hope that insurance costs will decrease if a simplified premium structure can be developed, with big pools of customers established so that their numbers can be used to bargain for more favorable rates. This seems to be the interim solution the 103rd Congress settled for. A second solution was to mandate that every individual must buy health insurance which, as he pointed out, cannot occur, because it will not be possible to make enough reforms in the insurance system for the uninsured and unemployed to participate. The third solution was a single-payer Canadian-type plan, which he rejected because of the difficulties and initial costs to implement, plus the loss of jobs that would occur in the insurance industry. The fourth solution, which he favored, was to introduce a mandate that employers must provide insurance for their employees, with a long phase-in period combined with a limitation on how large the premiums would be for small businesses operating on narrow profit margins. He favored this approach, because the system we have now works for most Americans, except for what he calls a "laundry list of problems," and expressed the belief that we can maintain the "world's finest health-care system." The statistics outlined here, regarding the costs and availability of health care in the United States indicate that the problems constitute more than a "laundry list"; they suggest that, instead of the "world's finest health-care system," we have an inadequate system edging toward catastrophic failure. It is unable to provide adequate care for many millions of our citizens, and even with its inadequacies, the costs could fatally cripple the U.S. economy.

HEALTH-CARE RATIONING SYSTEMS

Caplan (1986c) discussed the "no-fat thesis," which maintains that all of the technology that has entered medical practice in the past few decades is basically useful and beneficial. According to this thesis, the only way to cope with new technological advances is to "batten down the hatches" by immediately rationing access. He favored the counter-argument that there is a good deal of fat in the health-care system, and that those concerned with medical ethics should not consider establishing rationing criteria until the system is made as efficient and fat-free as possible. The statistics cited earlier, the lack of any coherent system to assess the effectiveness of health-care technologies, and the widespread use of expensive new technologies (as well as expensive older ones of questionable effectiveness) all argue that effective cost containment produced only by eliminating excesses is not likely to be effective. It is doubtful that these changes will have sufficient impact to make it possible for universal health care to become a reality. The rapid increase in the number of people over 65 suggests that the change in demographics alone could make it necessary to limit those treatments that only extend vital processes at the end of life.

Although Caplan (1989b) argued that every effort must be made to resist or delay implementation of any rationing policies, he subsequently suggested (Caplan, 1993) that we should examine the experience of other countries that deliver better care at less cost. Glaser (1993) also suggested that the United States needs to consider the characteristics of the more successful health-care systems of other countries. Caplan's examination of the situation led him to suggest that one problem is that the United States has too many costly specialists and not enough general and primary-care practitioners. It has been noted by Konner (1993) and several other authorities that the U.S. system relies on specialists so much that the poor usually have no one doctor corresponding to the British GP who serves to coordinate the efforts of specialists and to mediate between them and the patient. In Britain about one-half of all physicians are GPs, whereas many places in the United States, such as the inner cities and rural farm areas, have no primary-care physicians at all. In France, 58% of physicians were primary-care physicians in 1990, compared to 34% in the United States (Fielding & Lancry, 1993). Other countries adhere to fixed overall medical budgets, with the consequence that there is inevitable rationing of access to specialists and expensive technology, but general access to primary care and preventive medicine. Caplan seems to have moved toward a position that is more receptive to the idea of health-care rationing.

The data reviewed in this chapter suggest it is unlikely that adequate reform can be implemented easily in the near future, especially given the power of the lobbies representing the medical and insurance interests. Too much discussion has been centered on the financial considerations involved in health-care reform to the exclusion of moral or humanitarian concerns. Passell (1993), in an article dealing with the problems involved in developing a National Health Plan (NHP), raised a number of concerns that bedevil the process: the health-delivery industry is concerned about unemployment and a loss in profits; politicians are worried about the effects on the overall economy, on the process of deficit reduction, and their own reelection; those with medical insurance are concerned that their insurance rates will be increased; the public focuses on the "princely earnings" of physician specialists and worries about their own decrease in standard of living; the medical-technology industry worries about the loss of the earnings it gains from the export of drugs and machines; major industries worry that costs for medical insurance will make them less able to compete with foreign companies; small businesses worry that major industries will gain preferential advantages that will work against the profit margins of the smaller companies, and the major industries worry that small business will gain an advantage if the government bears part of the costs of their medical plans; the insurance industry is concerned with the profit margins for their stockholders; the media want lively and evocative real-life drama. Only in the last line of his article does Passell mention a concern for the lives of the less privileged, a concern the Clinton Administration emphasized in the hope that it would be possible to enhance the quality of existence in the United States.

The issue of health-care reform has been clouded by partisan politics, with some of the arguments representing legitimate conflicts of interest between different groups and others based on political posturing to set the agenda for national elections to come. Toner (1993) reported that Representative Dick Armey of Texas, at that time the third-ranking Republican in the House and Chairman of the Republican Conference, argued against the Clinton health-care plan even before it was announced. Armey, in his initial salvo, declared that the Clinton plan will destroy jobs, burden the economy with massive new taxes, and lead to health-care rationing. The plan would put the "best health-care system in the world" in jeopardy because it would be a "monolithic change to be affected in one fell swoop." The statement included the colorful rhetoric that the Clinton plan is a "Dr. Kevorkian plan for jobs," and characterized the managed-competition plan as really being managed coercion. He characterized the plan as bureaucratic and a step down the road toward a government-controlled system. President Clinton, who also characterized the U.S. health-care system to be "the world's finest," seems to agree with Rep. Armey regarding the overall quality of our present system, but to have different views regarding what should be done to it. Early in the proceedings, Senator Dole floated a trial balloon when he stated that there was no serious health-care problem, and even though there was little positive response by either politicians or the public, he returned to that line from time to time. The specter of "socialized medicine" was invoked, especially when mention of a single-payer plan was raised. The debates began with a divisive and partisan tone that was not centered on the basic needs of the uninsured or the quality of health-care available to the poor.

Existing U.S. Rationing Systems

It can be argued that health care should be rationed to limit extraordinary services to the terminally ill, and that unnecessary and expensive diagnostic tests for the insured should be eliminated if there are not enough resources to provide health care for all. The millions of people not protected by any health-care safety net and who receive marginal or no care at all must be protected. Moral concern should be centered more on the quality of health care and the relative costs and efficacy of treatments, and less on the economic well-being of the more privileged.

Statistics support the argument that the U.S. health-care system needs a complete overhaul, and that some plan to ration health care might have to be devised, unless the economic realities are altered. One such health-care rationing systems is called *triage* (the French word for sorting or choosing), and is used by military medical personnel in the field of combat. When there are large numbers of combat casualties and limited time and facilities to treat all the wounded, treatment is given preferentially to those whose likelihood of recovery is judged to be greater, given the severity and type of wound, even though these decisions mean some other wounded people will die from lack of surgical care.

Caplan (1988) noted that the concept of rationing is used presently with IVF. The government assumes no obligation to assure the availability of the elective IVF procedure to any person who might want it, any more than it has assumed the obligation to help find mates for unmarried people who desire to reproduce. Another instance of rationing occurs when decisions are made regarding who is to receive an organ transplant (Caplan, 1989c). There are (usually unstated) rules to determine who will receive an organ: Some transplant centers consider 55 to be too old for a liver transplant, and 65 for a heart transplant; there is usually a means test (those who can pay or are insured receive a higher priority); the overall health of the potential recipient is a factor; and "psychosocial" or moral factors are often brought in by considering such things as the quality of care that will be available to a patient who is a baby or child, or whether the baby being considered for a transplant is the product of "unmarried bliss." Both Caplan's and Konner's analyses led them to conclude that there has been rationing for some time in the case of organ transplants, artificial organs, rehabilitation procedures, and I would add, even basic health care.

Prottas (1993) examined issues involved in organ transplants, noting that every organ procured is needed and will be given to some patient. If a state does not pay for transplants under Medicaid, then the poor in that state will not receive a transplant. The rich can get a transplant, whereas the poor cannot, a clear instance of the *de facto* rationing that occurs for this and a large number of medical treatments. A number of people examined the criteria used (with varying degrees of formality) to ration health care (see Battin, 1987; Caplan, 1986c, 1989a, 1989b, 1989c; Fox & Swazey, 1992). Among them, are age, length of time on a waiting list, ability to pay, medical urgency, ease and speed of the procedure, prognosis (in terms of medical benefit and quality of life), antigen matching (between donor and recipient), expected future contribution to society, responsibility for others, random selection, regional availability of resources and facilities, preventive measures, quality of care available, personal responsibility for the health problem, and a variety of moral considerations, such as marital status and substance abuse.

Development of Formal Rationing Systems

Battin (1987) discussed an example of one type of rationing system that might be established:

> Ten units of medical care given to a ninety-two-year-old man with multiple chronic conditions might make it possible for him to live an additional two years, but ten units of care given to an eight-year-old girl in an acute episode might make it possible for her to live a normal life span, or about sixty-four additional years.

The elderly individual in this situation might be disgruntled by the decision to provide care only to the girl, but it can be argued that this decision could be a just one in the face of limited resources. The problem is to determine how to scale the "units" of medical care, determine the relative weight to be given to the quality and number of the additional years of life, and establish the rules by which to combine these weights to arrive at a priority score.

A point system could be established for each of the criteria, with people receiving care based on the priority points assigned to the various medical treatments until funds are no longer available. When the State of Oregon proposed such a system, it was challenged by special interest groups, such as the elderly and physically disabled. If any rationing system is to be considered, it would seem that the principles outlined when discussing the SJT procedures should be taken to heart: The value-laden criteria to guide the allocations should be agreed upon by the public; then the relevant economic, scientific, and medical experts should examine the data related to each of the criteria; and the final rankings that result from the agreed upon criteria and weighting system should be examined. If there are unanticipated or undesirable inequities, then the criteria based on these publicly established weightings can be revised after discussion among the involved parties. The GHPM

(Kaplan, 1994a) approach that was used with success in Oregon will be discussed in the next chapter. Typically, everyone will not be satisfied with any system that is developed, but it should be possible to develop a system that is satisfactory to a large majority who can agree that the interests of justice have been honored democratically, given the realities of the situation. At least, such a procedure is preferable to the haphazard and unformulated systems that now prevail.

Murray (1993), a bioethicist and Director of the Center for Biomedical Ethics at Case Western Reserve University, noted that it has been difficult to establish a set of basic moral premises that everyone can accept as a basis for a deductivist approach to apply to particular cases. He noted that the National Commission for the Protection of Subjects of Biomedical and Behavioral Research, established in 1974, was able to agree when considering policy in the context of specific actions concerning individual cases, no matter what difficulties the cases presented. There was no agreement, however, regarding the general moral principles that should, or did, regulate the decisions. Some have argued for what Murray called the "bioethics mantra," the chanting of which is presumed to solve moral dilemmas: autonomy, respect for human liberty, and beneficence (both an injunction to do good and avoid harming). The problem is that even if these are accepted as the proper moral principles, there is no agreement regarding the relative importance of each or how to apply and interpret them in specific instances.

This state of affairs led Murray to suggest that one should not proceed deductively, but use what he called the *method of casuistry*—case-centered moral reasoning employing the general procedures involved when the sorites-style argument is used (see Petrinovich, 1995). With this argument, one begins by considering cases on which all agree as to what should be done, moving to more and more difficult ones, until disagreement appears. When the point of disagreement is reached, then conceptual principles should be examined, and agreement should be sought regarding why the case is difficult. Discussion could then center on the general principles that might be invoked to resolve the case. This procedure is essentially the scientific method: Clearly state a set of general principles (a theory); apply the principles to a relevant set of cases (perform an experiment); identify where the principles are not adequate to provide guidelines to resolve the cases (analyze the data); revise the principles so that the difficulties are no longer present (modify the theory); examine a new set of cases to determine whether the revised principles are adequate (perform a new experiment); and follow this process continually (examine the predictions generated by the theory and subject them to further experimentation).

Rather than characterizing scientific procedures as deductive versus inductive, it is more reasonable to emphasize the continual interplay of each. There is a conceptual framework that leads scientists to observe some things rather than others, and to perform certain kinds of experiments to understand aspects of the world that appear to be promising places to look. This conceptual framework is developed on the basis of what has been observed before, or examination of the results of previous observations and experiments. These new experimental or observational data, in turn, lead to a modification of the conceptual framework, and this recursive process continues, it is hoped, to a more and more adequate vision of the nature of reality. A polarity between deductivism and casuistry does not characterize good scientific method, nor will it further sound public policy making.

The first task is to obtain public agreement concerning the criteria that should apply and to establish some relative priority for each of the criteria selected for inclusion. Once this step has been taken, panels of experts could examine the relevant data to evaluate the costs and benefits of each procedure. One serious problem, identified by Konner (1993) and Kaplan (1994a), is that there is little solid scientific data regarding the relative effectiveness of medical treatments. The benefits of surgery for prostate cancer have been questioned in light of the financial and mortality costs of the surgery, as well as the production of undesirable side effects, such as impotence and incontinence. There is concern regarding the wisdom of the treatment, because the increase in survival time following surgery is only months, and the side effects can make those months low quality for many of the patients.

Konner noted that a large number of different surgical procedures are used to reverse the atherosclerosis of the arteries that causes anginal pain, heart attack, and many strokes. He expressed amazement when he realized that the positive results of all of the operations invented to treat the symptoms of atherosclerosis can be attained just as well if the patient is put on an antiatherosclerotic diet. Konner's bottom line is that research evaluating surgical outcomes should be given the highest priority in medical science.

Some types of data, such as the cost of different procedures, will be relatively easy to gather. Caplan (1986c) listed the average costs of common surgical procedures and diagnostic devices, and costs to support individual types of patients (such as premature newborns, and children born with severe immune deficiency syndrome) who have specific medical problems.

The amount spent on health care in the United States staggers the imagination. Angell (1993a) estimated that in 1993, the United States will spend $900 billion, or $3,380 per citizen, and this $900 billion figure has been revised to $939.9 billion by the Commerce Department. The projected figures are of astounding magnitude in recent years, and almost all the projections have turned out to be underestimates.

Despite the enormity of these amounts, almost nothing is spent to determine whether much of it does any good. Konner (1993) noted that after hundreds of years, only a small percentage of medical practices have been evaluated properly. The federal government spends only about $100 million a year in research to study outcomes, to identify inappropriate utilization of medical procedures, and to develop guidelines for physicians (De Lew et al., 1992). They cited evidence of widespread inappropriate use of expensive and potentially dangerous procedures, such as an inappropriate use of coronary artery bypass surgery 20–35% of the time. Other studies have found about 15–30% of certain innovative medical procedures are inappropriate, unnecessary, or both. Their estimate was that these unnecessary expenditures, if applied as a percentage of all medical spending, would amount to between $99 billion and $198 billion in 1990. The highest priority for major research expenditures should be to evaluate the efficacy of existing procedures, with the development of new and exotic procedures having a lower priority.

It should be mandated that medical institutions and health-maintenance organizations make continual assessment of the efficacy of every procedure, and funding should be provided for high-quality evaluation research by qualified independent teams of evaluation researchers. No sensible rationing plan, or effective health-care system can be devised without including outcomes as a major component to direct the type of system that will be adequate to meet the needs of society.

Caplan (1989c) reminded us that the distribution of health-care resources occurs at three distinct levels: (1) between health-care and other societal expenditures; (2) within the health-care sector; and (3) among individual patients. Considering the first level, politicians and the public will have to establish overall priorities within the realities of governmental budgets. At this level is the arbitrariness that goes into constructing any personal budget—so much allocated for food, housing, clothing, transportation, medical care, entertainment, and what have you.

Once decisions have been made at this level, the medical profession, the public and its representatives, medical institutions (hospitals and laboratories), and insurers can decide how to allocate these resources within agreed-upon budget limits. At this second level, the questions refer to such things as how much should be allocated for hospices, how much to research, how much to prevention programs, how much to purchase the latest medical technology, and so forth.

After these decisions have been made, individual medical facilities and physicians (with the help of medical review boards) can deal with the issue of triage—how much of scarce resources will be used to support which patients. It is at this level that the medical profession needs assistance to establish rational policies based on agreed-upon criteria, and it is here that the principles of SJT and GHPM can be used to advantage.

In this chapter some of the basic problems involved in the development of an adequate health-care plan have been identified and some of the economic and moral implications outlined. Data were presented to argue that the health-care system of the United States is a sorry and immoral mess in terms of effectiveness, access, and cost. The following four chapters will deal with three plans that were proposed to resolve these problems and some of their practical, economic, and moral implications. The progress of the debate throughout 1993–1994 will be examined, and the likely future for health-care reform in the United States will be considered in Chapter 14.

CHAPTER 10

Two Proposed Health-Care Plans

Oregon Rationing and Managed Competition

THE OREGON RATIONING PLAN

Attempts have been made to devise rationing plans for health care, and they have met with varying degrees of acceptance. In 1987, the Oregon legislature voted to eliminate $1.1 billion in Medicaid expenditures for all organ transplants except kidneys and corneas and to use the funds to provide prenatal care for an estimated 2,000 medically indigent women. This decision was made with an explicit understanding that there was a trade-off between basic care for the many and expensive care for the few.

In 1991, about 450,000 Oregonians were uninsured, with another 230,000 underinsured. These people were ineligible for Medicaid because their income exceed the $5,700 per year limit for a family of three. People earning low wages but making over $5,700 could not afford insurance—a clear form of implicit rationing. The Oregon legislature proposed a bipartisan plan to ration explicitly the health care provided by Oregon's Medicaid program. The plan had to be approved by the U.S. government to qualify for the federal portion of Medicaid funds, and it was rejected by the Bush Administration in August 1992. The plan would have added thousands of poor people to the health-care system. To accomplish this goal, however, some benefits received by those covered under the Medicaid program would have been eliminated. The original proposal contained a ranking of 709 medical procedures according to their costs and benefits, with those ranking below 587 no longer financed by Medicaid. The rankings were done by Oregonians who were asked to estimate the value of a given treatment to contribute positively to the quality of life. According to *The Oregonian* (August 4, 1992), the plan would have made it possible to add 120 thousand uninsured people to the 231 thousand that were on the Medicaid rolls, and would have required employers to provide insurance coverage for another 300 thousand uninsured workers. Even with employer mandates, it was estimated that about 50 thousand Oregonians still would not have health-care coverage, because they work more than one part-time job, which put their total earnings above the eligibility level for Medicaid. The other reason that there still would be 50 thousand uninsured was that no one would be required to sign-up for health-care coverage.

The plan was opposed by the Roman Catholic Church, antiabortion groups, groups representing disabled people, the elderly, and the Children's Defense Fund. The Bush Administration concluded that the ranking process was tainted by discrimination on the basis of disability and was therefore unconstitutional.

The methods used to revise the plan have been described by Kaplan (1993, 1994a) and provide a model of informed decision making. As a first step, the Health Services Commission held 48 meetings attended by more than 1,000 people to learn about preferences for medical care in Oregon communities—the citizenry not willing to accept standards that had been developed by Californians. Thirteen community values were identified, including prevention, cost-effectiveness, quality of life, ability to function, and length of life, with a special emphasis on preventive medical services at the level of primary care. People consistently stated a willingness to forego expensive, extraordinary treatments for some in order to offer basic services for all.

To realize these goals, it was necessary to draw up a priority list. At this point the Commission had a medical committee of experts in a variety of specialities evaluate services using the QWB from the GHPM (described in Chapter 8). This step provided the expert testimony (just as the ballistics experts did in the Denver bullet instance) to estimate the expected benefit of the 709 condition-treatment pairs. The QWB was then used to obtain the opinions of 1,001 Oregon citizens in order to determine the subjective judgments required to score the desirability of the various health conditions. The commission placed greatest emphasis on problems that were acute and fatal, especially those in which treatment prevents death and there is full recovery. At the bottom of the list were treatments for fatal or nonfatal conditions that did not improve or extend the quality of life. The commission ignored information regarding costs when establishing the priority list, relying on subjective judgments instead. This was sensible, because it avoided raising the ire of many physicians and honored the concern that the medical value of treatments should be considered first, with financial ones being secondary. Obviously the financial realities will rear their ugly heads when the priority line is to be drawn.

The revised plan was resubmitted to the Clinton Administration, which approved it in March 1993, and it was implemented on February 1, 1994. In final form, the list of treatment–condition pairs was reduced to 688 by eliminating some redundancies, and a cutoff was set for those below 568. The conditions not covered were those in which treatment is ineffective and the condition will just run its course (e.g., the common cold), a home remedy is as effective as medical treatment (e.g., diaper rash), the treatment is cosmetic (e.g., premature graying of hair), or the treatment is futile (e.g., cancer that has spread throughout the body, and further medical treatment would not result in an estimated 5% chance of a 5-year survival).

Steinbrook and Lo (1992) discussed the Oregon Plan as it was initially proposed. They considered the Plan to represent a step in the right direction, but questioned whether the proposal represented a complete basic health-care package. One concern was that the process of setting priorities did not account for severity of illness among patients with the same diagnosis, a problem that will be addressed later. Another concern was that the original priority line was drawn on a financial as well as a medical basis, and they stated that it made little clinical sense to have the level of coverage rise and fall with budgetary pressures. (It is the budgetary reality, however, that forces the health-care system to consider a rationing plan at all.) They also were concerned that physicians would have to practice medicine according to a list, producing a conflict between a physician's ethical duties and responsibilities as defined by the state. These concerns should be kept in mind, as should the benefits to thousands of individuals who would receive essential coverage. The benefits to the uninsured should be accorded as much weight as the costs to the conscience and freedom of action accorded the members of the medical profession. The present threats to the autonomy of the medical profession are so great that it cannot afford to stick its collective head in the sand any longer. It must accept the reality that the financial stakes are so high that the market mentality is dictating the practice of medicine, that this mentality is submerging the interests of good health care, eroding the autonomy of physicians and the freedom of patients to select their own physicians.

Eddy (1991) evaluated the original Oregon Plan, concentrating his critical attention on the ranking system. He pointed out, as did Steinbrook and Lo, that the rankings are based on the assumption that all services are of equal value to all patients regardless of age, severity of symptoms, or specific indications for treatment. He also noted, as did

Konner (1993), that it is not possible to estimate either the benefits or costs of medical treatments without adequate outcome research. Given the lack of research evidence, any method of rationing must be based on crude guidelines. His bottom line was that attention should be focused on the methods used to establish priorities, concluding that these methods should be revised completely to reflect medical and financial cost-effectiveness more reasonably. If such revisions are not possible, he suggested the Oregon Plan should be abandoned in favor of some other approach. The Canadian Health Plan (CHP) might be a better approach and its adoption might well avoid the necessity of rationing health care at all.

In brief, the Oregon Health Plan covers all effective preventive medical care (including physicals, mammograms, and prenatal care), visits to physicians for diagnosis of any condition, treatment of most conditions (including hospitalization), psychological treatment for conditions ranging from depression to schizophrenia, treatment for drug and alcohol abuse, gynecological care (including tubal ligation and abortion), noncosmetic surgery (including most organ transplants), physical and occupational therapy, dental services, prescription drugs, and hospice care for the terminally ill. The Oregon legislature approved the 5-year Plan August 5, 1993, financing the first steps by a 10 cent per pack increase in the state cigarette tax. Further implementation of the plan awaits a battle with the business lobby, which has opposed proposals to require businesses to provide coverage for all of their employees in every session of the legislature since the plan was passed. The legislature voted to delay the employer mandate until March 1997 for businesses with 26 or more employees, and until January 1998 for those with fewer than 26 employees. New employers would not have to insure their employees for the first 18 months after they start their business.

The concern regarding the lack of a requirement to sign up for health coverage seemed not to be a problem. During the initial sign-up period, the rates of enrollment exceeded all expectations: It had been projected that 48 thousand Oregonians would sign up during the first 6 months of the program, but 81,084 had done so. The greater than expected enrollment occurred because more two-parent families—the working poor—signed up than expected, more rural Oregonians used the plan than had been projected, and a small number of uninsured residents of other states moved to Oregon to take advantage of the government-financed health care (about 2.5% of those enrolling in the plan indicated that they had lived in Oregon less than 3 months). This enrollment pattern supports the conclusion that there is a health-care crisis for those without health insurance, and that people will respond when a positive alternative is made available.

Indications are that the battle with business interests will continue in future legislative sessions. After the November 1994 elections, the newly elected Governor of Oregon, John Kitzhaber (a physician who had been the primary force in developing the Oregon Plan) decided that the idea of establishing an employer mandate was dead. In a November 16, 1994 editorial, *The Oregonian* expressed the opinion that without the mandate, true universal coverage probably cannot be reached. The editorial suggested that the 1995 legislature should consider forming a purchasing cooperative to buy insurance for Medicaid recipients and state employees, reform insurance laws to require some form of community rating, scale back the expectation of what minimal coverage should be ("Lean coverage is better than no coverage at all"), work for changes in the federal income-tax law to make health-insurance costs to individuals deductible and allow medical savings accounts, and "study" a targeted employer mandate that would affect only those businesses that offer no health coverage now, but not affect employers who cover their workers already.

Questions regarding employer and individual mandates have been raised regarding all health-care plans that have been introduced in the United States, with many businesses objecting to the former and some citizens rejecting the latter as an impingement on their freedom. There has been considerable experience with both kinds of mandate. There is an employer mandate with workers' compensation insurance and an individual mandate with auto insurance. Experience indicates that with an employer mandate, universal coverage is approached—for example, it has been estimated that only about 3% of Hawaiians do not have health insurance under their employer-mandated system. With the individual, mandated auto insurance system, there are an estimated 10% of drivers in the United States who are uninsured, with as many as 20% in those states that do not have strong enforcement policies. There is a question regarding the effectiveness of an individually mandated health-insurance system, even if it is established. The police can confiscate an uninsured car, but what can be done with an uninsured family should catastrophic illness strike?

Employer-mandated benefits tend to be shifted back to the employee in the form of lower long-run take-home pay, with little impact on the individual firms within a given industry (Reinhardt, 1994a). Employer mandates would have no impact on an industry that already offers employees health insurance, and a high proportion do at the present time. It seems that the strong opposition to employer mandates is mainly a diversionary tactic to sidetrack any reform to the health-care delivery system.

Scaling Medical Priorities

Several efforts have been made to develop priority systems for medical treatments, with medical economists, physicians, psychologists, and ethicists working to develop scales to rank treatments according to relative costs and benefits. Engelhardt and Rie (1986) argued that one place to begin attacking the problem of allocating scarce resources is with the hospital intensive care unit (ICU). Given the increase in the number of people over 65 in the world population and the improvements in life-prolonging medical treatments, the problem of allocating ICU resources is not trivial. In 1978–1979, Massachusetts General Hospital in Boston budgeted 18% of its budget to the ICU for approximately 7% of total hospital patient-days. Each ICU bed-day costs three to four times more than a routine hospital bed-day, and they argued that resources should be targeted for those patients most likely to benefit from them (Kapp, 1993). Research should be done to develop defensible criteria for providing life-sustaining technologies for some patients, but not for others, and these criteria should be based on medical efficacy.

Engelhardt and Rie (1986) considered problems involved in the allocation of ICU beds when further admissions to an ICU will jeopardize the standard of health for all currently in the ICU. This could occur whenever someone is eligible for admission and shows greater promise of benefiting from treatment than those already in the unit (requires moving the latter out and the former in). The problem is that the investment of resources in some patients becomes disproportionate to expected gains, because only marginal benefits are likely to be obtained for patients who are in the final stages of debilitation. Explicit general criteria should be developed to govern policies to admit, continue treatment, and discharge patients.

The criteria they recommended were based on indices of likelihood of success, quality of life if successful, and length of survival. These factors would be entered into an equation they call the ICU treatment entitlement index (ICU-EI). The values these factors assume would be considered along with permissible limits in terms of cost of the treatment option, and these costs would be used as one element to establish treatment policies. Because societal resources are finite, it is possible that there may not be sufficient funds to treat all those needing treatments that cost millions of dollars per person (even if the probability of success was 1.0 and the quality of life was unimpaired), especially in view of competing needs for the same medical resources. Their ICU-EI formula has four variables: P (probability of successful outcomes), multiplied by Q (quality of life), multiplied by L (length of life), with the value of that term divided by C (cost required for the treatment). Two comments are in order: (1) P and Q will be difficult, but not impossible, to estimate; and (2) if any one of the multiplicative variables is zero, the value of the equation becomes zero, and no treatment would be in order, no matter what the value of the other three variables.

Emanuel (1991) was troubled because the equation uses a mechanical system to assess benefits and lacks sensitivity to how people actually conceive of the situations they encounter. His concern was that such an equation does not respect the pluralism of values that people have in contemporary society. One person with laryngeal cancer, when considering surgery, may decline because speech is valued more, and there is a willingness to accept a shorter life with speech, while another person may value longevity over the ability to speak. It is difficult to capture such subtle difference in values and goals using any equation.

Emanuel (1991, pp. 114–154) examined some of the lists of basic medical services that have been proposed, noted some of the objections that have been raised to them, and discussed problems involved in legislating different health-care schemes. He concluded that there are serious problems and suggested that the best hope is to turn to democratic political procedures involving public debate regarding conceptions of the good life. Even though he recognized the difficulties involved and is critical of many of the quick fixes that have been suggested, he believes it is essential to change the current distribution of health-care resources. His solution is to work within the framework of an overall political philosophy—specifically his Liberal Communitarian Vision.

Although the problems involved in developing indices are numerous, attempts to create such scales have merit, if only because they add explicit, rational criteria that can be examined, compared, and discussed. Such procedures are preferable to the existing "first-come, first-serve" practice. It is possible to develop guidelines to decide whom to admit and discharge and whom to continue to treat, given competing demands and availability of resources. Any system necessarily is based on a subjective assignment of scale values, but the procedure is public and can lead to profitable debate regarding the relative importance of the different factors that enter into the equation and to a discussion of the moral principles that should regulate decisions. Having explicit criteria also makes it possible for physicians to communicate the reasons for decisions to patients and their families. It is important that the public understand what medical policies exist, how they will be used, and the basis on which any specific decision was made.

Engelhardt and Rie (1986) suggested a procedure that could help develop an empirically based treatment equation. They ask us to suppose that we have finite funds and are able to purchase insurance for ourselves to cover costs of treatment, with these treatments leading to different qualities and length of life. People could make a series of such decisions, and investigators could capture the rules used to decide when the purchase of insurance is a prudent investment and when it is not worth the expenditure. It might be found that people do not want to purchase insurance that would provide 3 additional months of life, if that life is to be extremely painful, if they will not be conscious, if they will exist in a demented state, or if the probability of the treatment's successful outcome is very low. A systematic set of hypothetical cases could be constructed using different values of the factors to be considered, and judgment policies could be captured for a representative sample of taxpayers, patients, medical staff, and whoever else should be involved in medical decision making. Such procedures would establish whether people have systematic policies at all, indicate how the policies differ for different types of people, and reveal any communal policies. Such understanding could focus discussions of the relative priorities on which a rationing system could be based, and develop a shared and explicit policy in areas where there is confusion.

There are two polar views regarding priority rankings. Some people reject outright any attempt to rank treatments, stating that this is a judgment that only patients, in consultation with their personal physician, can make. Others believe that economic realities have already forced the use of some such system to ensure all citizens the minimally satisfactory quality of life to which they are entitled, will continue to do so, and we should face that reality explicitly.

Difficult statistical-scaling problems are encountered whenever subjective judgments are to be converted into quantitative scores. A problem with many attempts is that the qualitative assessments on which quantitative scales have been based were made by healthy people asked to imagine how they would value such things as a year of life confined to bed, compared to one in which they are fully mobile, but in constant pain. The conjectural nature of this task could make the procedure questionable because the actual choices that would be made, given the reality of pain, might be weighted quite differently in the presence of real pain. Kaplan (1993) has found, however, that such preferences did not vary greatly across social or demographic groups, nor between those who have had direct experience with a disability and those who had not.

A variant is to ask respondents to imagine a hypothetical game of chance in which they would be asked to take a gamble that either would lead to perfect health or instant death. This could be another fantasy situation that might be too many levels removed from reality to be useful. The dilemma could be posed as one in which respondents could attain a specified health status with absolute certainty, but one that falls short of perfect health. Yet another variation could involve the purchase of insurance against the different eventualities described earlier. If there is input from communities of taxpayers, patient groups, and medical experts, then triangulating the outcome of these ratings might provide a converging set of communal opinions, focusing the discussion on the moral principles that are involved and those that should be.

There is no question that the establishment of adequate priority scales is a difficult task, and that much tinkering would be required to respond to political pressures and deal with intuitively perceived inequalities. The development of a priority scale might progress better if the underlying moral rationale on which the evaluations are based is agreed upon before the results of any scale are used to determine health-care priorities. If agreement can be reached on the applicable moral principles, as well as the duties and responsibilities they impose on society, then efforts made by medical experts, ethicists, taxpayers, and various groups of patients involved will be much more likely to result in satisfactory outcomes. The experience in Oregon provides a useful model to deal with the realities to be faced, given the current inequities in distributing limited health-care resources.

MANAGED COMPETITION

The Concept

The Clinton administration proposed modifications of the health-care system, using what it called managed competition (MC), which aimed to combine free-market competition with government management to control costs. An essential element of MC was to create consumer alliances to bring large numbers of businesses, communities, and occupational groups together to create a large-enough patient base to negotiate better terms with health-care insurers and to absorb individuals at high medical risk without imposing too much cost on other individuals in the alliance. Enthoven (1993a), Professor of Public and

Private Management at Stanford University and one of the developers of the MC concept, defined it as a purchasing strategy to obtain maximum value for the money for both employers and consumers. The trick was to design and administer rules of competition that would not reward health plans for selecting only good risks or segment markets, both of which defeat the goals of MC. MC is price competition focused on an annual premium for comprehensive health-care services rather than on the price for individual services.

One of the key elements in the MC system was what Enthoven called "sponsors"; purchasing agencies that contract with health-care plans on behalf of a large group of subscribers and continuously structure and adjust the market to overcome attempts by insurers and providers to avoid price competition. The sponsor system was an essential element of the plan to defeat the cost-increasing incentives of fee-for-service payments, especially given the lack of cost consciousness of insured patients who do not have information regarding the effectiveness of different treatment options, and who have what Reinhardt (1993) called a "free-lunch myth." This myth is that "the company" is paying for employee health care, so the costs to the patient do not matter. Reinhardt noted that this is not the case; workers have given up wage increases to gain health-care benefits, and the increased costs to business are passed on to consumers through price increases.

In an article in which they compared employee and individual mandates Krueger and Reinhardt (1994) remarked that other countries (Germany and France) that use a social insurance system to finance health care refer to it as a *contribution rate* rather than a payroll tax. They argued that this usage does not reflect a sloppy, stupid, or cynical use of language but is an appreciation of the significant economic, administrative, and political differences between contribution rates and genuine taxes. It is important to maintain this distinction, because it leads European workers to understand that the total amount contributed by both employers and employees is a percentage of gross wages—a percentage of their own money. This is quite different from being told that the company pays a certain percentage for the health-insurance premium—which leads to the free-lunch perception.

Enthoven (1993b) believes the use of sponsors would overcome insurers' attempts to group customers by expected medical costs and to charge those in each group a premium that reflects those expected costs. This underwriting practice makes insurance unaffordable for sick people, leads healthy people to ride free by underinsuring when healthy and gamble that they can get insurance if they become ill. These practices create a public burden to care for those under- and uninsured members of the community. Because of the costs of underwriting procedures, as well as high marketing costs to appeal to desirable groups of individuals, the costs to administer individual health-insurance policies are at least 40% of medical claims. Enthoven believes that insurers make health-insurance contracts extremely complicated deliberately, because this makes it difficult for consumers to compare prices in a market that is highly segmented. Group insurance with a sponsor is his solution to these problems.

Sponsors have several important functions in MC. Among them are to ensure that every eligible person is covered at a moderate cost, that everyone has subsidized access to the lowest priced plan meeting minimal national standards, that there is community rating whereby the same premium is paid for the same coverage regardless of health status, and that there are no exclusions or limitations on the basis of preexisting conditions.

Competition was to be assured, because it is always possible for the lowest priced plan to take business away from higher priced plans by cutting annual premiums. To ensure that plans do not stint on the quality of services, a national Outcomes Management Standards Board was proposed, that would set uniform standards for outcome reporting to enable consumers to shop for the best coverage at the lowest price. Because it is important that the sponsor be an impartial broker rather than a biased participant, the sponsor should not have its own plan.

Large Health Insurance Purchasing Cooperatives (HPICs) would serve as the sponsors. For large employers who have 10 thousand or more employees in one geographic area, there would be no difficulty forming them into a HPIC. Over 40% of the employee population are in groups of 100 or fewer workers, and these groups are too small to spread risks effectively. Enthoven found that administrative expenses are about 40% of claims in groups with one to four employees, 35% of claims in groups of five to nine, and 5.5% in groups of 10 thousand and more. For small companies, the self-employed, and in sparsely populated areas, there would have to be flexibility to incorporate them into established urban, comprehensive-care organizations that would establish and operate a network of primary-care outposts large enough to enjoy low administrative expenses.

Reinhardt (1993) believes that employers should be the "pumping stations" to deal with the flow of funds from U.S. households to health-insurance funds. Employers should collect payroll-based premiums and pass them on to HPICs. Such payroll taxes have proven worldwide to be the most popular way to finance health care and provide a reliable source of financing. He noted that no industrial country other than the United

States has employers directly and actively involved in the markets for health insurance and health care, and recommended that it would be wise to abandon this uniquely American tradition.

Reinhardt (1993, p. 183) summed up part of his argument as follows:

> Given the rather checkered history of private-sector employers in American health care, a good case can be made for leaving employers out altogether. Alternatively, if employers are to be engaged as pumping stations in financing health care, then one ought to limit their role strictly to those tasks they can do fairly well: the collection and transmission of payroll-based premiums.

He added that paychecks should note how much take-home pay was reduced to provide health insurance so employees have the information to understand they are paying directly.

Some Economic Realities

A major stumbling block concerns who will pay the estimated $30–50 billion required to cover the 41 million uninsured and the estimated 22 million underinsured in the population (Altman & Cohen, 1993). The administration released a trial balloon, suggesting that businesses would be expected to contribute toward these costs. The Kaiser Health Reform Project (1994) reported that in 1992, 60% of the 223.3 million insured Americans under age 65 had private health-insurance coverage, obtained through their employer, and 9% had individual private insurance. The suggestion that there be an employer mandate to provide health insurance produced concern in the small-business community and objections by various segments of the political community, even though employers presently provide coverage for the majority of Americans. Less than one-third of firms with fewer than 25 workers offer health benefits to their employees, in contrast to 95% of firms with 100–199 workers, and 98% with more than 1,000. From 1988 to 1993, the average family premium for employer-based group health insurance in the United States more than doubled, from $2,500 to $5,200. Private insurance is available but expensive. The Kaiser report noted that, in New York City, family coverage ranged from $6,000 to $11,000 per year, with high deductibles and coinsurance levels, and more limited benefit packages than those provided through employer group coverage.

Reinhardt (1993) suggested there should be an earmarked indigent-care tax, perhaps 1% of taxable income, to supplement the modest premiums collected from low-income families, and that the government should collect earmarked health taxes on alcohol, tobacco, and gasoline.

The idea of any increased taxes provoked screams of protest from members of Congress, executives in the industries involved, and taxpayers.

According to the Kaiser report, only 18% of the uninsured were from poor families (defined as having an income less than $11,570 for a family of three), 32% had low income (100–199% of poverty), 27% middle income (200–399%), and 13% high income (greater than 400% of the poverty level). Of these uninsured, 22% were 17 years old or younger, and by region of the United States, the largest number of uninsured were in the South (42%).

President Clinton's Proposal

President Clinton addressed the Congress on the issue of health care on September 22, 1993. In that address, he considered the general issues that must be resolved to fix what he called a badly broken health system. He outlined six principles that must be embodied in efforts to reform the health-care system. The most important principle was security, and this principle received support from almost everyone. What the principle means is that every American must have the security provided by guaranteed health coverage, no matter what their status in society, whether they have or change jobs, and without concern for the state of their health ("preexisting conditions"). Guaranteed health-care coverage is to provide security to those who are now uninsured and make sure they will never lose the security of that coverage.

He offered five other principles, about which there was less agreement that the proposed plan would meet the requirements of each. These principles were

1. Simplicity, aimed at eliminating the hundreds of different forms currently required by the over 1,500 insurers.
2. Savings, to stop the high rate of medical inflation by using MC.
3. Choice, by patients of physicians and by physicians of the setting within which they will pursue their practice.
4. Quality, creating "report cards" on health plans that are intelligible enough to enable consumers to make informed choices of health care, and to provide physicians with information regarding the effectiveness of treatment decisions.
5. Responsibility, to provide coverage for all members of society at a reasonable cost.

A key element of the original Clinton proposal was to provide universal health insurance for all people by the year 2000. With no changes in the current health-care system, it has been estimated that the cost of health care would be $1.63 trillion in 2000, whereas with the Clinton plan it was estimated to be $1.49 trillion. The President proposed to meet the costs of his plan by holding down the increase in Medicare and Medicaid spending by $238 billion, and saving $47 billion in other federal programs. He also proposed to increase taxes on tobacco (and possibly alcohol) by $105 billion, and to collect $51 billion in higher income taxes that would be levied on previously untaxed health-care benefits to workers. The administration's estimate was that there would be a total of $441 billion in savings and additional revenues by 2000. In that same period, increased spending would be necessary to provide a new long-term care program, at a cost of $80 billion, and a new prescription drug program, at $72 billion. A national health-care board was to be established to create and supervise the regional health alliances that would function within the 50 states, at a cost of $29 billion. The plan also called for $169 billion in subsidies for companies and low-income workers who cannot pay their share of insurance costs. The additional costs come to $350 billion dollars in new federal spending. He proposed to use the $91 billion difference between the $441 billion in cuts and the $350 billion in new spending to reduce the federal deficit, probably an unfortunate claim, because it introduced an extraneous economic issue into an already explosive set. The CBO supported the administration's estimates: By 2004, the plan would limit overall health-care spending to $2.07 trillion dollars, $150 billion less than otherwise would be spent.

These figures underwent intense debate regarding the reasonableness of the assumptions on which they were based. The administration proposed to set state-by-state budgets for health spending if the health alliances failed to reduce costs by a projected 15%. These proposed spending caps produced cries that they introduce price controls that would cause reductions in medical care, restrict access to new technologies, and threaten high-quality medical research. The details were changed continually, and the plan was the subject of massive legislative finagling.

Freudenheim, in a report published in *The New York Times* on October 8, 1993, noted the increasingly uneasy feelings that had developed. The Clinton administration, in order to make better estimates of the costs of the proposed MC plan, consulted a group of actuaries from several large accounting firms. This was done because the administration's original projections used incomplete and outdated information. It is critical to have accurate figures regarding such things as the per-capita costs for a guaranteed standard package of benefits to estimate the budgetary implications of the package. To provide better estimates, these actuaries were asked to construct a new set of economic models. The major problem was that the information available to construct the models was not comprehensive, and much of it was not current.

One of the actuaries decided it was not possible to collect appropriate data because of the multifaceted structure of the health-care industry and the many self-funded employer health plans. The actuaries noted that the information for the individual states was also outdated; the last complete figures being used were collected in 1982. Although data for the states have now been provided for 1992, the experts said that the information was pulled together from such a variety of sources, including suspect sources (i.e., hospital trade association statistics), it was not precise enough to construct quantitative models that could reliably estimate the cost of health-care premiums. They suggested that, given the incomplete nature of the data, it might be possible only to suggest the methodology that could be used to obtain the needed values whenever adequate data were available.

The difficulties encountered suggest that the market-driven private sector has had little interest in cost containment. Profit increases have been satisfactory, and all was well, until the system threatened to bring the rest of the economy to ruin. If an inordinate proportion of the economy is devoted to health care, then there are fewer resources available to support other segments, such as education and public works. This might not be so bad if our health care was the best in the world, but it is not. Perhaps the health-care delivery system should be freed from a market mentality.

Even with the Clinton proposal, health-care spending in the United States would exceed that of other countries: the 17% of the GDP projected to be spent by the United States in 2000 would be almost twice as large as the 10% projected for many other countries, all of which have controls to keep spending from rising higher. Some argue that the level of spending in the United States would be too high, that debate should center on reconsidering the use of expensive procedures and medical technologies, and that resources devoted to such activities as keeping extremely premature babies alive and providing extensive life-support systems to the terminally ill should be restricted.

The administration did not want to provoke a debate on such divisive questions, and Reinhardt defended the administration's implicit handling of these issues as a good solution, because it hid what he called *ad hoc* rationing by keeping it within the obscurity of the HMO. He pointed out that HMOs could avoid saying "We're over the budget" and, instead, could argue that "It's inhumane to continue treatment" (quoted by Eckholm, 1993b). Reinhardt commented that the concept of rationing is anathema to the American public, but it is essential to set priorities on treatment procedures based on their relative costs in order to provide basic care to everyone. He believes it is better economically to have decisions regarding medical care made in the privacy of the boardroom of the HMO than to have them exposed to public scrutiny. This paternalism and duplicity is disturbing, and one would hope that the public considers it intolerable.

Gaylin (1993), Professor of Psychiatry at Columbia University and cofounder and president of the Hastings Center for bioethical research, expressed the view that if we must have allocation (and he suggested it is a "cruel necessity"), then it should not be hidden from public view or determined by a small group of health-care professionals. It can be argued that it is even more objectionable if it is done by a small group of medical economists. Gaylin argued that any decisions regarding the propriety of certain medical treatments that involved the setting of limits and priorities on health care should be subjected to public scrutiny and debate. Any limits set on health care should be done using explicit principles regarding equity and justice, and these principles should be developed with the full exposure of the democratic process, much as Emanuel (1991) argued.

Instances have been described, in which rational policy making has been done by concerned, informed, and well-intentioned members of the public, using procedures such as those involved in SJT and GHPM, in the development of the Oregon rationing plan, and in devising priorities for organ transplants. Steps have been taken to establish rational health-care priority systems in other countries, and these will be discussed. Economics has been referred to as "the dismal science," and Reinhardt's proposal that the public be hoodwinked by elite policy makers, no matter how well intentioned, is a dismal one.

STATE LESSONS IN STRATEGY

Several states have been considering health-care reform for a number of years, and some have implemented changes in their health-care systems. Mashaw and Marmor (1993), Yale professors of law and politics, respectively, argued that the Clinton plan should permit a federalist option mandating uniform national standards for health insurance that demands permanent and universal coverage not be based on residence or employment. These standards would free the states to experiment with structural options that could allow them to realize their goals. There are great differences among the states in the characteristics of the populations, economic resources, structure of the economy, and fluidity of movement to and from the different states.

A report by the Health Care Financing Administration (Levit, Lazenby, Cowan, & Letsch, 1993) estimated state spending for health care in 1991—the first report of such estimates since 1985. Per-capita health spending for the United States was $1,877 (11.5% of per-capita income) and ranged from a high of $2,402 (10.5%) for Massachusetts to a low of $1,234 (8.0%) for Idaho. Differences in per-capita health spending between the states, and even parts of the same state, were found to be greater than the differences between the United States and some other countries. These large differences suggest that the most effective health-care plan might be one in which states (or even regions) assume the role that the provinces do in Canada, and that the government should establish budgets and spending caps to realize those budgetary limits. The most reasonable solution, given political realities, might be to have the states serve as the basic units. This would respect strong, entrenched interests and the historical commitment to states' rights, as well as the fact that our representative form of government uses the political structure of the states as its basic unit. Iglehart (1994) noted that one of the best features of the Clinton Plan was its recognition of American federalism.

It is not possible to compare the effectiveness of the various state reforms that have been proposed, given the unique circumstances that exist in the different states, as well as the effect of the 1994 elections, which brought the concept of employer mandates into question. Hawaii is often cited as a possible model, yet Hawaii is unique because it has a relatively healthy climate, and its major industries are clean. Hawaii is located in the middle of the Pacific Ocean, which makes it difficult for businesses to move easily, and there can be very little border crossing by residents of other states who might want to take advantage of benefits provided by the Hawaiian system. The problem of such border crossings can be extreme between Washington, D.C. and Maryland, for example.

Rogal and Helms (1993) pointed out that health-care reform generally falls into two categories: one addressing health-care financing mechanisms, and the other addressing the organization of the delivery system. The MC plan proposed that the federal government would be involved in financing and mandate participation in the National Health Plan. It would define the basic benefit package and develop provider networks. The states would be expected to develop and oversee the integrated delivery system and implement mechanisms to ensure compliance with budget limits.

Although it is not possible to compare the effectiveness of the reforms attempted by different states, it will be instructive to examine problems that have been encountered in reform attempts. The Oregon rationing plan was discussed earlier, and several problems surfaced when a formal and explicit rationing plan was suggested to replace the existing *de facto* rationing. The Oregon plan overcame many of the problems with the health-care delivery system, but there is still a struggle regarding financing, with special interest groups bringing intense pressure on behalf of their constituencies.

It will be useful to consider the experience in several other states. The experience in Florida can be used to analyze some of the political realities that will be encountered by any proposed health plan and to suggest ways to move beyond political bickering. Vermont made progress toward establishing a uniform package of health-care services, with cost control, a statewide global budget, and a community rating system, but it has run into problems at the final stages of implementation. The state of Washington has moved to a universal MC system with insurance premium caps, but the recently empowered Republican legislature has vowed to undo the reforms in the name of reducing the impact of government on people's lives. Hawaii made rapid progress, but has encountered difficulties recently. Minnesota made strides toward reform by expanding access, reforming health insurance, and facing problems of financing expanded access. A discussion of the experiences of each of these states will provide insights into the issues, and suggest ways a national program might be approached.

Florida

One of the earliest steps taken to establish an MC system was taken by the state of Florida. The Florida legislature acted because of soaring health-care spending—from $9.4 billion in 1980 to $38 billion in 1992 (Rohter, 1993). It will be instructive to examine the problems that Florida had to face and consider the tactics and strategies used to arrive at a health-care plan that enjoyed almost universal support.

The high health costs in Florida occurred because the population of 13.5 million people includes the highest proportion of elderly in the country, as well as large numbers of immigrants, and rural and urban poor. Businesses with 50 or fewer employees account for 95% of all jobs in the state, and they are the most affected by rising health-care costs.

Florida has been characterized as a state of extremes by Brown (1993), head of the Division of Health Policy and Management at the Columbia University School of Public Health. Its Medicare expenditures per eligible beneficiary are the highest in the nation, and it tends to be one of the worst states in terms of such things as infant mortality, low-birthweight babies, low eligibility for Medicaid, and low rates of payment and provider participation in Medicaid. Much of the Florida economy depends on agriculture, construction, tourism, and service industries that often do not provide health insurance for employees. In 1992, 24.2% of those under 65 did not have health insurance, and if the aged (who have Medicare) are removed from the base, the percentage that are uninsured would be even higher (Eckholm, 1994). It was estimated that 2.7 million Florida citizens were unable to buy health insurance (Applebome, 1994). The proposed health-care reform was intended to help insure about one million of these people.

The recession swelled the ranks of the unemployed, the uninsured, and those on welfare. Federal mandates required the state to add new groups of people to the Medicaid program, and utilization of health-care facilities increased as a result of an increase in the number of drug-abuse cases and AIDS patients. About 40% of Florida's hospitals are for-profit and, in order to survive, they sent uninsured patients to public hospitals, which had to assume the costs for those patients. This pattern of hospital utilization produced fiscal problems for both private and public hospitals, and they made a common case for fiscal relief.

The first step was to forge an agreement regarding the goals to be realized by 1996. These goals included access to universal health care and a means of paying for it; provision of primary health care for 95% of employees and their dependents or employer mandates would be adopted; increase of per capital health-care costs not to exceed the overall rate of inflation.

Agreement was obtained regarding these goals, just as President Clinton had initial success obtaining agreement with the spirit of his six basic principles. The next move was to establish a single state agency to reform the health-care system. This agency was established; it concentrated public attention on negotiations within the public and private sectors, as well as between the liberal and conservative elements in the state.

Two other unique factors helped enable the process. One was the presence of a large number of liberal Democratic, largely Jewish, ex-New Yorkers, who had retired and settled in and around Miami and Miami Beach. Many of these people had political attitudes that were formed in the Roosevelt New Deal era, including a tendency to be more accepting of government-centered solutions than were the younger people in the state. The attitudes of these Northeasterners are quite different from those indigenous to typical Southern conservative states such as Florida.

The other factor was the active support provided by a popular governor, Lawton Chiles. The problem was how to deal with the self-serving positions taken by business, the insurance industry, hospitals, the medical association, and groups representing the elderly, such as the American Association of Retired Persons (AARP). Each of these special interests agreed to abstract goals and joined the consensus regarding the need for change, but each wanted to settle the costs of reform everywhere but on itself. This is a preview of the problems encountered at the national level.

Everyone agreed (except the physicians) that a new public-sector entity, the Agency for Health-Care Administration, was required. Compromises were struck: The requirement that mandates would be applied to businesses if 95% coverage was not attained by a certain date was dropped; there was no reference to legislation to set physicians' fee schedules, which kept the image of the free market alive. The Medical Association still opposed the plan but made a tactical error when it opposed legislation to curb widely criticized physician referrals of patients to facilities in which physicians had financial interests. Brown (1993, p. 19) attributed the medical profession's lack of political effectiveness to the fact that, "it was busy shooting itself in the foot over the "black hat" issue of physicians self-referrals." The final Health Care Reform Act of 1992 passed the House 109–0, and the Senate 35–2.

The health-care plan used the MC framework favored by the Clinton administration. Participation in the plan was, however, purely voluntary, and it lacked the features of required universal coverage and standard benefits that Enthoven believes are essential to the success of MC.

On April 2, 1992, the Florida legislature created 11 Community Health Purchasing Alliances to assist employers to obtain the best care at the cheapest price. These alliances were to allow people who did not have health care to obtain it, and small businesses to cover previously uninsured employees. Some opponents expressed concern that it was but the first step on the slippery slope that leads to socialized medicine, but the existing need was so great that only 10 state senators voted against this stage of the plan, and several of these were liberals who were concerned that the overhaul did not go far enough.

The plan was killed by a legislative deadlock in June 1994. Although the plan was quite similar to what the Republicans nationally considered to be a Conservative alternative to the Clinton plan, and a broad coalition of groups supported it, the Senate Republicans blocked the implementation of the plan in strict party-line votes. The Democrats accused the Republicans of opposing it to hurt Governor Chiles in his fall reelection bid and claimed that pressure was being exerted by the Republican National Committee and Senator Dole to defeat any plan proposed by Democrats. The Republicans denied the allegations, saying they opposed it because it relied entirely on speculative savings and depended too heavily on a flawed agreement with the federal government. They attributed these problems to the political ineptitude of Governor Chiles. Whatever, it is clear that party politics dominated the process to kill the bill. The experience in Florida provides an interesting case history, identifies some of the problems that must be faced to establish any reasonable plan, and can be instructive to understand the political tactics and strategies that might be used to move any NHP forward, as well as political problems to be avoided.

The successful strategy that led to the health-reform bill had three aspects. The first was to appoint a number of study commissions to explore problems and seek solutions. One achievement of these commissions was to make an intense effort to educate prominent public and private players about the issues, what could be done, and who wanted what. Second was the use of what Brown called "policy clubs" by the politicians advocating reform. These clubs included the threat of a single-payer plan, a hospital rate-setting system, and a mandate that employers would have to cover employees if voluntary efforts did not result in guaranteed health care for 95% of the population. When alternatives were proposed, interest groups sought compromises and seriously considered a revenue assessment plan. The third aspect was the policy makers' understanding that they should blend benefits with costs, so that all major participants could take away some victory from the bargaining table. Another important element is the general agreement that was obtained at the outset regarding the nature of the moral and practical goals to be achieved (universal coverage, cost containment, and adequate financing). The experience of other states supports the view that these strategies generally have been successful to promote those reforms that have been accomplished. The political agenda always rears its ugly head at both the national and state levels, even though the public favors universal health-care coverage. Politicians should be reminded of this fact whenever elections roll around.

Vermont

In the spring of 1992, Vermont passed a health-care act that was to provide universal availability of a standard package of health-care services, control costs through a global budget, establish community rating for health insurance, reform medical malpractice laws, and place health care under one state authority. Leichter (1993b) attributed the success in

Vermont to some of the same factors that enabled the initial process in Florida. There was a broad and deep support for health-care reform in the state. As early as 1988, the legislature extended access to health care to children up to age 7 and to pregnant women whose family income was above the level to qualify for Medicaid eligibility but below 225% of the poverty level. This program was highly popular and considered successful.

Another element was to educate the public regarding the need for reform. In 1991, a bill had been approved requiring insurance companies to use community rating and guarantee acceptance for insurance in the small-group health-insurance market. This essentially Democratic bill had the support of both the Republican governor and an important state House committee with a Republican chair. As in Florida, the efforts had bipartisan support of some highly visible, popular, and skilled politicians, and a great deal of public attention and discussion had been devoted to health-care reform before the legislative battles began.

A third element was the "policy club" of the single-payer insurance plan, which was introduced in the legislature by a progressive Democrat who had the support of the 8,000-member Vermont National Education Association, as well as the backing of a politically astute Socialist U.S. Representative. The introduction of this bill led to the formation of a coalition that had the common goal of defeating the single-payer proposal, a proposal strongly supported by the people but with little support in the legislature.

Finally, after agreement was obtained regarding general principles, the politicians accepted a compromise that contained as many elements as could be attained without creating a great deal of acrimony. These principles included the establishment of a single health-care authority to define the nature of the universal health-care package and develop two alternative financing models—one a multipayer model and the other a single-payer. The Vermont Medical School was authorized to develop a plan to train primary-care physicians and encourage physicians to enter rural practices. To further this goal, the state allocated funds to support six additional family-practice residency positions at the medical school. Two difficult decisions were postponed, however. One involved malpractice reform: The state bar and the medical society were not willing to accept any compromise on the issue. The other involved the question of whether financing would use the multi- or single-payer model. The Vermont health-care system has been successful to the extent that the state had the seventh lowest level of uninsured (11.1%) in the nation in 1992 (Eckholm, 1994).

Unfortunately, the plan failed to gain final approval when the question of financing was faced. It collapsed when liberal Democrats (favoring the single-payer option) decided it accomplished too little, and Republicans (who opposed employer mandates) decided it accomplished too much (Pear, 1994c). A campaign of negative advertising that frightened many people was the first step in the unraveling. Local newspapers published tables showing that state income taxes would double for many people, but they omitted the crucial fact that the new taxes would replace health-insurance premiums that were already being paid. Neither the single-payer advocates nor employer-mandate opponents would back a compromise plan to introduce new taxes to finance universal coverage. Vermont made advances by requiring insurers to sell coverage at standard rates without considering preexisting medical conditions and setting annual spending goals for each hospital in the state. Several observers consider it to have been a fatal mistake not to have faced the critical questions regarding financing until the end of the process.

The entire process ground to a halt, with both the political left and right acting to block what they described as "bad legislation," and the center was neither well-enough defined nor organized to accomplish anything. The governor pledged that there would be no new income taxes to support the health plan, leaving it to be funded by taxes on tobacco, alcohol, and gasoline. He also proposed a 50–50 employer mandate, but employer mandates seem to have been effectively removed from consideration by the 1994 election results. The sizable number of single-payer advocates became dissatisfied and refused to accept anything other than the ideal, while the right wing was pleased to see no changes whatever and convinced many people that the single-payer plan was too radical. Leichter (1994) considered the major lesson to be that the parties interested in health-care reform failed to convince Vermonters of the need for comprehensive change, and people accepted the argument that government did not have the capacity to accomplish positive change. It should also be noted that Vermont has a small (560 thousand), relatively homogeneous population (mostly white, rural, and healthy). Eighty-five to 90% of Vermonters have health insurance, and 80% are satisfied with their health coverage, while the dissatisfaction with government is at an all-time high (80% rated the 1994 legislature to be fair or poor). A left and center coalition will be necessary to accomplish anything, because the right does not want major reform.

The lessons to be learned from Vermont are that it is critical to educate the public regarding the need for reform, to develop effective political leadership to guide the plan through the maze of special-interest groups, and to make enough visible progress that the reforms become an active reality. It is now necessary to begin a new round of discussions regarding those issues on which there is disagreement to start the whole process again. It helped, initially, to have the "Canadian System" lurking in the wings to provide the political club to keep the players searching for a viable alternative solution, but the specter of radicalism defeated the single-payer plan. A misinformation campaign raised people's fears that they could lose what they have and might pay more taxes. These fears were sufficient to bring the process to a halt and counteract the legitimate concern that all are not able to afford health insurance. The failure to consider the realities of financing proved fatal in the face of the negative advertising campaign and the coalition of two opposed political groups that formed an alliance strong enough to block any compromise.

Washington State

In April 1993, Washington State passed what Crittenden (1993) called the most extensive health-reform act on the books in the United States, setting a goal of universal coverage by 1999. As a result, there is a low number of uninsured citizens (12.4%)—the 13th lowest rate in the United States in 1992 (Eckholm, 1994). The plan incorporated MC and adopted insurance-premium caps to control costs. It completely restructured the health-care system, authorized a single state regulatory commission to operate it, created four regional HPICs, mandated that employers cover employees and their dependents for at least 50% of the lowest cost benefit package, placed an emphasis on preventive care, and imposed mandatory mediation in medical malpractice suits.

A familiar pattern of events made reform possible. In 1992, a bill was introduced in the legislature that heightened public awareness. This health-reform bill mandated a subsidized health-insurance plan and was passed by the Democratic House but defeated in the Republican Senate. The debate, however, led to public sentiment for health reform. The process was enabled when the Republicans lost the Senate, and the newly elected Governor and Insurance Commissioner both supported health reform. Also, the state medical and hospital associations, as well as the major HMOs, changed their positions from opposing to supporting reform.

The special interests fell to squabbling among themselves—this diffused opposition efforts as former allies became competitors. The major opposition included large and small businesses and insurers. Insurers devoted their concern to a proposed premium cap and were not interested in controlling costs, big business talked about its self-insured plans and expressed alarm about possible tax increases, and small business focused its opposition on the employer mandate. This fragmented opposition was not able to mobilize an effective effort sufficient to defeat the bill.

The plan was financed by taxes imposed on hospitals, insurers, and tobacco and alcohol. The public supported the sin taxes because of the relevance to health and a general disapproval of smoking and drinking. A voter's initiative to roll back the almost $2 billion in new taxes to finance the program was defeated on November 2, 1993. Hospitals agreed to tax themselves if the money were used for health-care reform, and insurers concentrated their efforts to kill the entire bill, rather than opposing the taxes that targeted insurance providers. There was popular support for the principle of health-care reform, education of the public regarding the issues, involved and effective politicians who were in favor of the plan, a factionalized opposition, and the threat of single-payer alternative, which was introduced early when the state commission recommended that either a single-payer or employer-based system be tried. The coalition in favor of reform held together, the opposition fell to bickering among themselves over special interests, and an informed public did not succumb to the fears that had effectively defeated attempts at reform in other states.

Hawaii

Some attention should be given to the experience in Hawaii, because it has been held up by Hilary Rodham Clinton and others to be an appropriate national model of MC. Neubauer (1993), Professor of Political Science at the University of Hawaii, considers Hawaii a pioneer in health-system reform. It initiated its plan in 1974, and created a State Health Insurance Program (SHIP) in 1989, which moved the state closer to universal care. While Hawaii had initial success, there are reasons to suspect that part of this success was due to unique characteristics of Hawaii itself, and that the future of the plan might not be all that promising.

A very high proportion of the Hawaiian population is covered by a uniform and extensive benefit package, emphasizing primary preventive care and financed by mandated employer coverage of at least 50% of the costs. The uninsured rate was estimated to be 8.1% in 1992—the lowest in the United States (Eckholm, 1994). All who work at least 20 hours per week are insured, and Hawaii is presently the closest approximation to universal coverage. There are two major providers of health care in Hawaii, the Blues and Kaiser-Permanente. Although the plan has been effective, there are indications that problems loom in the future. Health-care spending between 1980 and 1991 increased at an average annual per-capita rate of 9.8%, compared to 9.4% for the total United States (Levit et al., 1994). The population is younger than for the United States (in 1991, 11.4% were 65 or older, compared with 12.6% for the United States). As the population ages, costs will continue to rise. Hawaii's insurance rates have risen 14–18% over the past several years, and state Medicaid expenditures are also rising rapidly (Neubauer, 1993).

Dick (1994) agreed that the level of coverage is close to Eckholm's estimate of 91.9% and noted that those not covered include many part-time workers, seasonal agricultural workers, and dependents of employees. Many persons have not complied with the mandate to acquire insurance. It appears that the fault is not with the effectiveness of the employer mandate, but with the fact that the mandates were not drawn up and applied properly to meet the goal of universal coverage. Hawaii ranks among the best states in terms of low infant mortality, longevity, and low rates of early death from heart and lung disease and cancer (Clymer, 1994c). Eighty-two percent of Hawaiians (compared to 71% of Americans overall) were satisfied with their health-care services. There is no evidence that jobs were lost or companies were forced out of business due to the employer mandates.

The initial success of the planning can be attributed to the influence of a powerful state senator who endorsed general social-welfare liberalism for Hawaii. His plan, which resulted in the Health-Care Act of 1967, was prepared and publicized in the political context of the Nixon administration's intention (which was not realized) to develop an employer-based mandate on a national level, which seemed to be the logical extension to the enactment of Medicare and Medicaid in 1965. In 1989, SHIP was created and hailed as a partnership between government, individuals and families, and the private sector. Because the plan fell short of achieving the goal of universal coverage, the governor formed a Blue-Ribbon Panel on Health Care, whose members were a cross-section of Hawaii's economic and social leaders, including representatives of business, insurance providers, unions, academia, consumers, government agencies, and health-care providers. This panel met biweekly for more than 15 months, and in 1992 made 36 recommendations to control costs. It was recommended that a central "entity" be established to implement the panel's recommendations and to mandate community rating to control costs. Powerful politicians spearheaded the drive for health-care reform, and there was a representative panel that studied the issues and made recommendations to provide the basis for action and to educate the public.

The plan stalled at the point of implementation: Recommendations to establish community rating and a central agency both failed to be approved by legislative committees, the chief political supporter was removed from his committee chairmanship, and the new committee chairman did not form a new coalition to support health-care reform. The Director of Health, who had been assuming an effective leadership role, was increasingly out of the state, working with the Clinton health-care task force. In Neubauer's view, Hawaii is caught at midstream. Until 2 years ago, Hawaii led the nation in the development of universal coverage while keeping costs under control, but now the element of purpose and vision has disappeared. One of the primary causes of this failure was that the public was not kept adequately informed regarding ongoing realities, which allowed the legislature to lose interest in working collectively to establish a central commission and to push for community rating, with the process stalling in the face of political and financial realities. Although the process started well, it could not get past the critical stage of the legislative action required to continue. Hawaii does not seem to provide a compelling model for the power of MC, and the factors that existed in Hawaii are unique, due to its geographic location and the structure of the population and economy.

Minnesota

The process used to develop and pass a health-care plan in Minnesota can serve to summarize the preceding "lessons" from other states. The plan falls short of universal coverage, although it had the third lowest level of uninsured citizens in the nation (10.0%) in 1992 (Eckholm, 1994). There is no community rating of health insurance (only gender-based rating is prohibited), and there is no employer-mandated coverage. Health spending was controlled through medical malpractice reform, the imposition of an effective curb on high-cost technology, the development of practice guidelines, and reduction in administration costs (Iglehart, 1994).

A basic benefit package was defined (although scaled back from what was proposed originally), some cost control was exercised through the questionable tactic of requiring people to be uninsured for 1 year and to be without employer-provided insurance for 18 months before becoming eligible to enter the state plan, as well as a 6-month state residency requirement. The plan is financed by a five-cent-per-pack cigarette tax increase, a tax on provider and hospital revenues, and a tax on nonprofit health plans. Although the plan is not ideal, it moves toward reform; the process is continuing, and the public is back in the game.

There was recognition by the public that the prevailing situation was bad: Total public spending on medical assistance increased 41.4% between 1985 and 1990; Minnesota ranked ninth highest in the nation in terms of the average health payments per family, increasing from $2,936 in 1980 to $7,252 in 1991; 33% of uninsured citizens had unpaid medical bills averaging $826, and about 20% of individually insured citizens had outstanding medical bills averaging $1,207.

The Minnesota legislature created a commission to develop a health-care plan for all residents. The commission recommended universal health care with subsidized insurance, community rating, limits on health-care spending, reforms to control administrative costs, and creation of a centralized health-care department. The plan passed the legislature with no specified financing mechanism. The coalition of providers, employers, and workers survived, because no one was being asked to make any financial sacrifices. The governor vetoed the bill but invited legislative leaders to work with his administration to develop "an affordable solution to the health-care access problem." The public wanted health-care reform, the governor needed a political boost to revive his sagging popularity, and the entire Minnesota legislature was running for reelection in the following year (Leichter, 1993a).

At this juncture, two powerful legislators (a conservative Republican and a liberal Democrat) met, realized they had some views in common, and decided they might make progress by bringing together experienced lawmakers from influential health and appropriation committees—what they called "The Gang of Seven." The group met, and to reduce political posturing, they agreed to defend all aspects of whatever bill was agreed upon. Agreement was reached on the general issues essential for reform, and a new state agency was created to enforce whatever plan resulted.

Although each special-interest group put in its oar when the bill was pushed through committees, the committee work was done so quickly that no orchestrated opposition was possible. Following the bill's formal introduction on the floor of the legislature, an intense lobbying campaign was conducted by the insurance industry, the medical association, and the hospital association, all opposing those particulars that impacted them, but the large HMOs supported the bill. Some legislators reported being "fed up" with the arrogance of the self-serving medical association, and they began to resist pressure from that quadrant.

At this point the political club had its effect. The medical association expressed concern that potentially a single-payer system ultimately might be accepted and speculated that the intent of the legislation was not to fine-tune a private health-sector plan but to replace the current system with the Canadian-style, single-payer system. The experience in Minnesota involved all of the elements that have come into play in the other states, and they might be useful to remember in any further attempts to develop a NHP.

The 1994 MinnesotaCare Act moved the process toward cost containment and required that universal coverage be achieved by July 1, 1997 (Blewett, 1994). The key issue now is how to finance that universal coverage. One cost-containment mechanism is the establishment of integrated service networks, which are prepaid health plans that compete on price and quality, along with an all-payer system that sets rates to meet growth limits established by the legislature through a fee schedule for physicians. Statewide expenditure limits were established to reduce the estimated rate of growth in health-care spending by 10% per year for the years 1994–1998. The complex issues of cost-containment and universal coverage were solved because there was commitment to solve the problems equitably.

Blewett described the basic beliefs that drove Minnesota's efforts as follows: (1) There is waste and inefficiency in the current system; (2) managed care is the appropriate vehicle to realize increased efficiency; (3) the competitive approach to cost-containment will not work without additional regulatory oversight. He attributed the Minnesota success to the fact that MC already was the norm in the state: In the metropolitan area of Minneapolis and St. Paul, almost half of the population is enrolled in HMOs (21% statewide), with prices for medical services in the metropolitan area about 18% below the national level. As for Hawaii, special circumstances might have existed in Minnesota that could make it difficult to generalize to the national level.

The task being tackled at present in Minnesota is to collect adequate data on health-care spending for personal health services, research and education, construction, and other capital expenditures. Efforts are being made to achieve these goals, and they reveal the tremendous complexities involved, because existing databases are not adequate, nor are they available in any readily accessible form.

Blewett considered the major lesson for a NHP to be that reform is an iterative process. The impact of reforms to control health-care spending can be measured only if accurate and timely information is collected routinely and consistently for the different components of the health-care system.

CONCLUSION

The lack of an adequate database regarding essential aspects of the health-care delivery system is the result of the loose-cannon, market-based medical system that has relied on continual inflationary expansion to generate profits, with little accountability or concern regarding cost-containment. Blewett made the sensible point that national leadership should be provided technical assistance and financial support for system design, rate-setting methodologies, and data collection. There should be national guidelines to define the elements to be included in the health-spending ledger and to establish data-collection standards. With such assistance, states should be encouraged to collect appropriate data on spending at the state level, and these data should be used to develop policies appropriate to each state. An integrated data-collection and evaluation system should be established at the national level by statisticians and researchers who can develop reliable and valid guidelines and establish evaluation methods free from the political influences that have characterized so much of the health-care debate.

Examination of the experience in the different states indicates that public opinion supports health-care reform. The public wants medical and financial security for themselves, and they do express the humanitarian sentiment that everyone should have access to adequate medical care. The public should be kept informed regarding the realities of health-care reform, and their opinions should be brought to bear to influence elected members of government. To mount an effective reform campaign, there must be a strong group of skilled and committed politicians who will, as the debate progresses, educate the public regarding the issues and bring public pressure to bear on their colleagues. This pressure is necessary to counter the activities of special-interest groups that focus almost exclusively on their own financial interests and are listened to by elected officials because of their economic power. What has been most effective is to set the different special interests to squabbling among themselves over specific details of reform, making it apparent to the public that reform has become a secondary consideration to the financial implications important to special interests. When such squabbling begins, however, it is essential that the center consistently pursue its own agenda in a forceful and effective manner.

It is important to establish a single commission to guide the process, to bring pressure on politicians, and to educate the public. The political club of the single-payer system can be used whenever the process seems to stall. It will be argued in Chapters 12–13 that the single-payer system provides the best solution to health-care reform, but it might not be attainable in a single step. Finally, one should consolidate those changes that can be made with a minimal level of acrimony, implement and finance them, establish uniform data-collection and evaluation procedures, and keep the process moving until an adequate universal benefit package is made affordable. It clearly is a mistake to pass legislation to implement highly commendable principles while ignoring the question of how to finance the system until the end of the process.

CHAPTER 11

Problems in Achieving Health-Care Reform

THE PROBLEM OF ADMINISTRATIVE COSTS

A serious problem with the concept of MC is that similar arrangements have not contained costs; administrative costs in U.S. hospitals are exorbitant. Woolhandler, Himmelstein, and Lewontin (1993) studied administrative costs for virtually all acute-care hospitals in the United States in 1990. Hospital administration accounted for an average of 24.8% of total costs (twice as high as those in Canada), ranging from 20.5% to 30.6% for individual hospitals. A discouraging fact was that administrative costs were not lower, but were slightly higher in states in which health-care management organization (HMO) enrollments included more than 25% of the population (California, Massachusetts, Minnesota, and Oregon—administrative costs were 25.6%) as compared to those with low HMO enrollment (administrative costs were 24.6%). This suggests that introducing MC will not solve the problem of high administrative costs—at least it did not in those states. No state had administrative costs as low as the 9–11% reported for Canadian hospitals, or the 10–15% reported for France. The 24.8% estimate of administrative costs in the United States is too low, because it did not include most hospital advertising and marketing costs, which add another 1% to total hospital costs, nor did it include clinical expenses for clerical personnel in clinical units, such as ward clerks, receptionists, and secretaries.

What activities do these excessive administrative costs support? One is to maintain detailed reports by physicians on the progress of patients who are under treatment, in order to obtain approval from insurers to start, stop, or continue medical treatments. These reports are called *utilization reviews* and are at the heart of MC plans, because they provide the mechanism by which managers hope to contain costs. Such reports require large office staffs, including clerks and nurses, working at three levels: physicians, hospitals, and insurers. Aetna Life and Casualty hires 1,000 nurses and 50 doctors to staff their medical-review department, and none of these are hired to provide primary care (Rosenthal, 1993). Their function is to review requests by physicians to provide treatments for their patients. Hospital business offices also hire clerks and accountants to keep records of medical treatments and to determine which the insurer will pay, which will have to be paid from other sources (such as the government or the patient), and which must be absorbed by the hospital or other treatment units.

The high salaries drawn by chief executive officers (CEOs) of hospitals, HMOs, and drug and insurance companies also contribute to the high administrative costs. Hilts (1993b) reported that the CEO of the Bristol-Meyers Squibb drug company received an average of $7 million for each of the last 5 years ($13 million in the last year reported), and these costs must be absorbed by consumers. The head of the Washington state Blue Cross was paid $712,788 in 1991 and $584,133 in 1992 (Cabrera, 1993). The executive vice-president of that company made a total of $445,460 in 1991. Blue Cross administrative costs in 1992 were 10.5% of income. The CEO of the American Hospital Corporation was paid $127 million in the most recent year reported. The CEO of the HMO, Foundation Health Corporation received a salary and bonus of $3.3 million in 1994 and also a stock-option package valued at roughly $15 million (Metz, 1995). These figures suggest that instituting a system based on competing HMOs will not contain health-care costs, at least as far as containing the one-fourth of hospital costs contributed by hospital administration, the high costs of medical utilization reviews, and the high salaries of administrators.

Another set of administrative costs is at the level of the approximately 1,500 private health insurers in the United States, whose overhead in 1987 was estimated to be 8% (Woolhandler & Himmelstein, 1991). This statistic is interesting, in that it is commonly asserted that the government should not be involved in health-care administration because government bureaucracy is so grossly inefficient and badly managed. Yet, Iglehart (1992) reported that the administrative cost for Medicare was only 2.1% in 1991, and for the Canadian Health Plan only 0.7% in 1989, whereas the administrative costs for private insurers had risen to 14.2% by 1990.

Another set of administrative costs is at the physician level, where it is estimated that about 10% of the physician's gross income is spent on billing costs. Another factor contributing to high administrative costs is that most commercial insurers process claims by hand rather than by computer. Medicare processes 89% of its hospital bills and 47% of its physicians' claims electronically, whereas the Blues have corresponding figures of only 60% and 20%, respectively. It was estimated by Sullivan, Secretary of Health and Human Services in the Bush administration, that at least $8 billion would be saved if the 4 billion medical claims generated each year were processed electronically. The Health Insurance Association of America set the year 2000 as a target to standardize claim forms and enter them into a computer system, the only change requested by President Clinton that has been adopted (Freudenheim, 1994b).

HEALTH FOR PROFIT

The moral issue is that health care should be viewed as a service that is every human's entitlement in a free, just, and humane world and not be considered as a commodity that is served up for profit. When my status as a consumer of commodities is involved, the free-market principle can be applied, because I can decide whether to buy a new car, a new suit, or a new home, given my particular economic realities. The realities of health care are completely different. I cannot decide to postpone having an accident this year because I cannot afford the expense, or to forego surgery required to save my life, no matter what the economic realities. I must preserve my health and my life when the necessity arises and cannot decide to put off the purchase of life until next year.

The Pharmaceutical Industry

The federal government proposed to provide free vaccines for millions of children who are poor or uninsured (Pear, 1993c). Vaccine was to be purchased at discount prices from drug manufacturers and distributed to physicians, who would not charge patients for the vaccine or be permitted to turn away a child whose parents were unable to pay for the service. Drug companies objected to the program, arguing that it is parental negligence, rather than the cost of vaccine, that is the main reason for the nation's low immunization rate, which hardly is an argument against providing it to those who are poor and not negligent. The evidence on which the argument of the drug companies was based was not provided. The companies also argued that if the government buys vaccines at reduced prices, then the companies will have less money available for research and development (R&D) on vaccines, and the program would destroy the vaccine industry in this country.

Pear noted that Congress, in order to calm this disturbance, instructed the Department of Health and Human Services to consider the cost of research when negotiating vaccine contracts under the new law. Although the drug companies expect to spend more than $10 billion on R&D in the current year, they also will spend about $11 billion on promotion and marketing. It has been suggested that the days should be past when pharmaceutical companies could routinely expect the double-digit increases in profits every 3 months that they have been enjoying. One sensible plan would be for NIH to take complete responsibility to pursue such research, with drug companies licensed to manufacture vaccines at agreed-upon prices, freeing them from what they seem to regard as an onerous burden (as well as being a conflict of interest) to develop new vaccines in the public interest.

The idea that the government might want to conduct the necessary R&D and license the production of new drugs, which would help to eliminate the exorbitant profits the pharmaceutical industry has come to expect, is supported by an article by Hilts in *The New York Times* of November 11, 1993. This article presented evidence that all is not as it should be with the pharmaceutical industry. The last major new contraceptive device since the development of the birth control pill 30 years ago is Norplant, the contraceptive implant that is effective for a period of 5 years. The price of the drug in the United States is $365, with another $350 required in physician fees over the 5-year period. The cost to make and market Norplant is estimated at $16 per device; the price charged in some developing countries is $23, with the top price outside the United States only $120.

The industry justified its high prices with the rather bizarre logic that keeping the price high prevents the drug from becoming known as a poor woman's drug, which would perhaps cause it to be shunned by middle-class women! Another justification was that if the industry was not allowed to enjoy the fruits of such high pricing, then they might be disinclined to develop innovative products. In the next breath, they acknowledged that the high prices would prevent many women from getting the device and, for this reason, had given it free to 13 thousand poor women in the United States. They seem to enjoy the privilege of having it both ways.

Representative Ron Wyden of Oregon brought to the attention of industry spokesman that 17 million taxpayer dollars were spent to develop the device, with another $25 million contributed by foundations. Wyden suggested that, given these realities, the price should be lower, at least at public clinics. That there is a need for greater public distribution is supported by an estimation that 100–400 thousand women have been prevented from getting the device because of its high cost.

It was reported by Neergaard, in the July 12, 1994 issue of *The Oregonian*, that Representative Wyden revealed the alarming fact that medicine developed with tax dollars costs the public a great deal more than industry-funded drugs. He found that the median wholesale price for company-financed drugs (either for a year's supply or a complete course of treatment) was $1,626, whereas taxpayer-funded drugs cost $4,584. Other examples are the cost of Taxol, the breast cancer drug discovered by NIH, which costs 25 cents per milligram to produce but is sold by a drug company for $4.87 per milligram, and by the drug Levamisole, a pill to treat colon cancer, which NIH provided $11 million to develop, and which is sold for $6 per pill to patients, but for only 6 cents per pill when used to deworm sheep.

One reason this situation has developed is that NIH allows outside scientists to voluntarily report inventions resulting from the almost $8 billion in grants it awards annually, but it has had only *one* employee to ensure that researchers comply. The government has legal rights to any drug or medical device developed with tax dollars. It can revoke a patent that is not used in the public interest, or license a product to a competitor if the price is too high. The NIH deputy director responded that the agency now has *two* people on the job and hopes to improve the program by allowing scientists to report inventions more easily using computers. Wyden considered that an inadequate response and insisted on better procedures to safeguard taxpayer's investments in medical research.

Another interesting report in *The New York Times* (August 15, 1993) was that the Pfizer pharmaceutical company has decided to provide 11 of its top-selling drugs free of charge to as many as 1 million poor Americans who have no health insurance, at a cost of about $11 million in the first year. The gesture is generous, even though the $11 million represents but a tiny portion of Pfizer's worldwide sales of $4.5 billion last year. Pfizer proposed to distribute the drugs indefinitely until a national health-care plan takes effect that would, presumably, provide health insurance for the 41 million Americans without it. This proposal indicates that the health-care industries, at least initially, took the health-care reform issue seriously and positioned themselves not just to maintain profit margins, but to enhance their public image as well.

Problems involved in trying to hold the drug industry accountable were emphasized in a civil lawsuit filed by drugstore chains against drug makers. The lawsuit charged that seven of the largest drug makers sold drugs to pharmacies at rates that were as much as 1,200% higher than were charged to hospitals and HMOs. The suit alleged that this variable pricing costs the American consumer hundreds of millions of dollars annually. Drug companies responded with a denial of price fixing of any kind. It has been estimated that about 2 billion outpatient prescriptions are filled annually by druggists and that, of the $56 billion spent on prescription drugs in 1992, about $35 billion are spent in retail drugstores. The implications for the financing of health care are extensive if this suit is found to have merit.

A study of the prices of America's 20 top-selling prescription drugs was done by Families USA. It was found that prices rose faster than inflation, according to a report in the March 16, 1995 *Boston Globe*. The prices of these drugs increased 4.3% from January 1993 to January 1994, whereas general inflation was only 2.7%. Between 1989 and 1994, cumulative general inflation was 22%, whereas the prices of 11 drugs increased at a rate that was more than double that figure, and 3 rose more than triple the general inflation rate. Officials of the Pharmacy Research and Manufacturers Association denounced the study because it was based on wholesale prices that are higher than those paid by insurance companies that have negotiated price discounts. The Association noted that, when those discounts are accounted for, the rise for the 20 drugs was only 3.1% (which is still higher than general inflation).

The Association justified the higher rates on the grounds that the companies are doing research to develop better drugs, and it complained that people should be told how to shop for prescription drugs rather than being exposed to quibbling over trivial percentage points. It should be remembered, however, that these trivial percentage points represent a great deal of money, given that an estimated $56 billion was spent on prescription drugs in 1992. It should also be remembered that the uninsured pay for medicine out-of-pocket, and whenever insurance includes high copayments and deductibles, uninsured people are impacted by high drug prices.

Hutton et al. (1994) considered the lessons that could be learned from Europe, where different countries have tried to reform the organization of their health-care systems and explored methods to regulate the pharmaceutical market in order to contain overall costs. Reforms have been directed toward creating a more efficient R&D process and more cost-effective innovations. The authors identified different ways to set the price of drugs. On the supply side, one way was to set the initial price of a new drug by using direct cost calculation on a product-by-product basis. In Italy, pricing was done by estimating the therapeutic value, the research effort, and the manufacturer's expenses. One problem with this process is that it is difficult to quantify the quality of therapeutic value.

Another way was to use a reference price system, whereby the prices of comparable products were used to determine the appropriate price level for a new product. Yet another was the one used in Germany since 1989 to price drugs that had gone beyond the patent-protection period. The generic prices of all competing drugs were taken as the reference price for drugs in the category, and if a drug was prescribed that costs more than that reference price, the surplus cost must be borne by the physician or patient.

Another approach was used in the United Kingdom and Spain. Overall company profitability was used to set profit rates for the sale of drugs to the national health service. Firms are free to set prices for specific drugs to achieve an overall profit rate, with some upper limits set for the general categories of allowable costs, such as promotion, production, R&D, distribution, and product information. One advantage of an overall company-based profit-regulation plan is that it avoids the need to identify costs for each individual product, recognizing the characteristics of the innovative process whereby many products are being developed but very few actually contribute to a company's overall profit at any one time. Such an approach might meet some of the objections of companies that do not want to reveal "trade secrets" to their competitors. All that need be provided are aggregate figures, leaving the item-by-item accounting a private matter, open only to inspection by appropriate monitoring authorities.

On the demand side, positive and negative lists of drugs have been developed. In France, Belgium, Denmark, Spain, Greece, and Italy, a positive list specifies which drugs will be reimbursed through the public health system. In the United Kingdom, Germany, and the Netherlands, a negative list specifies drugs that will not be reimbursed by the public health system, but may be financed privately (with Germany intending to move to a positive list approach in 1995). The purpose of this digression is, once again, to indicate that the debate could move profitably from the level of conjecture regarding the harmful outcomes of regulation and oversight, to a consideration of specific steps that could be taken to bring costs under control, while allowing reasonable profits.

The Health Insurance Industry

The premise on which a private insurance system is based is not compatible with the view that health care should be guaranteed to all. An insurance company not only has an obligation to provide service to those it insures, but also it is a business, and as such has an obligation to show a profit for its stockholders—and private insurance certainly is big business in America. Of the $186.2 billion spent on health by private businesses in 1990, $139.1 billion was for insurance premiums and direct payment of medical claims (Iglehart, 1992). In 1991, private insurance premiums totaled $244.4 billion (an increase of 10% from 1990) and the difference between premiums earned ($244.4 billion) and benefits paid ($209.3 billion) was 14.36%, which includes administration costs, additions to reserves, and profits (Letsch, 1993). The insurance industry collected about $265 billion in premiums in 1993 and, after paying hospitals and doctors, had $50 billion left for profits, marketing, and administrative expenses (Freudenheim, 1994b). One way insurers have increased profits in the United States is to use community- or experience-based rating procedures, and some communities and individuals at risk are not insured unless they pay high premiums. They often are denied insurance altogether, thereby shifting those individuals from the private to the public sector if they are to have any medical coverage.

Originally the Blues used a community-rating system in which premiums were calculated on the basis of the expected costs for all policyholders in the relevant community. With such a system, people who need more expensive services must be subsidized by other policyholders if the premiums the company receives are to keep pace with the cost of services. Private insurers used experience rating in which the rate for each group of employees was calculated using the historic overall costs for those in that occupation. Both systems have now moved to the experience-rating procedure, which has resulted in many individuals, businesses, and occupations being blacklisted. Iglehart (1992) noted that two of the most frequently blacklisted occupations are lawyers (too litigious) and physicians (high users of medical care)!

It is in the interests of commercial insurers to avoid financial risks, and in the interests of the public to avoid a system that jeopardizes receiving continual health care or produces a substantial risk of financial ruin. When procedures such as experience rating and blacklisting members of certain occupations are used, then heavy expenses are incurred by taxpayers, who must pick up the tab for those who are denied insurance, for the indigent sick who are not insurable, and for the unemployed. The poor and uneducated suffer the woes of poor health much more than do the privileged members of society. Senator Wofford, Democrat of Pennsylvania, argued during his 1991 election campaign that if criminals have the right to a lawyer, then working Americans should have the right to a doctor. This argument proved to be the turning point in his election campaign, and Wofford's defeat of the favored Thornburgh, a former Attorney General, was credited to his position on health-care reform.

Pear (1994a) analyzed the findings of the Census Bureau to determine the makeup of the estimated 25% of Americans who were without health insurance for at least 1 month during the 32-month period between February 1990 and September 1992. Young adults were more vulnerable (almost 50% of those 18–24 were uninsured for 1 month) than those 65 or older (only 1% uninsured). This difference occurred because most of the elderly are covered by Medicare. Hispanics were more likely to be uninsured (48%) than blacks (36%), and they more than whites (24%). Among people below the official poverty level, 49% were uninsured for 1 month or more, compared to only 9% of those with incomes at least four times the poverty level. There is clearly a serious problem of inequality in our health-care delivery system that almost any view of morality would insist must be addressed and eliminated in a just society.

If high-risk people are not shifted to public programs, such as Medicaid, then their high medical costs must be assumed by other individuals in the insurance pool, whose premiums will be raised to ensure a satisfactory financial picture for the insurance company, or copayments will be required, along with higher deductible levels, all of which will increase the number of people who cannot afford insurance and, therefore, must rely on programs funded by the public. A severe problem looms because the leaders of the Republican majority in the 104th Congress have proposed to reduce Medicaid and Medicare in order to balance the federal budget by 2002. The federal government expects to spend about $1 trillion on Medicaid between 1995 and 2002 (Gosselin, 1995b). Various proposals have been floated regarding the amount that Medicaid must be reduced to provide the politically motivated middle-class tax cuts proposed and to balance the federal budget at the same time. It also has been suggested that Medicare will have to be cut if the promise of a balanced budget by 2002 is to be honored. Indications are that the number of people likely to require assistance, given the present health-care system will only increase (as will inflation in health-care costs). One indication that the problem is likely to become more serious is the fact that, although 62% of people had private, employer-provided insurance in 1989, only 57% did in 1993. The Medicaid program is the major source available to support medical care for those people who lose employer-provided insurance. Another fact suggesting the problems will only become more severe is that the median household income of more and more families has been falling below the poverty line. The number of families below the poverty line increased from 13.1% in 1989 to 15.1% in 1993 (DeParle, 1994). This means that support for high-risk people, those in poverty, and those who cannot afford health insurance will be less available without major reforms in the health-insurance system.

Many individuals are spending large amounts out-of-pocket (in the form of deductibles and copayments, as well as direct payments by the uninsured) to pay for health care. In 1991, consumers spent a total of $144.3 billion out-of-pocket, an increase of 5.7% over 1990, and this was 19.2% of all health-care spending (Letsch, 1993). The Health Insurance Association of America reported that the cost of family coverage went from $235 per month in 1988 to $436 per month in 1992, an average increase of 17% (Clymer, 1994b). In addition to the cost of health-care insurance premiums, per-capita out-of-pocket expenses (adjusted for inflation) rose from about $300 in 1985 to almost $600 in 1991 (Rosenthal, 1994).

Insurers have had little interest in controlling costs as long as either individuals or businesses can continue to pay premiums. If an insurance company is receiving a set profit margin on premiums received, then it is to the company's advantage to let the gross-claims payments increase because, if profit ratios are kept constant, that increase will result in increased premium rates, an ever-increasing dollar profit for the stockholders, and justify hefty salaries and bonuses for the CEOs. Letsch (1993) agreed with the position of Reinhardt (discussed in Chapter 10) that consumers become insulated from the true prices of health care, thereby compromising the ability of the marketplace to set prices that reflect societal value. It has been argued that a system with third-party insurance payers weakens any incentive the insured consumer might have to make economical health-care purchases. This lack of contact with financial realities, combined with the technological and medical complexity of health-care systems, renders it almost impossible for consumers to make informed treatment decisions. If profits are calculated as a percentage of costs, then it is in the interest of the insurers to increase technology (and hence, costs), even though these technologies have not been evaluated in comparison to lower priced alternatives. A major problem is that the data do not exist to permit rational evaluation of even standard medical procedures, let alone of exotic new ones. It is not only the consumer who is in a mist regarding the wisdom of choices, but also the same problems confront and confound medical practitioners and utilization review boards, as well as medical economists and politicians. The result of all of this confusion has been that private profit rather than public health has come to be the bottom line.

HMOs

One of the primary purposes and concerns of HMOs is to control costs. Forty-four million Americans were enrolled in HMOs in 1992—a sharp increase from 9 million in 1980 and 19 million in 1985 (Enthoven, 1993a). It was estimated that more than 66 million people were covered by some form of provider coverage by the end of 1992. Yet, during that period, there has been a calamitous rise in medical costs. The controls intended to result in a cost-effective system are regulated by a battery of clerks working for HMOs, or are negotiated with insurance providers, and it is clerks who decide whether a treatment should be authorized for a given patient. Some HMOs, which cover more than 50 million people, are using a consulting firm that judges which services should be covered and which should not (Myerson, 1995). The firm was characterized by one observer as "the supreme court of medical insurance," employing nine doctors and nine nurses who have "plenty of clinical and administrative experience, especially at HMOs" to develop guidelines regarding treatment policies. The firm's revenues were reported to be $150 million, and it has issued four volumes of guidelines, with two more on the way. The AMA has fought the use of consulting firms but has not developed its own guidelines, and it was reported to have an interest in buying the consulting firm—a case of, if you can't beat 'em, join 'em.

The system is one in which, depending on the specific organizational structure, clerks working for HMOs are negotiating with clerks working for insurance companies in order to make critical decisions regarding medical treatment, and there is even a consulting firm to regulate the services that HMOs offer to their patients. This all requires layers of administrators at the levels of insurance providers, hospitals, consultants, and physicians. Many patients would prefer that their personal, medically trained and responsible primary-care physician make treatment decisions affecting them.

All HMOs must be cost-effective and their interest in cost-containment can lead them to restrict services to customers in order to remain within capitated payments. The counterforce is to offer a reasonable quality of care to be attractive to customers in order to maintain a large number of subscribers. The degree of oversight that such economic needs demand, however, might require considerable micromanagement of physicians' activities by personnel who have no contact with patients and, in many cases, have no medical training. The additional problem is that it is difficult for the public to make informed decisions regarding the relative quality of different health-care systems, given the complexity, the lack of data regarding the value of various treatments, and the difficulty in sifting through the advertising claims that are aimed to attract customers as much as they are to inform the consuming public.

Shortly after the Clinton administration indicated it would propose the MC system, there were rapid adjustments by politicians, insurers, drug companies, and the owners of hospitals. Before details of the Clinton health plan had been announced, hospitals, doctors, and insurance companies began to form health-care networks that would be the foundation of any medical system involving MC (Pear, 1993d). It has been estimated that the major beneficiaries of the MC system will be the largest insurers and HMOs.

Dr. Sidney Wolfe and Sara Nichols, in a letter to *The New York Times* (November 7, 1993), noted that the five largest health-insurance companies would have been major players under the proposed Clinton MC plan. They cited a study that found 21 of the 25 fastest-growing HMOs were for-profit, and what they called the "big 8," owned 251 of the country's 562 HMOs (45%), with the size of these holdings increasing steadily. They also remarked that the MC plan hands the deal of the century to the huge companies who got that way by profiteering at consumer expense.

Health-Care Mergers. The consolidation of large segments of the health-care industry is proceeding apace. For example, two of the largest Connecticut-based competitors proposed a merger that will create the largest provider of managed mental-health care in the United States (Ringer, 1993). These companies establish networks rather than provide treatment, and the CEO of one of them justified the merger on the grounds that the two companies provide similar services. It is anticipated that this conglomerate will be able to provide treatment at lower cost than many insurance companies can, and that additional revenues will be generated by their information services and prescription-drug units. The stock of the parent company immediately increased by almost 33% at the announcement. The stockholders must expect a reasonable return on this valuable investment.

Eli Lilly bought PCS Health Systems, the country's largest manager of drug-benefit programs (providing drug benefits for about 50 million people), for $4 billion cash; SmithKline Beecham and Merck & Co. (the world's largest drugmaker—which bought Medco Containment Services for $6.6 billion) acquired Pharmaceutical Services for $2.3 billion (Freudenheim, 1994a); in a drug company merger, American Home Products bought American Cyanamid for $9.7 billion, and the two companies were estimated (by *The New York Times*, August 21, 1994) to have combined sales of $12.6 billion. We are talking dollars followed by a lot of zeros here!

On October 3, 1993, a merger of two of the largest U.S. hospital chains, Columbia/HCA Health Care and Healthtrust was announced. The new company would include 311 hospitals and 125 outpatient centers in 26 states, mainly in the South (Feder, 1994b). The company made deals to take over 137 hospitals in 1994 and had $8.2 billion in revenues and $403 million in after-tax profits for the first 9 months of 1994 (Freudenheim, 1995a). National Medical Enterprises and American Medical Holdings announced a $3.3 billion merger, involving 84 hospitals in 13 states and four foreign countries. These two mergers will account for about 75% of the 110 thousand beds owned by for-profit hospitals (Hofmeister, 1994).

The new CEO of the merged Columbia/Healthtrust company estimated that the company will save $138 million per year by eliminating duplication in hospital administration and other costs that do not affect patient care, and executives of the National Medical/American Medical company said they expected the merger to save $60 million in costs for the first year. The physician who was chairman of the new company's board received $127 million in compensation as CEO of one of the merging companies, mainly through options to buy and sell stock at "steep profits," a system of compensation that has resulted in the good doctor owning an estimated $545 million in company shares, at the prices quoted on October 1, 1993.

There is a pattern of generosity in HMO executive paychecks that can only serve to raise the overhead costs to consumers (Freudenheim, 1995e). The CEOs of the seven biggest for-profit HMOs averaged $7 million in cash and stock awards in 1994, ranging from $2.8 million to $15.5 million. An executive compensation expert was quoted to have regarded these figures as "monstrously large" and among the highest seen for any industry. Some critics have expressed the opinion that the HMOs should direct their spending to improving health care, rather than producing large executive-compensation packages. The conclusion must be that these HMOs are big-time businesses, rather than quasi-charitable organizations. An indication that this statement is reasonable is that HMO stocks rose 32.7% in 1994, while the Standard and Poor (S&P) index of 500 companies fell 1.5%, and HMO stocks continued to rise 12% in the first quarter of 1995 (S&P rose 3%).

Freudenheim quoted Reinhardt, saying that all of this is being paid for by the doctors, nurses, and pharmaceutical companies who have suffered the squeeze after the government, for political reasons, failed to clean things up. Reinhardt offered the opinion that "It took certain people with guts to go after doctors in this brutal fashion." One pundit expressed the worry that these numbers obviously are fodder for groups advocating a NHP. I wonder why they should not be so regarded, because it is the consumer and taxpayer who are going to pay in the long run.

The Columbia/Healthcare executives expressed an intention to get a jump on the changes in the health-care system and be in a position to dominate the local markets in Miami, Houston, and El Paso. It was announced that the new company would probably buy 10 or 20 hospitals in the next 2–5 years, which they would shut down because the bed occupancy rate of the hospitals they now have is less than 50%. This policy has the possibility to either drive prices down due to increased efficiency or to remove any possibility of meaningful competition to provide the market incentive required for cost-containment to occur.

One of the vice-presidents in the operation explicitly expressed a view that causes concern when he noted that such mergers will produce conditions to lower the costs suppliers charge, offering the opinion that health-care reform is about consolidating in order to reduce costs (Caprino, 1993). One might hope that health-care reform is also about producing adequate health care to the public, with the economic issues assuming a secondary role. This corporation is moving quickly, as signaled by a report in *The New York Times*, October 8, 1993, that they had reached an agreement to jointly provide services with the nation's largest outpatient surgery center. This alliance is another step in the consolidation of the health-care industry such that former competitors are joining forces.

A novel approach was taken by Baxter International, the world's biggest hospital supply company, which signed a 5-year supply contract with the Duke University Medical Center that will generate at least $200 million in revenue for Baxter (Feder, 1994a). Baxter agreed to pay part of Duke's expenses for surgical supplies if costs are higher than projected, but it will collect a bonus based on the amount saved if Duke spends less than budgeted. The chief executive of the Duke University Hospital was quoted, "If we do this right, we believe it will be a national model." Baxter entered into this deal because the specter of MC has put pressure on hospitals to release patients sooner and to avoid unnecessary treatment, which leads the hospitals to press their suppliers to cut prices. The result of these changes has hurt the demand for Baxter's products, and has led to "sagging earnings." I will make one observation and ask two questions. It could be the case that meaningful cost-containment will be the result of such a novel agreement. But will the creation of a joint board of hospital personnel and Baxter representatives to establish standards and decide how materials are used seriously erode the autonomy of practicing physicians? And if such global budgeting of hospital operations between a private for-profit company and a public not-for-profit hospital is possible, then why not eliminate the for-profit company altogether and have the budget negotiations take place between a single, government payer and the public hospital?

Fraud is a major concern when such vast amounts of money are involved. There have been numerous cases in which kickbacks and bribes have been used to gain referrals. One of the largest psychiatric hospitals in the nation pleaded guilty to seven charges, agreeing to pay $362.7 million in fines to the Government (Myerson, 1994). These charges covered a period from 1985 to 1990, included false statements on Medicare expense reports, and a conspiracy to pay kickbacks to more than 50 physicians. Federal officials estimated that fraud consumes about 10% of the nation's health-care expenses, amounting to between $80–100 billion a year! This all reminds one of the savings-and-loans fiasco, which cost the taxpayer hundreds of billions of dollars, and leads to even greater pessimism regarding the likelihood that the MC plan will control costs and provide more adequate services.

A MORAL PERSPECTIVE

Two principles that can frame the issues involved in attaining the ideal of justice in health-care delivery were discussed by Dworkin (1994b) in a review of the Clinton Plan. The first principle is the "rescue principle" and the second is the "prudent insurance" ideal. The rescue principle assumes that life and health are chief among all goods, with everything else to be sacrificed until they are assured; health care must be distributed to achieve equality at the level of this basic distribution of goods. An implication of the rescue principle is that health-care rationing should not be based on money an individual has, as it now is in the U.S. The major difficulty with the principle is that wealth is so unfairly distributed in the United States that many people are unable to buy health insurance at market rates, most have inadequate information about health risks or the effectiveness of medical technology, and given an unregulated market, insurance companies charge higher rates whenever a person has greater health risks (which involve greater future health needs). He concluded that the rescue principle is almost useless to arrive at proper standards to distribute health care.

The prudent insurance ideal assumes that resources should be allocated between health and other social needs (such as shelter, food, and education), as well as between people, by making three conceptual transformations. First, assume a hypothetical situation in which wealth is distributed more equitably than it is at present—the great disparities between rich and poor have been eliminated. Second, assume that the public is informed about the value, cost, and side effects of medical procedures. Third, assume that no one—including insurance companies—has information about the medical status of any person. Given these assumptions, one can further assume that physicians, hospitals, and drug companies are able to use the free market to charge whatever they wish. It is assumed that all people have sufficient resources to buy the minimal medical care they decide is appropriate. The question then becomes, what kind of health care system would develop in such a hypothetical community, and what decisions would the members of the community be likely to make?

Dworkin proposed this model to enable speculation regarding what the members of this ideal community would spend on health care. He suggested that the spending, based on these assumptions, could provide an estimate of the morally appropriate amounts for a just community to guarantee its members. The manner in which this ideal could be used can be illustrated using a hypothetical case. A 25-year-old woman with average wealth and prospects can choose among a wide variety of possible arrangements to purchase health care under various contingencies over the course of her life. She could buy insurance that provides every form of treatment that might conceivably be beneficial under any circumstance. The cost of this insurance, however, would be prohibitively expensive, leaving insufficient resources to obtain many of the other goods required to sustain her conception of a satisfactory life. This woman might choose coverage that could change year by year to ensure her health in ways that enhance her ongoing conscious life, foregoing life-sustaining treatment if she falls into an irreversible coma. The choice would be dictated by her preference to use resources to enhance her remaining years of conscious life, but not to buy a longer, unconscious state.

The point of considering this case is that, through the use of a series of such hypothetical cases, it might be that, when given a choice, people preferentially would buy insurance covering ordinary medical care, hospitalization when necessary, prenatal and child care for their young, and regular physical examinations as a preventive measure. If these are the values that people have, then a universal health-care system should include these services if the principle of justice is to prevail. Dworkin argued that if few people would insure themselves to provide for heroic life-prolonging care at the end of life, then it is a disservice to force everyone to have such insurance through a mandatory scheme.

Using this inquiry procedure, it would be possible to set reasonable limits on the package of universal coverage for which funding should be assured. These decisions are not compromises with the principle of justice, but are required by it. People would be provided those things that the aggregate taxpayer wants a national health system to make available, and the decisions are based on what they would insure themselves for if they were funding it. Because we do not have an equitable distribution of wealth in our society, it is acceptable for those who can afford it to insure themselves against almost any eventuality they care to, as long as they can pay for it personally.

The rescue principle insists that society provide all treatments if there is a remote chance it will save a life, whereas the prudent insurance principle balances the anticipated value of medical treatments against other goods and risks. The latter approach honors the free-market system, and because the free market is near and dear to conservative approaches to economics, that principle should be supported by advocates of the free market. In Dworkin's view, the debate surrounding the NHP progressed badly, because too many people and special interest groups were unwilling to make any significant sacrifice to produce a more just society. Using the prudent insurance model might help to end the national disgrace of the United States being the only prosperous nation to cheat citizens of an adequate quality of life and health.

Problems with Managed Competition: Economics

The huge mergers described here produce anxieties that are increased by statements made by the captains of the health-care industry that the industry today is comparable to the oil industry of the 1980s, at which time they were perceived as "strong growth companies" (Gould, 1993). One disquieting statement is that the strategy should be to buy shares in health companies that dominate local markets "because you have to control the market to maintain pricing power." The exorbitant returns health-care funds have enjoyed between 1989 and 1991 cause concern over what the concept of "maintaining pricing power" would mean to the health-care consumer and the taxpayer.

It is difficult to consider MC to be viable unless there are enough competitors offering similar services. it is important that there be enough competing HMOs in a region, that they be convenient, that it be possible to evaluate their relative quality using publicly available and standardized information regarding treatment outcomes, and that all competitors provide similar benefits to prospective patients (Enthoven, 1993b). Will the costs for the newly merged companies be lower than those insurance companies are able to offer, given such things as the aforementioned high levels of compensation of CEOs relative to estimated administrative savings?

Many of the developments discussed here began shortly after the MC plan was announced. This raises the concern that true market competition between large HMOs is unlikely because they are carving out exclusive territories to control pricing. Such managed noncompetition could further restrict the control a personal physician has of a patient's treatment program, given the emphasis on cost-containment to maximize profits. A distressing development is that activities are being directed principally to control the structure of the health-care system, especially the flow of money, with little emphasis placed on the quality of care, especially to the poor.

Enthoven (1993b) argued that it should not be assumed that competition has failed, because it has not yet been tried; when he uses the word *competition*, he is referring to price competition. One problem is that HMOs have not been driven to cut prices to subscribers to the lowest possible levels. HMOs would not lose subscribers by raising prices as long as they set rates just below those of fee-for-service costs. This policy allows HMOs to concentrate on offering expanded coverage and improved service to attract more customers, rather than focusing on cost-containment. In Enthoven's view, one critical element for competition—better care at lower cost—has not been an element in the game as it has been played.

More industries are using HMOs to provide health-care coverage for their employees (Freudenheim, 1995b). In 1991, 19% of manufacturers used HMOs, while in 1993, 50% did; for the health care industry, 1991—26%, 1993—49%; utilities, 1991—26%, 1993—34%; colleges and universities, 1991—20%, 1993—40%; retailers, 1991—18%, 1993—67%.

Not only are HMOs playing a larger role in employer health plans, but also the Medicare program is being affected as well (Eckholm, 1995a). There have been intense marketing efforts by HMOs to enroll Medicare patients, and several politicians have emphasized that this should be the future structure of Medicare. The percentage of people eligible for Medicare who are enrolled in HMOs has increased steadily from 3.0% in 1987 to 5.7% in 1994, with a projection of 6.6% for 1995. In California, Oregon, and Arizona, 25% of the elderly have joined HMOs, with 14% in Florida and about 8% in New York. There is evidence that the government may be losing money through these increases, because the HMOs are enrolling the healthier among the elderly (called "cherry-picking"), which leaves the sicker ones to depend on pay-for-service (with the higher costs for these patients raising the Medicare payment schedules). Cost factors force physicians to be more cost-effective, which could lead them to deny treatments whenever there are questions. Because of the movement from fee-for-service payments to a fee-per-person-enrolled model, health plans face a risk if they have a greater than anticipated number of patients with catastrophic and costly illness. A new enterprise has developed to guard against such risks: The health plans can buy insurance to protect against this eventuality (Quinn, 1995).

Kaplan (1994a) discussed the possibility that the fee structure has been conceived improperly. Instead of a fee-for-service model, perhaps there should be a fee-for-benefit. He proposed that the value of a treatment be estimated in terms of the QALYs produced for individual patients, with physicians receiving maximum reimbursement only if they select the option most likely to benefit the patient. In this way, physicians could be reimbursed for selecting the appropriate option for those patients who will benefit most from it. Something of this sort might lead physicians to use available resources to produce more benefits than are now realized through the fee-for-service system, which places too much reliance on billing for procedures rather than on the quality of outcomes.

Problems with Managed Competition: Health Care

Glaser (1993) criticized the U.S. health-care system, noting that every developed country except the United States has a comprehensive health system that provides universal coverage, and these other systems have more effectively contained the costs of health care. The solution in the United States has been to move toward a requirement that all employers provide health insurance to their workers, with an overall emphasis on making the free market work. White (1993), of the Brookings Institute, noted that there are some aspects peculiar to health care; the cost of most other social functions is not out of control, whereas health care costs have been wildly increasing for several years. He suggested that the market approach is not a viable one, especially if it poses the issue in terms of free markets versus bureaucracy. The move should be to neither of these alternatives, but should deal with problems involved in managing the system, the best solution being one that creates a system embracing medical professionalism.

When people buy health insurance, they do so to avoid risk. Because the incidence of illness and the cost of treatments are uncertain, people want insurance, and they tend to be willing to insure for more care than they would buy if they were paying for the treatment directly. Enthoven (1993b) pointed out that the packages of health care that different health insurers offer make it difficult even for an expert to make side-by-side, value-for-money comparisons. The consumer is led to focus on features that can be sold rather than on the cost-effectiveness of a plan to meet their needs.

White concluded that price constraints do not work in the health-care area and argued that MC is another way to dress up bureaucratic control in market garb. He believes that MC makes it easier to displace blame from politicians to managers of the health-care or insurance systems. The result of third-party managed care is to reduce the medical profession to rules administered by insurers, and such a system is rarely successful, as evidenced by the excesses found in such operations as Pentagon procurement regulations. White worried that the problems introduced by MC become how to impose limits on physicians and get them to ration care. The appropriate solution is to negotiate budgets for entire hospitals, to set HMO premium rates, and to cap prescription totals per physician, as the Germans have done successfully. The medical profession should be given budgetary limits within which it must operate, leaving it free to meet those limits without any specification of how it must be done.

Patients are unable to judge whether they should request or permit a treatment; they have neither the data nor the training to make such decisions and must rely almost totally on the decisions of the physician. White believes that the best way to control costs, without interfering with professionalism, is to regulate fees rather than to continue to regulate treatment decisions. The basic principle would be to have decisions made by those with appropriate medical knowledge, and in this way, rely on the professionalism of physicians. It is seldom possible to evaluate outcomes because the research to understand what treatments are appropriate, when, and for whom is not available. The lack of data makes it impossible to decide whether capital expenditures or new technology will be cost-effective, or which of a range of new expenditures are the most desirable.

Glaser (1993) noted that the market approach has not been satisfactory, and that California, which comes close to total MC, has had the fastest growth in spending in the United States for a decade, now being the second most expensive state, with still an estimated 20% of those under 65 having no medical insurance (Quinn, 1994). Glaser also identified a disturbing tendency in Minnesota, where competitive pressures are causing mergers among independent provider networks and previously independent purchasers. His concern is that the MC plan may be farcical if there are a few large regional HMOs in place, providing little possibility for market competition.

No modern health-care system in the world is organized on market lines. The closest approximation is the U.S. system, and it seems to have failed magnificently (Barer & Evans, 1992). There have been a number of attempts to fix our ailing market-oriented system, and all have sustained that tradition of magnificent failure, accompanied by decreased access and skyrocketing costs. One reason for the failure is that the payment game is too difficult for most patients to play against health-care providers. Patients cannot control the state of their health; it is difficult for the public to obtain or understand the information necessary to make informed decisions, and the individual lacks the economic resources to play against formidable corporate opponents, just as it is almost impossible for an individual with limited resources to break the bank at Las Vegas, even given an unlikely long run of good luck at the casino. Private-sector health plans have not proven satisfactory; almost all other countries have abandoned them. Can it be, as Barer and Evans remarked, that Winston Churchill was correct when he said that one can always count on the Americans to do the right thing, after having exhausted all possible alternatives?

The new networks being developed are accumulating hospitals, clinics, and doctor's practices through mergers, acquisitions, and consolidations, and they will be in a position to regulate the delivery and management of care as well as the finances. It seems doubtful that the consumer will realize many beneficial fruits of competition in terms of either higher quality care or financial relief. HMO networks are large corporate entities, interested primarily in finances and profits, and only secondarily in the quality of health care. The financial stakes that physicians have in hospital networks could well conflict with good medical decisions, leaving the poor and uninsured in the dust as the race for profits continues.

Conflict-of-interest rules (which set regulations to prevent doctors from sending patients to businesses they own) were established to ensure that professional judgments concerning a primary interest (such as a patient's welfare and the validity of research) would not be influenced unduly by a secondary interest (such as financial gain). Although ethics cannot be legislated, law and morality overlap and interact in mutually reinforcing ways. Legislation should be passed to establish enforceable laws protecting patients from excesses due to physicians' secondary interests.

Physician's interests in merged health-care plans include keeping their "income streams up": they receive cash and stock, a promise of steady income from the growing number of patients the company hopes to recruit, and the companies often foot the bill to construct new medical buildings, acquire new technology and computers, and provide malpractice insurance (Freudenheim, 1993b). The companies also take responsibility to employ nurses, technicians, and clerks, and to provide facilities to analyze medical results. The General Accounting Office reported that physicians were three to five times more likely to order CAT scans, MRI, ultrasound procedures, and other diagnostic imaging services when the equipment was owned by the physicians or by groups with which they were affiliated (Pear, 1994b). All these particulars lead to the question of whether the medical profession can balance their primary and secondary interest without a little help from its friends (to paraphrase a popular Beatles' song).

Freudenheim cited a disturbing statistic: Each family doctor at a Southern California clinic that was acquired by a medical corporation sees 25–30 patients per day, which would allow less than 15 minutes per patient, assuming the physician saw patients in the office for a solid 8 hours each day (which is unlikely). The medical director of the clinic noted that physicians earned $10,000–11,000 per month last year, and in addition received benefits that included an allowance for continuing medical education, malpractice, life, and health insurance, stock options, and a year-end bonus of $5,000 or more if their unit's profit margins exceeded 7%. This is a good financial deal for the physicians, but one wonders about the deal in terms of the quality of patient health care.

Concerns have been expressed by members of the medical community regarding potential conflicts between medical values (emphasizing the importance of doing everything possible for the patient) and company shareholder's insistence on seeing quarterly profits grow. Perhaps medicine should not be organized around a profit-and-loss center, and there is reason to believe that stock-exchange companies put dollar profits ahead of the quality of health care. Health care is different from other corporate activities, because human health care deals with an essential that is necessary to maintain a minimally satisfactory level of life.

Corporate executives emphasized that physicians would have the clinical autonomy to decide the best way to practice and would have half the seats on the joint policy committee of the clinic, while acknowledging that under the pay system of salaries and shared bonuses, the physicians would be "as at risk as we are" keeping costs down. An economic analyst forecast that profit margins from the operations of some HMOs would be 5% to 10%. Another analyst noted that "managing physicians is like herding cats"—a remark that would give me concern were I a physician—or a cat!

It does not seem likely that costs will be diminished much, if at all, if large HMOs are involved as major players in the MC system. The same kinds of health utilization boards will be used, and it will be insurance and hospital industry boards that will continue to approve medical decisions made by physicians. Physicians have complained that hospitals are pressuring them to join affiliated networks, and that private physicians are intimidated by the fear of not having hospital privileges unless they join a network. The proposed networks restrict the ability of patients to choose their personal physician, as well as curtail the physician's freedom to make medical decisions or choose a preferred practice setting. Patients in HMOs have complained that they have little time with a physician on visits, have difficulties contacting physicians, experience long periods in waiting rooms, and have an impersonal relationship with the primary-care physician. These complaints all seem unlikely to be addressed, given the realities of MC.

Concern has been expressed regarding the effects of managed care on issues affecting physician–patient relationships (Emanuel & Dubler, 1995) and the attendant ethical issues (American Medical Association, 1995). Emanuel and Dubler based their observations on the assumption that major health-care legislation "died—or was killed" by the 103rd Congress, but that changes in the health-care system are continuing at an accelerated pace. There has been "woefully little discussion" of consequences of the effects of managed care in several areas: restrictions on clinical research, reduced funding for physician-training programs, and closing hospitals in rural areas and small communities. The strongest concerns expressed by the public and politicians regarding the Clinton plan were those concerning choice of physicians: a principal concern of a vast majority of citizens.

The issues involved in the ideal physician–patient relationship in terms of what Emanuel and Dubler called the six Cs: choice of physician by the patient; competence of the physician; communication so that a physician understands the patient's need to know (or preference not to know) the prognosis of any symptoms; compassion in terms of support during times of great stress; continuity, so that a competent and compassionate physician providing the care should be the same one over time; and (no) conflict of interest in the financial aspects of treatment options.

They pointed out that 40% of patients in HMOs had to change doctors when they joined the HMO, making continuity an elusive ideal for many. Another problem is that fee-for-service practitioners are almost disappearing in some regions, such as Northern California and

Minnesota, where managed care has expanded rapidly, reducing patients' choice of practice setting. The restriction by employers of their employees' choices is a serious problem. An increasing number of employers offer only one health-care plan; others require their workers to enroll in a particular managed-care plan or select a physician from a precertified list; others are requiring employees to pay substantially more to see a physician outside their managed-care panel, and others discourage employees from selecting higher priced health plans, some even reverting to the old practice of hiring their own "company" physicians.

In a price-competitive marketplace, employers are likely to switch plans from year to year, forcing patients to choose between continuing with their current physician at a higher price or switching to a cheaper plan and a different physician. Managed-care plans are introducing various mechanisms to reduce physician use of health-care resources for their patients, such as providing bonuses for physicians who order few tests and basing a percentage of physician's salaries on test volume and ordering standards. Emanuel and Dubler concluded that these developments are undermining the "six Cs" that they consider to be fundamental elements of the ideal physician–patient relationship.

The AMA Council on Ethical and Judicial Affairs (1995) highlighted many of the same concerns as those discussed by Emanuel and Dubler. The AMA Council referred to the Hippocratic Oath, in which trust is a central element in almost all ethical obligations of physicians. They discussed the importance of confidentiality, the necessity to "avoid mischief" and sexual misconduct, to give no harmful or death causing agent, and to put their own health at risk to do everything they can to help their patients.

The Council identified two conflicting loyalties for physicians involved in managed care: first, to balance the interests of their patients with those of other patients to conserve the plan's resources, and second, the conflict between medical needs of patients and the financial interests of physicians whenever there are bonuses and test-ordering or prescription quotas.

The Council believes that primary-care physicians should not be gatekeepers in the environment of managed care. To protect physicians' ethical obligations, they recommended that allocation decisions should not be determined by individual physicians at the bedside, but by guidelines established at higher policy-making levels. The physician should consider it a duty to recommend and advocate the use of an expensive procedure, based on the likelihood it will benefit a particular patient. Managed-care organizations should be required to establish a medical-staff structure with a governing board, including at least three physician members representing the participating physicians, as well as a medical board composed entirely of participating physicians. Those on the medical board would have the responsibility to review restrictions on services, the quality of care and physicians' credentialing, and to disclose review criteria to subscribers.

The governing board would "be ultimately responsible" for activities of the organization, but participating physicians would have formal mechanisms for input, as well as responsibilities regarding crucial medical issues. There should be a well-structured appeals process through which physicians and patients can challenge denials of a particular diagnostic test or therapeutic procedure.

The Council implicitly endorsed some aspects of the Oregon Health Plan by suggesting that the public participate in the formulation of benefits packages, and that patient autonomy does not guarantee funding for all treatment choices, arguing (p. 334), "Some limits on personal freedom are inevitable in a society that tries to provide all of its members with adequate health care." The Council issued guidelines to implement these changes, to assure the rights of physicians and patients, and to specify the duties and obligations of the medical profession.

LACK OF OUTCOME DATA

HMOs do not keep any standard statistics to make it possible to evaluate their performance (Freudenheim, 1993a). When a corporation shopping for an HMO asked different ones to provide data regarding how many days employees served by the HMOs had been hospitalized, the answers provided could not be used to arrive at comparative estimates: Some were based on claims for payment, some on medical charts, and some on the number of days approved before admission.

A dozen or so HMOs have developed a standardized reporting plan to make it possible to compile statistics regarding performance in some 60 categories, such as percentage of low-birth-weight infants, percentage of individuals at high risk for elevated cholesterol levels, number of minutes in waiting rooms for routine office visits, dropout rates, premiums charged, and so forth. These statistics would make it easier for a consumer, be it a corporation or an individual, to choose among available plans for the one that meets the group's or individual's specific needs.

The Group Health Association of America, a trade group for the nation's HMOs, proposed that nationwide uniform standards be developed for health plans, as reported by Pear in *The New York Times*, February 15, 1994. The Association devoted most of its attention to financial concerns in terms of capitalization. President Clinton had proposed that all health plans have a capitalization of at least $500,000 to be included as a player in the health care networks that would be the foundation of his MC plan. The Association stated that this amount was too small to protect consumers or to guarantee the stability of health plans. Less attention was devoted to issues concerning the quality of care.

The alarming implications of Freudenheim's (1993a) report concern the amount of time the HMO business has continued without gathering evaluations, the immense sums of money involved, the millions of Americans enrolled in HMOs, and the dismal overall quality of medical services available to many members of society. Some observers have opposed gathering data, because it is not possible to measure quality, the data might be difficult to interpret, doctors and HMOs would concentrate on getting a "good grade" rather than improving care, and consumers are not able to evaluate the medical information they might be provided. It was reported that, when asked about government proposals to monitor quality, Dr. James Todd, the AMA executive vice-president, remarked (Meier, 1994), "They are talking about setting up a whole new bureaucracy to collect data that people in the profession are already collecting. All we are seeing now is a lot of overkill." There are very little data available and very little is being done by the medical profession to gather them. Dr. David Eddy, Professor of Health Policy and Management at Duke University, noted that there have been rigorous clinical trials for only about 20% of the medical procedures. The Rand Corporation found that 25% to 33% of some common procedures are unnecessary because benefits probably do not exceed the risks.

Kaiser-Permanente is now computerizing patient records to establish detailed, running studies of a wide range of medical procedures, a process that will require an estimated $1 billion and take over a decade to accomplish. An innovative attempt to measure quality was begun in the mid-1980s in the LDS Hospital in Salt Lake City, Utah (Brinkley, 1994). Through a reexamination of existing treatment records, it was found to be advantageous to compare each patient's medical history with the literature regarding the effectiveness of drugs before deciding which ones to prescribe—an advantage that produced an estimated saving of $450,000 a year by reducing the number of prolonged hospital stays required because of adverse reactions to drugs.

When surgical records of 2,847 patients were examined, it was found that surgeons were administering antibiotics almost at random times before or after surgery, ranging from 24 hours before to 24 hours after surgery. When the timing of antibiotic administration was related to outcome, it was found that patients given antibiotics in the 2 hours before surgery had far smaller rates of infection. When surgeons began following the recommended treatment regime, the frequency of infection fell from 1.8% to 0.4%, and the number of deaths decreased as well. It was estimated that this relatively minor change was saving about $500,000 a year.

Dr. Brent Jones, a physician who runs an innovative quality evaluation program for a chain of hospitals in Utah, was quoted to have said, "When someone imposes guidelines from outside, what happens is the doctors will find 1,000 ways to fight you and end up doing things the other way." When quality-improvement programs are begun, however, it becomes possible to identify physicians whose records indicate they are not performing adequately. Some in the profession have expressed concern that medical boards are reluctant to identify incompetent physicians and less likely to prosecute, even though patients suffer increased levels of morbidity and mortality. This is a dilemma that health reformers should ponder. Such concerns typify the paternalism toward patients that is evident in the attitudes of many in medicine and that contributes to the public's negative attitudes toward the medical profession and the health-care industry.

The complexity of the problems involved in attempts to evaluate outcomes, as well as the primitive level of existing procedures, are attested to in a report by Steinwachs, Wu, and Skinner (1994). They defined the goal of outcomes management (p. 153) as the development

> of a common patient-understood language of health care outcomes: a national data base containing information on clinical, financial, and health outcomes that establishes ... the relation between medical intervention and health outcomes, as well as the relation between health outcomes and money.

They described a study in which 11 members of the Managed Health Care Association joined in an effort to use standardized data-collection methods; the data were pooled, and the feasibility and usefulness of outcome management systems were analyzed. Two patient populations were studied for 2 years: those with asthma and those undergoing coronary angiography.

Severe sampling problems were encountered; response rates were so highly variable that the question was raised whether such studies were feasible, and some of the participating organizations decided they were not staffed adequately to undertake outcome management. Although the authors identified several problems and suggested ways to improve the quality of outcomes studies, it was concluded that there was much to learn regarding the conduct of evaluation studies if they are to provide the information needed to manage the care of patients to assure the best health outcomes. The problems identified in this report are serious and complex, and the evaluative procedures are at a very primitive stage of development within the medical community.

Most reporting measures now used throughout the nation relate to access or ease of getting care, rather than the quality of the care delivered. A nationwide study of 17,671 patients who rated their satisfaction with the treatments used by their health provider was reported by Rubin et al. (1993). Some patients were served by doctors working in their own offices, some by small, single-specialty groups, some by large multispecialty medical practices, and some by HMOs with salaried doctors. The sample of 367 clinicians included internists, family practitioners, cardiologists, endocrinologists, and a few nurse practitioners, and they were located in three major metropolitan areas.

Dissatisfaction was expressed with the treatment provided by HMOs; the complaints were that long hours were spent in the waiting room, visits with physicians were too short, and it was difficult to schedule office appointments on short notice or to contact physicians by telephone. Patients responded that independent physicians showed more interest in patients' well-being. It was found that the patients of physicians who were rated among the highest one-fifth in terms of patient satisfaction showed a significantly higher tendency to stay with their physicians over the next 6 months than did patients of those rated in the lowest one-fifth, which suggests that the results of the survey were related to patients' behavior as well as attitude. The results of this study support the conclusion that patients tend not to be satisfied with the operation of the large HMO networks that are being established, and that patients will have increased difficulty gaining ready access to services.

A telephone survey was conducted to determine physicians' satisfaction with MC practices (Baker & Cantor, 1993). Data were obtained from 4,257 physicians under 45 years of age, who had been in practice between 2 and 9 years. The physician groups of interest here are those who were self-employed in solo or group practice, and who did not engage in any managed care ($N = 558$), those who were employed by a HMO ($N = 198$), and those who were employed by government agencies ($N = 298$). HMO physicians were significantly more likely to be female and a member of a racial or ethnic minority, more likely to be generalists, to spend a comparatively greater portion of their time providing primary care, and had fewer poor patients. HMO physicians were less likely to agree that they could spend sufficient time with patients or control their schedules. Those employed by the government were least satisfied with their current practice, more planned to leave that practice within 2 years, and believed their salary was below what they perceived as adequate. Those employed by HMOs did not express career regrets any more than the self-employed.

Why should the health-care system move in the direction of immense HMO bureaucracies, given that they have structural aspects that leave patients unhappy with quality of care and physicians unhappy with the lack of freedom to control medical decisions? Rubin et al. interpreted their results to mean that small, single-specialty practices, even when prepaid on a capitated basis, provided care that patients consider to be superior to that provided by large HMOs.

A serious effort should be made to develop databases, and this should be done by independent investigators who are trained and experienced in the intricacies of statistics and understand the complexities involved in evaluations research. The knowledge and skills of such research groups as the Hammond and Kaplan groups could be used to develop outcome measures. They have solved some of the difficult problems involved in the developing measurement scales and have addressed questions of patient values. The profit-oriented, free-market approach has not led to investments of the time, money, and effort required to evaluate outcomes that represent more than a gesture, nor is it likely that such research programs will be developed under the present system. Some attempts to measure outcomes have been reported in medical journals and some of them amount to little more than reinventing the wheel, given the primitive nature of the methods.

CONCLUSIONS

The MC plan has little to recommend it in terms of producing economies: It is not likely to improve the quality of health care delivery; the general public would have to assume much of the burden to cover the uninsured; and there would be even less patient satisfaction with the nature of the relationship with their physician. Even if businesses are required to absorb the costs of health-care plans (a requirement they have resisted successfully), it is inevitable that these costs would be passed on to consumers in the form of higher prices for goods and services. The creation of huge HMOs will place physicians under even greater strictures to be cost-effective; patients will have a lessened ability to choose their physicians; the quality and efficiency of individual treatment procedures will not be improved; and the amount of micromanagement by outside review boards of medical practice will increase. Utilization reviews result in extremely high administrative costs, and there are few incentives for private insurers to institute efficient accounting and billing procedures (which can be passed on to the consumer). These facts suggest that neither insurance companies nor large HMOs should be the major primary players in the public health-care system.

At least one social ogre would be exterminated if the MC system is adopted. The "free-market" machinations by HMOs, physicians, political action committees, and insurers will remove the threat of creeping socialism, although some might wonder about the reappearance of the robber barons of yore.

Although the CEOs of some of the largest U.S. corporations expressed a concern that they see "a real, genuine risk here of worsening health care," the specific items discussed tended to concern such things as the slow sales of MRI equipment to those hospitals that had been hurt by the current debate over health care and the possible effect of reforms on the budget deficit.

CHAPTER 12

A Single-Payer National Health Plan

THE CANADIAN HEALTH PLAN

Of all health plans, many have argued that the Canadian Health Plan (CHP) is the one that should be considered for the United States. The Physicians for a National Health Program (Himmelstein et al., 1989; Grumbach et al., 1991) endorsed the CHP and suggested it be implemented with adaptations to meet circumstances that exist in the United States.

Structure of the Canadian Health Plan

An excellent summary of single-source financing systems, describing both the tax-based Canadian system and European social-insurance plans was published by Saltman (1992). The CHP generates 37% of its revenues at the national level through personal income, corporate, and other taxes. The remaining amounts are obtained through direct taxation at the level of the provinces, predominantly through personal income taxes that vary across the different provinces. This revenue system establishes a single "financial spigot" for all revenues paid to hospitals and physicians, taking a public regulatory approach to health-care spending. The national government controls aggregate expenditure levels and regulates the quality and composition of the services of health-care providers. These controls are exercised through bilateral negotiations between the universal financing agency and independently owned hospitals, and by direct bilateral negotiations between the agency and physicians, nurses, and other health-care workers. Annual negotiations are conducted at the provincial level, and physician's associations are organized at the provincial level to establish fee-for-service schedules. Negotiations with hospitals determine the size of prospective global budgets between each provincial government and independent not-for-profit hospitals. Capital funds are set in advance as a proportion of the total budget and are then allocated through competitive proposals. It is necessary for hospitals to negotiate with the provincial government for large capital expenditures, with funds allocated to a hospital after a project is approved.

The CHP provides a stable funding base for the provision of health services; this base is linked to the aggregate earned income of the province's inhabitants. Providers are allowed to plan service offerings based on a relatively stable income, which frees hospitals and physicians from worries regarding reimbursement for services. The global-budget approach caps health-related expenditures, thereby providing a mechanism for cost-containment. The most important aspect of the CHP is that it provides universal coverage, with all citizens receiving basic health services. There is no consideration of preexisting medical conditions, no waiting periods for insurance coverage to be effective, coverage is not tied to jobs or residence, and financing does not require patient-paid deductibles or cost-sharing. The system is simple and relatively inexpensive to administer, which is an important advantage when administrative costs are compared to those in the United States.

Central control of financing makes it easier to implement policy objectives to encourage greater primary and preventive care for pregnant women and for children. The single-source system makes it easier to generate a uniform national database to support epidemiological, clinical, managerial, and preventive studies.

Proposals for the United States

The McDermott–Wellstone Proposal. A proposal for a single-source financing system, sponsored by 92 Representatives and 5 Senators, was introduced in the 103rd Congress in January, 1994 by Representative Jim McDermott of the state of Washington and Senator Paul Wellstone of Minnesota. They proposed that health care be paid by an 8.4% payroll tax on most employers and a 2.1% income tax on workers. McDermott estimated that under this proposal, 75% of Americans would pay substantially less for health insurance than at present. Wellstone noted that the 8.4% payroll tax is well below the 12% payroll tax that most employers presently pay for employee's health insurance. Employers with more than 75 full-time employees would be taxed the full 8.4% of their payroll and those with fewer than 75, only 4%. The Congressional Joint Committee on Taxation calculated that the proposed taxes would raise $290 billion in fiscal 1997, rising to $479 billion by 1999. The bill proposed to increase the current cigarette tax of 24 cents a pack to $2, raising $21 billion by 1999, and to impose a 50% excise tax on retail prices of handguns and ammunition, raising about $350 million a year. About $370 billion a year would be rechanneled (from Medicare and Medicaid, which would no longer be necessary because their services would be included under the single-payer umbrella), and it was estimated by the sponsors that total spending for health care would be about $1.47 trillion in 2000.

The McDermott–Wellstone bill proposed annual global budgets for health care and to limit growth in expenditures to that of the GDP (Kaiser Health Reform Project, 1994). States would set physician fees, hospital and nursing-home budgets, with a National Health Board negotiating prescription drug prices with drug companies. Separate budgets would be negotiated for capital projects, such as the purchase of new equipment, and considerable cost containment would be realized through administrative savings.

The proposal would have provided universal mandatory coverage of all legal residents by 1995, eliminated private health insurance for benefits that are included in the national package, but private insurance could be obtained for supplemental coverage, and HMOs could operate within the system. The plan would have eliminated government spending controls over pricing, be administered at both the federal and state levels, and incorporate Medicare and Medicaid within the single plan. It had a comprehensive and explicitly defined benefit package covering primary, acute, and long-term care benefits, with no cost-sharing by patients. The benefits included inpatient and outpatient services, community-based primary health services, home dialysis, emergency ambulance services, prosthetic devices, prescription drugs, prenatal, and well-child care. Family planning services were included, as were preventive service, mental health services, and chemical dependency treatment. Long-term care benefits included home and community-based care, nursing facility services, home health services, and hospice care. Abortion was not excluded, and states and employers could provide additional benefits at their expense.

Geographic distribution of the population would be used to calculate the budgets and for reimbursements. The bill proposed a 1:1 target of primary-care providers to specialists 5 years after enactment, would reduce payments to states that failed to meet national goals for graduate medical education, increase funding to support the education of health professionals, and establish centers of excellence.

Physicians for a National Health Program. A proposal made by the Physicians for a National Health Program (Grumbach et al., 1991) was based on the premise that our health-care system is failing because it denies access to many in need; is expensive, inefficient, and bureaucratic; and pressures for cost control, competition, and profit threaten the traditional tenets of medical practice. The UN Universal Declaration of Human Rights asserted that a just society must guarantee all people the basic necessities to support a minimally satisfactory life at a level that recognizes the inherent dignity of persons: People should not be allowed to starve; they should have adequate shelter and clothing; and should enjoy a level of health care to permit them to live a life free from the onslaughts of preventable disease. All are entitled to a level of support giving them the freedom to reach the reasonable limits of their autobiographical life with dignity.

Among the cherished tenets considered to be of paramount importance by the medical profession (and the public agrees) are the personal physician–patient relationship and the ability of patients to choose a physician (see Emanuel & Dubler, 1995). Konner (1993) pointed out that many critics in the United States castigate the British system, using the phrase that it is "socialized medicine," and many have criticized the CHP on these grounds as well, although the CHP is organized along totally different lines. Over one-half of the physicians in Britain, Canada, and France are GPs who deal with their patients as primary health-care providers and mediate between primary patients and specialists. The reason the British system is not the envy of the world, Konner argued, is because Britain spends a relatively small proportion of its GDP to pay for health care, rather than due to a fault in the organization of the system. This is a telling point, especially because the United States spends twice as much as Britain, does not have universal coverage and has less personalized care, and the trend toward large HMOs is eroding the level of personalized care and choice of physicians.

Differences between the Canadian Health Plan and Current U.S. Practices

The CHP is a single-payer system providing universal coverage. Patients have free choice of physicians (who have an average net income that is about 26% less than that paid to physicians in the United States), care is rationed according to medical appropriateness (rather than by income as in the United States), and those who are economically able to buy additional private care can, either by purchase of private insurance or on a fee-for-service basis. Another important aspect is that private insurers are replaced by the single-payer system, reducing enormous administrative costs and, in Konner's phrase (1993, p. 235), "freeing doctors to do what they love best: doctoring."

The Physician's Group (Himmelstein et al., 1989) emphasized several essential concerns that any changes in the U.S. health-care plan must address. First, all people should be included in a single public plan that covers all medically necessary services. Because of the severe deficiencies in health care that have existed for so many years, some health service priorities might have to be established for the United States, and if so, this priority ranking might exclude some services from coverage. There should be no patient copayment or deductibles.

Billions of dollars are spent each year to administer private insurance claims. In 1987, administrative costs consumed about 20% of total medical revenues in California (Grumbach et al., 1991). The report by Woolhandler et al. (1993) found the national average for hospital administrative costs to be 24.8% (Canadian administrative costs were only 9%). Private health insurers in the United States use about 8% of their revenues for overhead, whereas both the Medicare program and the CHP have overhead costs of only 2–3% (Himmelstein et al., 1989). The insurance industry, which would be severely reduced in size and importance with this type of NHP, would have a minor role in health-care financing or in determining desirable medical treatments. (The varying estimates cited regarding administrative costs, although comparable within the context of the specific comparisons, used different components to define *administrative* and *overhead*.)

Woolhandler and Himmelstein (1991, p. 1253) wrote,

> Medicine is increasingly becoming a spectator sport. Doctors, patients, and nurses perform before an enlarging audience of utilization reviewers, efficiency experts, and cost managers.

They examined four components of administrative costs in the United States and Canada for the fiscal year 1987: private-insurance overhead, hospital administration, nursing-home administration, and physician's overhead and billing expenses. The administrative costs of the U.S. system were far in excess of those for Canada for all components: Private insurance overhead was 11.9% of premiums ($106 per capita) in the United States, but only 0.9% for Canada's provincial insurance plan ($17 per capita in U.S. adjusted dollars); hospital administration was $162 per capita in the United States and $50 per capita in Canada; nursing home administration was $26 per capita in the United States and $9 per capita in Canada. Physicians' billing expenses were estimated in two ways: By one method they accounted for $203 per capita in the United States and $80 per capita in Canada; by the other method (based on the number of clerical and managerial workers employed by physicians) they were $106 per capita in the United States and $41 per capita in Canada.

If U.S. health-care administration had been as efficient as Canada's, it was estimated there would have been a saving of between $69.0 billion to $83.2 billion in 1987. The differences between the U.S. and Canadian administration costs steadily increased through 1987, as did the total amount spent on health care in both systems. Woolhandler and Himmelstein (1991, p. 1256) concluded, "Reducing our administrative costs to Canadian levels would save enough money to fund coverage for all uninsured and underinsured Americans." This conclusion did not occupy much of the Clinton administration's attention, nor was it mentioned much by the media or those suggesting various alternative plans.

It has been argued that a system such as the Canadian one will not work in the United States, because the two different systems differ in some unexplained, critical ways. Critics of global health-care budgets asserted that negotiating structures and cultural norms that enable spending limits to work in other countries are absent in the United States, thus rendering such efforts unenforceable and ineffective when applied to the U.S. system, but they do not spell out any particulars to support that assertion (Altman & Cohen, 1993).

Woolhandler and Himmelstein considered data regarding the relative effectiveness of comparable systems throughout the world: (1) the small private-insurance sectors of Canada, the United Kingdom, and Germany have overheads of 10.9%, 16%, and 15.7%, respectively—percentages not too dissimilar to those for U.S. private insurers (11.9%), and much higher than the CHP overhead of 0.9%; (2) the Blues of Massachusetts cover 2.7 million subscribers and employ 6,682 workers. This number of workers is greater than that involved for all of Canada's provincial health plans, which cover more than 25 million people; (3) although the Blues employed 6,682 to serve their 2.7 million subscribers, a comparable size segment of the Canadian system (British Columbia) employed 435 workers to serve more than 3 million people; (4) the Shriner's hospitals in the United States bill neither patients nor third parties and devote only 2% of their revenues to administration; (5) the administrative costs of British hospitals (which are assigned global budgets, as in the Canadian Plan) are 6.9%, while those British hospitals that were paid on a per-patient basis have administrative costs of 18%. These figures indicate that administrative costs are determined by payment mechanisms rather that the cultural or political milieu. It is incumbent on those who argue that the U.S. and Canadian societies are so different that it is impossible for the United States to adapt the CHP to identify the critical differences and argue why they preclude the use of a single-payer system.

A comparative study was done by Redelmeier and Fuchs (1993) of hospital costs in 1987 at the national level (United States with Canada), the regional level (California, the largest state, with Ontario, the largest province), and institutional level (two hospitals in California with two in Ontario). They noted that an American child was nearly 50% less likely to enter a hospital than was a Canadian child, expenditures for administration were 39% higher in the United States than in Canada, and 63% higher in California than in Ontario. Yet, the U.S. hospitals used 24% more resources per admission than Canadian hospitals, and California used 46% more resources than Ontario. These differences in costs occurred even though occupancy rates were lower in California hospitals (62%) than in Ontario (89%), the length of stay was longer in Canadian hospitals (11.2 days) than in the United States (7.2 days), and hospital wages in the United States were lower than those in Canada. The longer stay in Canadian hospitals implies that Canadian patients would have more frequent clinical evaluations of their vital signs, more dressing changes, and other services that are provided on a daily basis while in hospital, which could mean that the Canadians receive superior medical care as compared to U.S. patients.

One source of excessive expenditures in the United States in 1987 is that there were three times as many hospitals with units providing open-heart surgery in California as in Ontario, five times as many with MRI machines, and ten times as many with lithotriptors. The greater cost for less service is the result of greater administration costs in the United States and a fuller use of capacity in Canada, especially when expensive technologies are involved. The Canadian centralization, reliance on referrals, and establishment of waiting lists result in less idle time for centralized high-cost equipment and associated personnel. The savings realized by more efficient utilization of medical technology is quite significant: Medical experts estimated that the overuse of medical technology could account for as much as $135 billion in unnecessary health-care spending (Freudenheim, 1993c). It has been estimated that hospitals would be relieved of bad debts from unpaid charges if the 59 million under- or uninsured people were covered—it is estimated that unpaid hospital charges are about 6% of the total charges each year.

Detsky (1993), in an editorial comment on this study, emphasized that Canada provides more care for children, performs fewer coronary-artery catheterizations, and has a longer wait for the use of MRI, whereas Americans with Medicaid wait for such things as basic prenatal care, with many Americans having no prenatal care at all. He noted that overall government reimbursement of hospital budgets in Ontario increased only 3.6% in 1992–1993, and it was announced that there would be no increases in base budgets for the next 2 years. Detsky (1993, p. 806) concluded

> There are some sectors in which markets simply will not work. In these sectors, the price signals and incentives are so severely distorted that the resulting allocation is neither efficient nor equitable. The market for health care may be one of these.

An editorial rebuttal to Angell's (1993a) editorial endorsing a single-payer plan was published in the *New England Journal of Medicine* by Morse (1993), President of the Massachusetts Medical Society, which owns and publishes *The Journal.* Morse challenged Angell's editorial position, referring to political and moral issues, as well as economic realities. He alluded to the profoundly different political culture of the United States as compared to Canada. The basis for this allusion was that the U.S. Declaration of Independence holds as inalienable rights "life, liberty, and the pursuit of happiness," whereas the goals of Canada's Fathers of Confederation were "peace, order, and good government." What difference these statements of principle make in terms of fundamental human freedom and dignity are not clear. He believes that any viable reform proposal for the United States has to respect "self-reliance, local decisions, private action, pluralism, multiple competing systems, and individual choices" (p. 804)—although he does not really tell us on what tablets those principles are inscribed and how we honor them at present. He concluded that the fundamental differences between our cultures make centralized decision making acceptable to a larger proportion of Canadians than Americans. I wonder if Morse has had occasion to visit Quebec of late to discuss the use of the English language in that province?

In any event, an appeal to the differences in the Canadian and U.S. cultural heritage seems to be quite beside the point. Of course, the United States and Canada are different, as are the United States and England, France, Germany, Japan, and all other countries with low-cost universal health care. Yet, people in each of these countries have their basic health needs ministered to, no matter what their cultural heritage. Perhaps we should attempt to overcome whatever cultural differences we might have with other countries to provide health care at a level comparable to that provided by those countries. A modified CHP might move us in that direction.

Morse cited a study by Torrey and Jacobs (1993), which he interpreted to indicate that the net personal consumption of health care is similar in the United States and Canada. Examination of that report reveals that the conclusion is not as straightforward as Morse indicated. Torrey and Jacobs found that Canadian households spend about half of what American households spend for health coverage. They also stated that the personal tax burden is almost twice that of the American, hence providing a similar level of what they called "personal consumption."

An interesting aspect of their study is that health-care spending as a percentage of current household consumption is 5.6% for the United States and 2.2% for Canada, but there is a major increase in the United States as age increases from 55 to 75 and over. In Canada, when the head of a household is 55–64, the health-care percentage of household consumption is 2.6% (United States, 7.17%), when 65–74, in Canada it is 2.7% (United States, 10.9%), when 75 or older, in Canada it is 3.2% (United States, 17.1%). The increase in the United States is due to increased expenses for physicians, health insurance, and hospital care. There is no evidence that the United States provides better care for these increased costs: Life expectancy in Canada is slightly higher for both males and females than in the United States.

Morse's bottom line is that it costs money to serve and to offer the best health care. He should remember not only the per-capita dollars involved, but take into account that Canadians, for less or the same amount of money, have universal health care, whereas the United States has about 62 million under- or uninsured! Morse frets that a health-care system will prove too costly if it comes at the cost of independence. The array of statistics presented in the last two chapters provide a basis on which to judge whether the United States offers the best health care, especially to its least well-off, and question whether the independence of health-care providers justifies the sacrifice of the lives and well-being of untreated people to secure that independence.

Several doctors responded to Angell's editorial in the same issue of *The Journal* that contained Morse's response. Most of them insisted on the importance of maintaining the free-market system for the United States. It has been under the structure of the free-market system, however, that health-care delivery has deteriorated so much that it is generally agreed that a massive overhaul is needed. Angell suggested, in response, that if we continue on our present course, we are likely to bankrupt the country and increase the *de facto* rationing that now prevails. She noted that the Canadian system has basic universal coverage, no third-party interference with physicians' primary-care decisions, low administrative costs, enjoys the strong support of the public, and has more support from its doctors than is the case in the United States. She emphasized that it is difficult to see how a libertarian capitalism can be appropriately applied to the health-care needs of people who are both poor and sick. Society grants physicians a virtual monopoly for their services and subsidizes their training, as well as the research on which they depend. She suggested that physicians owe something to society in return, and that medical costs have tended to rise as the number of providers has increased, leading her to ask why our system—the most market-driven in the world—is the most expensive in the world.

High administrative costs in the United States are the result of such things as the greater time required to process each health-care claim than that required to process medical claims in Canada, the army of bureaucrats required to scrutinize clinical decisions to eliminate the costs of unnecessary care to contain overall costs, and the existence of hundreds of claims-processing units in the United States, rather than a single one, as in each province in Canada. Woolhandler and Himmelstein concluded that competition among insurers leads to marketing and cost-shifting strategies that benefit individual insurance companies but raise the systemwide costs that must be borne by either patients or taxpayers.

HOW THE NATIONAL HEALTH PLAN COULD WORK

The most reasonable proposals for the NHP involve a program that would be federally mandated, mainly funded by the federal government, and administered largely at the state and local level (Himmelstein et al., 1989). Coverage would be universal, and if it is not economically feasible for all care to be covered, local boards of experts and community representatives would determine what services would be excluded, in much the same way as accomplished in Oregon. Private insurance for those specific services included in the NHP would be eliminated, as would patient copayments and deductibles, both of which endanger the health of the indigent sick.

Each hospital would negotiate a budget and receive an annual lump-sum payment to cover all operating expenses. This budget would be negotiated with a state NHP payment board and would be based on past expenditures, previous financial and clinical performance, changes projected by the hospital staff in levels of services and wages, and proposed new and innovative programs the physicians decide to pursue. The members of the Physician's Group argued that such budgeting would simplify hospital administration and virtually eliminate current billing procedures, with a savings of many billions of dollars. They proposed that, in the interests of minimizing disruptions to existing patterns of care, the NHP should include two basic payment options to physicians and other practitioners: one a fee-for-service payment; the other, to create salaried positions in medical care institutions or HMOs that receive capitated payments.

Funding for the NHP would be based on a progressive income tax and employer contributions. The increased cost t⁻ taxpayers would be recovered, because no private insurance fees would have to be paid by consumers, and the present fees paid for Medicaire and Medicaid would be included in the financing of the NHP. It is estimated, based on the Canadian experience, that employer contributions would decrease for most firms that now provide health insurance. The Physician's Group anticipated that the average physician's income would change little, again based on the Canadian experience. The income of a physician could be increased if the physician provided additional care, on a fee-for-service basis, to private patients who desired it, and the simplifications of billing procedures would save thousands of dollars per practitioner in reduced office expenses, due to simplified administrative bookkeeping (which has been estimated to amount to about 10% of physician's gross income).

Woolhandler et al. (1993) noted that reducing hospital bureaucracy to the Canadian level should save about $50 billion annually, with another $50 billion saved on insurance overhead and physicians' paperwork. To resurrect and inflate the old saw: $50 billion dollars here and $50 billion dollars there, and pretty soon you're talking about real money.

One concern is that if health care is free, then people will overutilize the system until it collapses—the "free lunch" syndrome. There was an initial increased demand for acute care in Canada when the CHP began, but that demand decreased after the initial surge. Improvements in health planning, preventive medicine, and lowered administrative costs also slowed the escalation of health-care costs. This concern appears to have no foundation, given the Canadian experience.

Grumbach et al. (1991) examined the budgetary implications of the proposed NHP and concluded that the administrative-cost reductions and elimination of insurance-board reviews would result in immediate savings and contribute toward containing future costs. These economies would be realized with no loss of clinical freedom by physicians or patients, and many of the high-technology services we now have could still be provided if the level of U.S. health spending is kept higher than that of Canada. The Canadians get more health care for their health-care dollar, even given their lower levels of expenditure. The improved performance is the result of lower insurance overhead and less administrative paper-pushing. The major benefit is universal health insurance for all Canadians, with no copayments or deductibles.

Woolhandler and Himmelstein (1991) estimated that the savings produced by moving to the single-payer system could cover all of the costs of providing care for the under- and uninsured. Because potential costs seem to be the major concern of both the business community and the middle-income taxpayer, there should be a study of the mechanisms whereby a single-payer system can be put into place as soon as possible. Quickly moving to such a system would meet the economic demands that threaten us and respond to the moral concerns that have become increasingly worrisome.

Grumbach et al. (1991), using an estimate that $602 billion would be the total medical budget for 1991, projected two budgets: one using current policies, and the other assuming their version of the NHP. The total health-care budget of $602 billion estimated for 1991 was too low (it was $675.0 billion in 1990, and $751.8 billion in 1991, as reported by Letsch, 1993), but the basic argument seems reasonable and on the conservative side. The proposed NHP allowed the same amount for hospital costs as estimated under current policies. Physician costs were increased 12.25% to allow for the increased patient load that would be produced when the under- and uninsured are included in the system, and to allow physician's a substantial increase in their income.

The provision of increased income for physicians could be questioned, given the fact that the average net income of U.S. physicians is 26% higher than in Canada, partly due to a larger ratio of specialists to primary-care physicians in the United States (Detsky, 1993), but never mind. The AMA reported that the mean income for internists in 1991 was $149,600 and for general surgeons, $223,800. For physicians in HMOs, the average income, before bonuses, was $101,267 for internists and $140,899 for general surgeons. These income levels seem to be generous.

A more recent survey of physician's incomes was reported by Eckholm (1993c). He examined the 1992 incomes of the one-third of all physicians who have private group practices that include three or more physicians. The incomes of these physicians were somewhat lower than for those in solo or academic practice. At the high end were cardiac surgeons, who made an average of $525,000, and neurosurgeons, who averaged $449,000. Orthopedic and hand surgeons averaged well over $300,000, whereas diagnostic radiologists had average incomes of $310,000. At the low end were physicians in family practice ($119,000), pediatrics ($124,000), and internal medicine ($130,000). In 1993, the average HMO salary of orthopedic surgeons was $207,000 plus bonuses averaging $12,000, and for pediatricians it was $98,000, plus $7,000 in bonuses. Eckholm concluded that these figures indicate that physicians have largely evaded normal economic forces; that supply and demand economics never applied to them. He quoted Eisenberg, chairman of medicine at Georgetown University, who described "the trick" being to start as high as you could when adopting a new service, and then others will tend to follow that lead.

Even given increases in the handsome income of physicians, a savings would be realized under the NHP, because insurance administration and profits should be less than one-fourth of what they are now: Grumbach et al. estimated this savings to be the difference between the current costs of $35 billion and the $8 billion that would be expected under the NHP, a total saving of $27 billion. The amount saved would be realized through reductions in administrative costs and elimination of profits for private insurers. The insurance industry collected $265 billion in premiums in 1993 and had $50 billion in profits and for marketing and administrative expenses. These savings could cover NHP start-up costs, job training and placement programs for displaced administrative and insurance personnel, and revitalized public-health programs. The latter are important, not only because they could reduce future costs of primary care, but also because they will allow people to live a more adequate, healthful life.

The amount of money saved by adopting the NHP could exceed the aforementioned figures. Much of the 1991 budget overrun was due to a disproportionate increase in administrative costs. Everything else being constant, eliminating these administrative costs would produce an even greater proportional reduction in the total budget. The reduced budgets should make it possible to decrease the overall federal deficit (as Clinton projected under his plan), which would produce large savings in interest payments required to service the federal indebtedness. The same quality of health care could be attained by those of us who are insured now, if funding levels stay the same, with the under- and uninsured included in the system as well. Universal coverage could be realized at no greater cost, and it is possible that these benefits will be provided at a lesser cost than at present.

Although the budget figures available from different sources vary somewhat (because different bases are used to express percentages, and different items are included in the different estimates), three things stand out:

1. The amount of money spent for health care in the United States is increasing rapidly (by 2000, it is projected to reach 17% of the GDP with the projected Clinton plan and about 19% without the plan), and actual expenditures have always exceeded estimates.
2. More and more people (including more and more of the middle class) are under- or uninsured each year.
3. There is a lot of money involved, both in terms of absolute dollars and in proportion of the total U.S. budget.

Glaser (1993) stated that the U.S. debate regarding care has been misled by the "red herring" of the public-financing method used in Canada, which he considered to be the only reasonable one that had been presented as an alternative to MC. Devotees of the free market, who argue that the CHP is a "governmental takeover" and that the Canadian model is unique with its full-treasury financing and private-service provision, are only creating a diversion. Given the strength of the opposition that has been expressed, Glaser concluded that the CHP is impossible to enact in the U.S. legislature, making it a "straw man" that performs the function of making MC seem plausible and inescapable. He presented no arguments against the CHP, except those based on public relations, and I believe he misinterprets the motivations of economic interests. They full well realize possible negative implications of the CHP for their profit margins and will oppose anything not resembling the current system under which they have flourished.

Given the negative perceptions of the CHP, Glaser examined how other developed countries organize and operate their statutory health-insurance programs. These countries include France, Germany, the Netherlands, Belgium, Switzerland, Austria, and Japan. All faced the problems the United States is now facing, and they solved them to some extent. Some of Glazer's conclusions, based on his examination of the health plans of these countries, were the following: Health-care financing at the national level must be part of a centralized system, and this financing should be done through a percentage payroll tax that would go directly to the IRS, with the plan absorbing both Medicare and Medicaid; coverage should be for all persons in an occupation, and not according to type of employer; coverage should include the entire family; all licensed physicians should be able to treat any insured patients; patients should have freedom to choose physicians; physicians should have the freedom to choose patients and compete to attract patients. He recommended that medical associations should become crucial comanagers of the system, because a health plan that enacted a national reform cannot possibly succeed if the medical profession opposes it. It would seem that this latter concern is one that physicians should face and fight out within their conservative, business-oriented professional associations. The huge, conglomerate HMOs being formed threaten the autonomy and financial security of the medical profession. It does not seem to be in their interests to attempt to position themselves above the fray.

Glaser argued that hospital operations should be scrutinized to evaluate staffing levels, the proliferation of technologies, and waste. If abuses are found, then effective challenges could be made by appropriate oversight agencies, and budgetary controls could be used to guarantee compliance to effect desired economies.

Morals and Profits

Glaser concluded that health is a field that involves so many weak, dependent, and poor people that it cannot be a system geared to conventional consumer markets, but requires redistributive financing. By all means, if the CHP is a "red-herring" that diverts the debate from the important issues, then let's oppose it for that reason. The CHP, however, embodies all of the elements Glaser has identified as crucial, and something like it would fill the bill. Beyond the sound-bite level, it is not apparent what crucial differences exist between the United States and Canada that make the CHP system inappropriate. Glaser's analysis of the health-care system of other countries is sound, and the MC proposal would accomplish little toward resolving the dilemmas that face the U.S. health-care system.

Most other advanced countries, according to Konner (1993), have either abolished or stringently regulated the private-insurance business, because it has no direct function in health-care delivery, a point emphasized by Glaser. Konner pointed out that some form of NHP is successfully in place in every other industrialized nation in the world, with the exception of South Africa, which he suggested is not good company to keep.

Although the issues involved in health-care reform are complicated, it is clear there must be reform of the health-care system, whether it be some form of rationing, managed competition, or single-payer coverage. Morally, it is demanded that reform occur if we are to have a just society, even though there are difficult decisions that may impose costs to the few in order to produce benefits to the many.

Konner (1993, p. xx) neatly summarized his conclusions as follows:

> I prefer personal medicine to the impersonal kind that has become common; low-tech primary care to high-tech specialized care wherever safely possible ...; prevention and early intervention to late-stage crisis management; patient and family involvement in decision making to medical authoritarianism; a dignified death a little sooner to an overmedicalized one later; a true system of accountability for doctors and compensation for injured patients to chaotic malpractice litigation; careful measurement of the outcomes of medical and surgical procedures to the haphazard system of trial and error we have now; rational and continuous monitoring of the health of a population and outreach for treatment and prevention to the passive stance of a medical system that waits for an illness to present itself; not-for-profit medicine to the cash-register variety; and universal health insurance to the ragged health safety net that exists in the United States.

Angell (1993b) argued that there are three reasons for the financial predicament created in the United States: (1) The United States, unlike other Western countries, treats health care as a market commodity, not as a social good; (2) the United States does not have a health-care system at all, but a multitude of arrangements that often work at cross purposes; (3) the United States does not limit the total amount spent on health care by establishing a global cap.

She made three recommendations: (1) Congress should set a global cap on health spending as a percentage of GDP and allocate funds according to relevant demographic variables; (2) the funding of health-care delivery should be done through an adaptation of the single-payer CHP; (3) price competition should be minimized, because experience has shown that when prices for health care are determined by the market, costs nearly always rise and the types of care provided reflect financial incentives rather than human needs. If the United States continues to spend far more per citizen than any other country, adoption of a single-source system should result in American citizens having the best medical care in the world, with no need for the queues or *de facto* rationing that some other countries have experienced because they spend much less on medical care.

Saltman (1992) reviewed the pros and cons of a single-source financing system for the United States. After careful consideration of the advantages and disadvantages of the single-source system he, too, came to the conclusion that such a system should be established. He recommended that policy-framing decisions and regional budgets should be established at the national level, and there should be regional and local control over the specific decisions regarding resource allocations.

Konner (1993) concluded that he did not see any positive role for private-insurance companies to provide a safety net for the needy and that, if we continue to consume too many available resources to prolong life at the expense of denying others the quality of life that we ourselves have enjoyed, we should be prepared to face a rationing of resources.

The arguments made by such experts as Angell, Konner, Saltman, and the Physicians for a National Health Program are convincing, and their conclusions, compelling. MC does not provide a viable or sensible alternative to cure the ills that afflict the U.S. health-care system. At its best, it provides a band-aid to treat a cancer, and at its worst, it continues to let an economic and moral malignancy grow that diminishes the United States as a just society. A single-source payer health plan similar to the CHP should be the immediate goal for the U.S. health system to provide for the basic health needs of the American public.

CHAPTER 13

Moral, Medical, and Financial Issues

A number of medical, economic, and moral issues involved in the health-care system have been considered. Some of the major proposals for health-care reform that were presented during 1993–1994 were discussed in Chapters 10 and 12, and it was argued that the McDermott–Wellstone single-payer proposal best meets the needs of the country. Unfortunately, a great deal of energy and rhetoric has been devoted to obscure the facts and issues through negative and misleading advertising, political posturing, and slogan-mongering. In this chapter, an attempt will be made to reduce the confusion to focus on the major moral, medical, and financial issues.

Public figures are fond of platitudes that serve to assure themselves, and those to whom the pronouncements are directed, that the United States is the best there is—land of the free and home of the brave—no matter the number of underprivileged. This bias is apparent when assertions are made that the U.S. health-care system is the best in the world, and that only a minor amount of tinkering with insurance financing is required. The Republican leadership of the 103rd Congress consistently denied the seriousness of any problems. Senator Dole stated, "Our country has health-care problems, but no health-care crisis." Representative Bill Gooding of Pennsylvania, ranking Republican on the House Education and Labor Committee in the 103rd Congress, was quoted to have said that all the nation's health-care system needed was fine-tuning. Senator Dole threatened to filibuster any plan including employer mandates and was unwilling to compromise that position during the course of the Senate debate.

Senator Orrin Hatch of Utah called President Clinton's plan "nothing more than a pasteurized version of Clinton's blueprint for socialized medicine." Representative Newt Gingrich of Georgia, the number-two

House Republican in the 103rd Congress and Majority leader in the 104th, likened President Clinton's plan to Soviet Communism, stating on one occasion that guaranteed health coverage is "socialism, now or later, and a dictatorship on health care," and on another that "you cannot get to universal coverage without a police state." Representative Dick Armey of Texas, the third-ranking Republican in the House and Chairman of the Republican Conference in the 103rd Congress, declared (before the Clinton Plan had been presented) that he was in opposition to whatever it is the President was going to propose because it would destroy jobs, burden the economy with massive new taxes, and lead to health-care rationing. The battle of rhetoric was joined at a rather low intellectual level that should have, and did, make reasoned debate difficult, and resulted in the defeat of any significant bipartisan accommodation, or even a bill to vote on.

HOW TO PROCEED

The factors involved in any revision of the U.S. health-care system are numerous and complex. Revisions will affect every individual's potential well-being, have an impact on the financial security of many, and have implications for almost every sector of the U.S. economy. Because of this complexity it will be useful to decompose the problem into its different aspects. The moral issues should be addressed first to understand them and decide what aspects of the present system should be changed to fulfill the moral obligations that society has to its citizens.

Once the moral issues have been analyzed (and there probably will not be much disagreement over the ideal) then questions regarding implementation can be approached. Practical questions involve how to deliver health care to patients so that the benefits of medical knowledge and technology can be realized, using existing facilities with maximum effectiveness. After these moral, medical, and practical realities have been considered, it should not be difficult to identify the problems to be resolved.

Only after these basic issues have been considered should financial realities be introduced. The financial questions involve such things as how much it will all cost, who is to pay how much for what, and in what form payments will be made. When agreement has been reached regarding the financial realities, it is appropriate for the special interests of the various economic factions in society to contribute to the discussion. Considerations of economics bring to the fore the actions of pressure groups, with their special pleading and political posturing. These considerations embody many of the ugly realities that characterize the modern-day democratic process and force compromises and accommodations. The problem is that economic questions have taken precedence and have driven the entire process of revising U.S. health-care policy.

Special interest groups staked out their territories at the outset, with a focus on how to minimize the burdens each had to accept to get as much as they could for themselves. The question asked was who—rather than we—should pay, and who of you should get less care when funds are limited. There was a great deal of misinformation, with the brandishing of scare-words at the outset, all of which tended to cloud the essential moral, medical, and financial issues. Those with access to the media and the financial capabilities to mount campaigns confounded issues to further their own particular agenda and urged the public to beware of tyrannization by "big government." Another underlying theme is that one should not proceed too quickly but should do what can be done, given existing circumstances, and certainly not disturb any economic interests.

Moral Issues

The basic moral issue regarding the system needed is not difficult to state, nor is it difficult to obtain. It was affirmed, in the UN Declaration of Human Rights, that all people in the world should be entitled to adequate health care as a guaranteed right. Polling data indicate that the U.S. public agrees that all who are citizens, or who are attempting to qualify for citizenship, are entitled to adequate medical care. There should be universal health coverage, at least at a level sufficient to guarantee a minimally satisfactory life. A poll reported July 20, 1994 in *The New York Times* indicated that 96% of Americans considered universal coverage to be important (79% responded *Very Important*, and 17% *Somewhat Important*). An employer mandate was favored by 49%. Agreement with the principle of universal coverage implies that current inequities between the social classes must be eliminated so that the lower socioeconomic classes are not held below a minimally acceptable level of health care.

Security should also prevail; people should not live with the fear that they will be unable to afford health care if they become ill. They should not have their insurance canceled because they lose a job and, with it, their health insurance. They should not fear for their future when they are elderly, or when they or their family members develop an expensive ailment that would lead to the loss of affordable insurance. The public and most bioethicists agree that continuous, affordable, and universal health care should be available to everyone.

President Clinton, in his address to Congress on September 22, 1993, identified the major philosophical concern regarding universal coverage. He phrased this concern in terms of "security"—that all Americans should have a guaranteed minimal level of health care with no fear of losing it because they switch jobs or retire early, that they should not be denied health insurance because of preexisting medical problems, or be unable to obtain insurance because it is not affordable. The principle of security received strong endorsement from both the public and politicians. Senator Dole remarked, "Who could be against security and savings and responsibility?" President Clinton seized the high ground and achieved agreement regarding the major moral issue. He moved the debate to questions regarding administrative and economic structures to achieve the goal, and to a consideration of the costs and risks involved using different approaches. Unfortunately, in the course of the ugly realities of the debate, there was a retreat from that moral high ground.

It is not agreed widely that noncitizens—be they legal, foreign migrant workers or illegal aliens living in the United States—should be covered by the U.S. health system. Because these noncitizens are human beings living in our community, it can be argued that they should have the same minimal care as do citizens. The problem is that someone may be in our community illegally, but that fact should not justify restrictions on the level of health care provided to any human. The issue of who is permitted to come to and remain in the United States should be addressed in forums concerned with immigration, naturalization, and law enforcement—not within the arena of health-care policy. All will agree that noncitizens suffer from illness the same as American citizens, and that such needless suffering is unjust.

Although it is easy to get agreement by most regarding the basic necessity for universal health care, how quickly this universal health care should be made available becomes more problematic. Representative McDermott's goal was universal coverage by 1995; President Clinton's was 1998. Senator Chafee of Rhode Island proposed a plan to provide coverage by 2005. The plan proposed by Representative Cooper of Tennessee did not have any mandate to guarantee universal coverage. Senator Dole declared the President's proposals for universal coverage and employer mandates "dead" on the Senate floor at the time they were proposed.

Those who would lack health insurance if universal coverage is not realized are not just the very poor, many of whom are covered by

Medicaid. Nor are they the elderly, who have Medicare. Many of the uninsured are those in working households who are fully employed, but whose income is too high to qualify for Medicaid. There was talk of introducing employer mandates through a "trigger mechanism" that would be activated by some date if a specified level of coverage had not been attained, but the details varied in the different plans, or were not specified at all, with only the vague assurance that something would be done if a specified proportion of the population did not have medical coverage.

Some argued that any changes made in the health-care system should be approached with all due deliberation, taking several years to reach the desirable goal of universal care. These suggestions were made by those who already have far more than the minimal care proposed for all, by the way. One would hope that those arguing for slow progress could assume the Rawlsian veil of ignorance and view the situation from the standpoint of those people who cannot qualify for or afford health care, but who suffer illness and the risk of premature death within their family. Even though we might decide not to rush to universal care, at least we should be able to imagine what our situation would be if we lost all medical insurance and became ill. Would we still want society to take several years to fade in universal coverage if we were included in that group without insurance?

Principles of justice require that everything possible be done to introduce universal coverage immediately, even though this might produce dislocations and economic inconveniences for those of us who have adequate medical protection. These changes might result in less coverage or flexibility at even greater costs than we experience at present. If it costs those of us who are privileged some medical luxuries and options, we should be willing to accept those burdens in the interest of providing for the worst-off—those who have nothing approaching minimal medical care. It is morally preferable for the privileged to be less well-off in order to raise the least well-off to the minimally satisfactory level, and almost any moral position, except that of the most hidebound Social Darwinists, would accept that argument. Even those who argue a strict libertarian position, that individuals should be free to pursue their own self-interest, using whatever resources they have to their own advantage, might agree that it is to their own personal benefit not to have infectious disease in their midst, or to permit the quality of people's lives to become so dismal that they are led to violent criminal activities or rebellion. Injustice could rebound to the individual disadvantage of the privileged, and this will not be in their ultimate reproductive interests, using the language of evolution.

One finds strange statements appearing, such as that attributed to Jack Faris, the president of the small business lobbying group, National Federation of Independent Business (Greenhouse, 1993a). He was quoted by Greenhouse as saying that he had checked the Constitution and couldn't see where it gave the right to universal health insurance. I will quote him in the interest of accuracy and fairness, not wanting to be accused of unfair paraphrasing:

> The Constitution says we have the right to life, liberty, and the pursuit of happiness. That's what small business is about. Well, this [requiring small business to provide health insurance to workers] is reducing our pursuit of happiness.

Surely, you jest, Mr. Faris; for one thing the phrase "life, liberty, and the pursuit of happiness" is in the Declaration of Independence, not the Constitution. One reasonable interpretation of this phrase in the Declaration is that the right to life might include guaranteed health care when it is required, and that universal health insurance could be one way to ensure that all people have adequate access to medical care. The Constitution does, in the 14th Amendment, note that persons cannot be deprived of life or liberty by the States, and that could be interpreted to mean that health is a Constitutional right. Greenhouse characterized Faris as "the federation's silver-haired, acid-tongued president," and he certainly is a wild and crazy guy, who should be commended for providing some comic relief to the discussion of a weighty issue.

Hopefully, this small-business National Federation will consider seriously the issues involved in society's attempts to resolve the dilemmas posed by health care. They also should be concerned when they consider the position expressed by the Business Roundtable, a group of executives from more than 200 of the nation's largest businesses, who announced their opposition to any system that would give breaks to small businesses, as every bill that has been proposed does by providing subsidies and tax breaks for any business with fewer than some specified number of employees. The Roundtable insisted that new costs hurt large employers as much as they do small ones. Therefore, assistance to small business would create an unfair competitive situation.

The U.S. public had reached a reasonable consensus that the U.S. health care system does not provide adequate care or security to the public, and agreed that the levels of medical expenditures were excessive, especially given the millions of people at risk. There was also a strong consensus that all citizens should receive adequate medical care. There is a moral imperative to achieve a minimal level of care immediately, and it should be extended to all people in the U.S. community.

With these understandings (and perhaps agreements to disagree with the latter belief) we can move to the next set of questions.

Structural Problems to Be Solved

It is not morally acceptable to consider health care as a market commodity. Human freedom depends on adequate health care; it is not the privilege that the medical associations argued it was not too many years ago. Adequate health care is necessary if persons are to have the ability and freedom to pursue life, liberty, and happiness with a sense of fulfillment, or to have any likelihood of enjoying a minimally satisfactory life. In the market-economy model, everything is available to the well-off; any scarce item goes to the highest bidder, and that means the least well-off are excluded at almost every move of the game.

Medical services can be scaled rationally along a dimension ranging from essentials to luxuries, and this can be done publicly and rationally by the public, with the assistance of social psychologists who have experience in SJT and GHPM (discussed in Chapter 8), using the input of physicians, medical researchers, medical economists, and moral philosophers.

Treatments to ease pain so that one can earn a living should have high priority, whereas the same treatments to allow me to enjoy the pleasure of tennis would have a low priority. If I can pay for it, I should be able to purchase any treatment on a fee-for-service basis, as long as that does not deny the less privileged access to any scarce treatment facilities they need to maintain an adequate life. Organ transplants probably should be considered luxuries rather than routine procedures, and would have to be evaluated in terms of the years of quality life that could be expected as a result of the transplant. IVF might not be available through a federally funded program, using a fee-for-service procedure available to those who want and can afford it. Those who can afford to pay for treatments that have a low probability of success should have that choice, but not at public expense.

The dreaded "R"-word—*rationing*—has not been used, although that is what is being talked about (some evade it by using the ugly word "prioritization"). The semantics should not be important, but they are, because words such as *rationing* are used as scare-words. The negative quality of the slogan blurs the intention behind a rationing system, the way it would work, and the fact that it exists presently. An explicit rationing system is preferable to the present situation in which the well-off and well-connected have first crack at everything, while others depend on first-come, first-serve systems, or on a lottery among the have-nots that relies on a random draw. It is enough that life involves the natural lottery of physical and psychological attributes, without imposing an arbitrary social lottery to compound misfortune and further empower the fortunate.

One problem comes to the fore when one considers rationing systems; when it is your health that is involved, you want every procedure done, no matter how expensive or how low the likelihood the treatment will help. Although these desires are understandable, the public might not be able to provide procedures for a person if economic realities deny others essential treatments. The basic needs of all must be met, and when those have been attained, then providing more of everything to everybody can be considered. Unfortunately, given present day economic realities, it seems that only the rich will be able to purchase exotic medical treatments for the immediate future. This inequality can be tolerated as long as the poorer members of society receive an adequate level of care. Given the rate at which expensive new medical technologies are being developed, the problems of making every procedure available to all who want them will only become worse. Members of the medical profession should face such realities instead of insisting that it is their duty to do everything possible for every patient, no matter what the circumstances. The progress of medical science makes the professional ethic now being used as outmoded as the Hippocratic Oath, and that point was emphasized in the statement of the AMA Council on Ethical and Judicial Affairs regarding ethical issues in managed care (1995).

Butler (1993), vice-president and director of domestic policy studies at the Heritage Foundation, expressed skepticism regarding the possibility that the Clinton Plan could succeed. He favored the creation of a "real health-care market" with families having control of health dollars (rather than businesses), along with the freedom to seek medical plans or services that offer the best value for the money. He considered this a better option than the Clinton Plan and rejected the single-payer option, although he considered it to be a clear and honest choice. His rejection of the single-payer plan was based on the belief that it would mean explicit rationing, "which most of us vehemently oppose." One can be just as vehement in opposing the implicit rationing we now have, because it subverts the ability to have open and honest discussion. Explicit rationing is superior to the alternatives. Butler is correct in his conclusion that business should not control the health dollar. Health financing should be controlled by a governmental single-payer entity, which will have the economic and political clout to ensure that services are provided—a clout that is unlikely as long as interests of the individual family must conflict with the interests of those who control the commercial and professional enterprises.

Another set of priorities concerns the cessation of treatment and the denial of access to facilities. To reach affordable cost levels, treatments should not be done at public expense whenever there is little or no possibility that the individual can pursue a biographical life. This includes continuing life support to maintain anencephalic neonates (babies that are so premature that they have almost no possibility of surviving for any more than a short period of time) or to prolong the life of terminally ill, comatose adults. ICU facilities should be allocated on the basis of the likelihood that treatment will allow a patient to return to a life of satisfactory quality.

These concerns should be considered in the the moral terms that reflect the values and beliefs of the public, the informed opinion of the medical profession, competing needs of other persons, with cost being factored in last and playing the smallest role in the equation. Decisions should be based on public and democratic consensus rather than on paternalistic and secretive decisions made by elite policy makers, shielded from public scrutiny.

Special Interests: Tobacco

The positions of special interest groups are based on an insistence that they should profit, while not losing any of their current privileges or incurring any greater costs. These interests include organizations that represent the elderly, large industry, small business, organized labor, the self-employed, the disabled, hospitals, physicians, psychologists, medical associations, HMOs, teachers, veterans, insurance companies, drug companies, pharmacists, and those commercial interests that would be affected if "sin taxes" are imposed.

The arguments by one of the commercial interests can be used to illustrate the problems that exist in many sectors of society. It is widely agreed that the tobacco industry should contribute to health-care costs, because of the harmful effects of tobacco and the burdens tobacco users place on society (although all of these negative aspects for many years were adamantly denied by the tobacco industry, and any evidence that there might be other negatives, due to such things as secondhand smoke, is challenged even before studies have been made public).

It was estimated by Congress's Office of Technology Assessment that the nation could save a large amount of the $68 billion lost each year in lower worker productivity and higher health-care costs for smokers if smoking levels were curtailed (Luther, 1993). The Centers for Disease Control and Prevention of the Department of Health and Human Services estimated that direct costs for smoking-related illness were $50 billion in 1993, about 7% of all health-care costs in the United States. These costs involved $26.9 billion for hospitalization and $15.5 billion for physicians' bills (Hilts, 1994). An article in *The New York Times*, February 14, 1995, reported that smoking and drug abuse will cost the federal government $77.6 billion in health and benefit payments in 1995, amounting to 20% of Medicare, Medicaid, and Social Security disability payments. This estimate led the administration to recommend that Medicare premiums be higher for smokers.

The human costs of tobacco use were considered in depth in a two-part article (Bartecchi, MacKenzie, & Schrier, 1994; MacKenzie, Bartecchi, & Schrier, 1994). Medical evidence and economic factors were discussed, and these articles should be examined carefully to appreciate the strength of the documentation regarding the negative effects of tobacco use, as well as the duplicity that has characterized the tobacco industry's campaigns to promote smoking. About 46.3 million adults in the United States smoke, and tobacco was the leading preventable cause of death in 1990 (400,000 deaths; 19% of total preventable deaths; 20 times those associated with drug use and 16 times those associated with auto crashes). The World Health Organization (WHO) estimated that there will be 10 million tobacco-related deaths worldwide per year by 2010 (Kaplan, 1994b). Smoking contributed heavily to cardiovascular disease, cancer, lung disease, spontaneous abortion of fetuses, and low birth weight of neonates. Tobacco sales are enormously profitable; the industry is subsidized by the U.S. government (through crop subsidies); tobacco advertising (especially targeted at the young) has increased from $500 million in 1975 to $3.9 billion in 1990 (a threefold increase in constant 1975 dollars); the United States exported 194 billion cigarettes in 1991 (more than three times as many as any other country in the world), and the tobacco lobby is considered to be one of the most pervasive groups in politics today. The cigarette is the only consumer product sold legally in the United States that is unequivocally carcinogenic when used as directed. There is strong evidence (disputed by the tobacco industry) that smoking becomes an addiction that is among the worst known.

Advertising campaigns conducted by the tobacco industry have been, and continue to be, aimed to systematically distort the scientific evidence and to confuse the public, even though the scientific debate has been settled for nearly 30 years. It has been demonstrated beyond a reasonable doubt that cigarettes cause lethal disease in humans. Even though the executives for the tobacco industry maintain steadfastly that cigarette smoking has not been shown to cause illness or death, it was reported that three insurance firms owned by tobacco companies charge smokers nearly double for term insurance because smokers are about twice as likely as nonsmokers to die at any given age!

The tobacco industry seemed to have accepted the fact that a cigarette tax was inevitable and concentrated its efforts to keep the amount to a minimum, as is expected in a market economy. The alcohol industry, which similarly contributes to the nation's health-care costs, agrees that tobacco should be taxed, but differences have surfaced between beer and hard-liquor interests. Hard-liquor distillers argue that their taxes already are excessive compared to the amount that beer makers pay, so all taxes should be levied on the brewing segment of the industry. The brewers argue that taxation on beer hurts the lower- and middle-class consumers, because they are the ones who drink the most beer. No mention is made, by either the brewer or the distiller, of the harmful effects their products have on the health of consumers. Some alcohol interests argued that alcohol should not be taxed at all because there is evidence that alcohol, used in moderation, might have beneficial effects on the health of drinkers, whereas tobacco is all bad. The tobacco industry responded that alcohol should be taxed heavily, because alcohol generally is harmful to health. A tobacco executive, testifying before a Congressional committee, suggested that tobacco poses no greater hazard to be addictive than is the case for Twinkies. The object of these debates scarcely seems to be to enlighten the public.

Some Southern Democrats stated they will oppose any health plan as long as tobacco stands alone as the only product taxed for health care. Tobacco interests have used the disingenuous argument that tobacco should not be taxed heavily or people will quit smoking and even less tax revenues will be collected (a risk, I think, the health community would be willing to take, given the costs in human health and misery that smoking exacts).

The revenues that tobacco taxes produce are considerable: California raises nearly $600 million per year with only a 25 cent per pack tax. The figures provided by WHO indicate that there is a linear relationship between the price of cigarettes and consumption; countries that charge less per pack have higher smoking rates. The cost of cigarettes varies quite widely: from $4.17 per pack in Norway (which has the lowest smoking rate in the world) to less than 50 cents per pack in Hungary, Romania, and Greece (which have the highest smoking rates). Smoking had been declining 0.7% per year in California prior to the introduction of the new state tax (Kaplan, 1994b). Since the tax was levied, the decline has accelerated to 1.27% per year, and the rate of change is significantly greater for socioeconomically disadvantaged groups. If the motivation for health-care reform is to save lives and improve health, one way to accomplish those goals is to introduce a high tobacco excise tax to enhance health status through lowering the prevalence of smoking and to raise revenues that can be used to finance health care. The tax would benefit society by reducing the burden of disease and disability, could directly benefit smokers by providing an increased incentive to quite earlier, and would provide funds to care for those who continue to smoke. The data from California surveys indicate that the majority of smokers want to quit, and about 50% make an active attempt to quit each year.

One incredible suggestion was that smokers already pay their way because they do not live long enough to enjoy their share of the Social Security and pension benefits for which they have paid. According to MacKenzie et al. (1994), the fact that there are premature deaths due to smoking hardly provides a humane means to control health-care costs.

A scare tactic is to scold that jobs will be lost in tobacco-producing states (feel free to insert the name of whatever special interest you like—liquor, insurance, hospitals, drug manufacturers, small business, etc.) if there is any burden placed on their operations and profits. And so, the lobbyists lobby, the politicians pose, the profiteers profit, and the poor perish.

There are reports that medical associations are gearing up their legal staffs to challenge any fee cap that the government might initiate. Insurance companies will sue to stop any attempts to limit insurance rates, and associations representing the elderly and the disabled will challenge the legality of any changes in Medicaid or Medicare. The threatened court battles should make the legal profession happy, but there have not been many concerted legal activities on behalf of the indigent sick.

It is not going to be simple to chart a course around the shoals of those selfish societal interests that will be affected by changes in health-care delivery and financing. All of these interests, however, should be compelled to assume a fair share of the burden. Although most accept the principles of fairness, disputes revolve around the question of what portion of the burden (if any) is their own fair share in each specific instance.

Arguments Concerning the Single-Payer Plan

The NHP could be similar to the government-mandated plans for universal health care that several countries have adopted successfully, and it could be modeled on the CHP. Some have stated that the government should not mandate any universal health coverage, but that position is immoral, given the suffering of so many who will receive no health care under the present system. There have been arguments that the CHP has so many problems that it should not be used as a model. There are at least 10 objections, and each will be examined briefly, although some are of questionable merit on the surface.

1. *What is okay for Canada is not appropriate for us, because this is America, and we should do things our way.* To be sure, Canada is not the United States, but no reasonable specifics have been presented regarding the essential differences between Canadian and U.S. citizens, between the economies of the two countries, or in the nature of the provinces compared to the states (except that there are more of the latter) that would produce the essential differences that mandate different health-care systems. One suggested critical difference was that when the U.S. system is compared with that found in Britain and other Western European countries, the U.S. social order is relatively classless (Stevens, 1993). The importance of this difference is that, although the professional and managerial classes in Britain are less willing to roll over and accept decisions with which they do not approve, the "other" social classes will. The per-capita expenditures by the British National Health Services were 41% higher for members of the upper two socioeconomic groups (professionals, employers, and managers), than for members of the "lowest" two classes, according to Stevens. His argument continues that patients in the United States are apt to behave as do the British upper socioeconomic groups—namely be less willing to accept "no" for an answer—making budget caps difficult to administer in the United States. He concluded, for this reason, that any budget caps for the United States would be ill-advised. This is a less than compelling argument, as well as a bit insulting to the British people. Even if there was any merit to the argument that the British lower classes are willing to settle for less than adequate treatment, that does not provide a justification for the practice. There is greater inequality between the rich and poor in terms of the availability of health care in the United States than in Britain, because Britain has universal medical coverage and the United States has millions who do not receive adequate care. The crucial differences that would make a difference must be specified more adequately than through such speculative arguments.

2. *A single-payer plan is socialized medicine.* This has been the rallying cry of opponents of health-care reform for many years and was the official position of medical associations throughout much of the latter half of this century, with the additional slippery-slope caveat (issued by several Republican leaders of the 103rd Congress when global

health-care reform was proposed) that socialized medicine is the first step to socialism, and that step will be followed by the inevitable slide to communism, with the ultimate loss of all of our cherished American ideals. In the 1930s, Republicans fought the creation of the Social Security program because it was an inevitable step on the slippery slope leading to socialism. The California Medical Association opposed the highly successful Kaiser-Permanente health-care system, which has been a prototype for the HMOs, whose framework they now enjoy for their practices. At the outset, the California Medical Association imposed sanctions on doctors who participated in the Kaiser system, exploiting the fear of socialized medicine and lack of free choice of physicians.

In the 1960s, the AMA fought the creation of Medicare because they considered it an irreversible step toward complete socialization of medical care. Medicare now accounts for a substantial portion of doctors' income and 40% of hospital revenues. The AMA now considers reduction in Medicare reimbursements to be unacceptable, because it would threaten access to care for elderly people, which, incidentally would reduce the income of physicians. Officials of the AMA were opposed to any national health-spending limits, asserting that controls of private fees and private expenditures are something new to this country.

The argument about creeping socialism was being made less often, but it resurfaced during the health-care debate. Meyers (1993) quoted Haislmaier, senior policy analyst for health care at the conservative Heritage Foundation: "I hate to say it, because it makes me sound like a lunatic right-winger, but this is textbook Marxism. This is central planning." Haislmaier is the one who characterized what that statement sounds like, but it seems he was compelled to make the statement anyway. Representative Dick Armey of Texas was quoted by Clymer (1993) to have said in June, 1993, that, "Her [Hillary Rodham Clinton] thoughts sound a lot like Karl Marx. She hangs around a lot of Marxists. All her friends are Marxists." Shades of the days when one could make the subversive lists for having a copy of *Das Kapital* on the bookshelf.

Cutting through the right-wing rhetoric, it is possible that a government-mandated health-care plan could deliver universal health care and security to the millions who have been cast aside by the present free-market system. One would hope that providing for the less well-off in society represents an American style of freedom and equality for all, rather than foreign, creeping socialism. It is distasteful to think that socialism has a monopoly on morality and that the United States resists efforts to provide for the poor in the interest of opposing the specter of socialism. If that is the case, we should examine the meaning of freedom and equality. Universal health care is in the spirit of the American ideal, and we should refuse to surrender that ideal to socialism.

3. The single-payer plan creates a huge government bureaucracy, and we all know about the inefficiency of big government. This argument needs to be examined in several ways. First, many who oppose what they see as government intrusion into the affairs of American citizens support some of the most bureaucratic and centralized aspects of our government, such as the Defense Department, Department of Agriculture, and NASA. They tend to support federal funds for police and prisons, for centralized research in nuclear physics and to understand and treat AIDS, for federal relief to disaster victims, and for numerous subsidy programs to farmers, ranchers, and various industries. There seem to be few right-wing arguments to eliminate such centralized government programs as the Veterans Administration or to close VA hospitals, which provide "socialized" care for American veterans.

The one standard government agency that is subjected to routine derisive comments is the U.S. postal system. But, as I heard Himmelstein remark on TV, the post office does not deny mail delivery to 40 million Americans because they live too far away. Yet, the free-market health-care system fails the 41 million people who cannot afford to pay for care. The thrust of the attack seems to speak to an interest by business to transfer the Postal Service to the private sector as a delivery-for-profit enterprise, as much as a comment on the quality and cost of the service it provides.

The special nature of the argument that big government should not be involved in the affairs of the people is used in selected instances. People do not argue that states should contribute an equal share to the defense budget or risk being removed from the protection provided by the U.S. defense system. Poorer states often receive massive amounts of federal funding to maintain military bases, even though other states contribute more of the funds to pay for those military facilities. Many choose to support this or that aspect of bureaucracy that benefits them, attack those entities of the government that provide benefits for others, and in general insist that government is no good because "they are all crooks"—but don't take away my entitlements or any of my share of the government, be it Medicare, Social Security, veterans benefits, agricultural supports, disaster assistance, space stations, military installations, or grazing and crop subsidies. All of the government interventions that benefit "me" seem to be acceptable and justifiable in the name of preserving free enterprise, even with an admission that although it is "pork," it is important that we get our fair share of that pork.

A more compelling bit of evidence, specifically involving health care, exists in the statistics regarding the administrative costs of different systems. In this book, a number of the comparative statistics presented indicate that in every country for which statistics are available (including the United States), the administrative costs of government programs are less than those for the private sector that has been driven by the market. Biderman, of the School of Public Affairs at American University in Washington, D.C. noted in a letter to *The New York Times* (November 7, 1993), that in 1990, the number of people employed by the U.S. Government in all legislative, judicial and nondefense executive agencies (including the Postal Service and 111 thousand temporary Census workers) was 2.173 million. In the same year, the number of people employed by private insurance carriers and insurance agents was 2.389 million. Such statistics, along with those for administrative costs, give the lie to the claim that government-administered health programs necessarily are more cumbersome and expensive than those in the private sector.

4. *If the government is involved, patients will have less choice.* Patients have more choice of primary-care physicians in Canada (and Britain) than they do in the United States, because all physicians are involved in the CHP. Patients can choose to pay on a fee-for-service basis for procedures not covered by the CHP, and they are free to choose any physician they wish to perform those procedures, just as in the United States, for those who can afford it. In Canada, access to specialists is controlled by primary-care physicians, who refer their own patients to specialists, and that seems neither undesirable nor any different from the way HMOs operate in the United States. Under the CHP, all referrals for diagnostic work, specialist care, and hospital services are under the control of the primary-care physician, who has a continuing contact with the patient. Rationing is done by qualified physicians who know the patient, rather than by utilization review boards, run by insurance companies or HMOs. These review boards are often staffed by clerks who have neither extensive medical training nor firsthand knowledge of the patient. In effect, the primary-care physicians in Canada run an internal utilization review and management system, with an eye to the needs of the patient and the available facilities. In Canada, the capacity and use of facilities are matched through a primary medical review rather than through lists and queues (Barer & Evans, 1992).

Choice of physicians in the United States is limited in HMOs. Patients can only use physicians who participate in the plan, and the huge health alliances that have formed in the United States make it probable that private practitioners and small-group providers will be frozen out of the system, because they will have difficulty gaining privileges at hospitals owned and controlled by HMOs. As Himmelstein and Woolhandler noted in a letter to *The New York Times* (September 26, 1993), in Canada everyone is covered, the quality of care is high, patients can freely choose and change physicians, and physicians are free to choose among a variety of practice settings. Much of the cost-containment in the CHP is due to the streamlining of bureaucracy, produced by "evicting" insurance companies from the system and developing a simple, centralized budgeting system.

5. *The adoption of the single-payer plan will result in long lines to see physicians and long delays to receive treatment from specialists, especially treatments that involve the use of high technology.* This is a frequent characterization of the CHP, but it does not represent the situation in Canada. Waiting lists do not exist for primary care in Canada. The utilization reviews by primary-care physicians avoid waiting of the first-come, first-serve kind, and evidence indicates that serious medical problems receive prompt treatment because cost is not a factor in the system (Barer & Evans, 1992). There are waiting lists for certain high-technology procedures, such as the use of MRI facilities and open-heart surgery, but the available facilities are used more fully and economically than in the United States, and priorities are established, based on medical indications rather than on the ability to pay, as is the case in the United States.

6. *People will swamp the system, because it is free.* In Canada there was an initial surge in the number of people seeking medical assistance. Those who had not had diagnoses or treatment for minor problems, because they could not afford the cost, were able to receive medical attention when the system began. An even greater surge probably will occur in the United States, because the health-care system (as evidenced by the surge in initial enrollment for the Oregon health plan) has been inadequate for such a long time, and the backlog of untreated medical problems is so large. In Canada, however, there was no long-term overuse of the CHP system, even though it is universal and carries no deductibles or copayments. An increase in utilization in the United States will occur, because needed medical care is provided to those who have not been able to afford it. The Canadian experience does not support the offensive view that the poor must pay dearly for services or they will abuse anything that is provided for their benefit.

7. *The single-payer system will lead to low-quality medical care.* This complaint usually is voiced by those who receive the best care available in the United States, who have access to their own private physicians, and whose expenses are paid by a good insurance plan. Their fear is that they might have to sacrifice some privileged benefits or

even pay more to keep them. Hillary Rodham Clinton, in her initial testimony before Congressional committees on September 28, 1993, estimated that under the Clinton Plan, insurance costs would increase for 10–12% of the insured, who would pay more for the same benefits they have now. However, 63–65% would pay the same or less for better benefits, with the remaining people paying a little more for better benefits. Even those who pay more now would save in the long run because, when they grow older, their benefits would be cheaper than will be the case if the system is not changed. It must not be forgotten that millions of people who have no coverage at all will receive it at modest or no cost to them.

An interesting example of the problems involved was provided by jousting that occurred between two sets of Democratic legislators. Wellstone of Minnesota and Stark of California are opposed to the Clinton Plan, favoring the single-payer plan. Both of these legislators wanted members of Congress to go on record that they would take the minimum coverage of whatever health plan might come out of Congress. Two Senators, Boxer of California and Murray of Washington, objected that they and their families should have the right that other Americans would have to select the best option for themselves. This skirmish exposes a delicate issue that one hopes Congress would explore regarding justice and equity. If we are not willing to live within the limits of the proposed minimal coverage, then should that be the coverage that is the entitlement for people in the United States?

There is little for many millions of people to worry about in terms of the quality of their medical care—they have little or none. The balance, then, is between a possible decrease in breadth of coverage for many in order to make it possible to provide some coverage for another sizable proportion of the population. The evidence is that the quality of care in the alliances and HMOs envisaged within the Clinton plan would involve the risk of lowering the quality for the already insured to a much greater degree than would occur with the single-payer plan.

The current system is badly broken, and Americans have expressed dissatisfaction with their health-care system. Compared to 10 Western nations, American citizens were less satisfied with their health-care system (even though the U.S. per-capita cost was the highest, at $2,051), and Canadians were the most satisfied (per-capita cost = $1,483 in U.S. dollars; Blendon, Leitman, Morrison, & Donelan, 1990). The American dissatisfaction was equaled only by Italians (per-capita cost = $841). There was a tendency in most countries for a relationship to exist between per-capita spending and satisfaction, with the United States being the major outlier—it had the highest per-capita spending and the lowest level of satisfaction.

These authors rejected a suggested interpretation that the greater satisfaction by the Canadians is due to their more positive general attitude toward life. In comparable surveys, American have expressed more general optimism than have Canadians regarding their personal futures and the direction the government is taking. Canadians are more satisfied than Americans *specifically* in regard to their health system. It was suggested that American dissatisfaction arises from sharply rising health-care costs and inadequate financial protection provided by the U.S. insurance system. Twenty-eight percent of Americans surveyed in 1989 by the U.S. Census Bureau reported they were without health insurance coverage for some time in the preceding 28 months; a 1990 survey found that 18% of adults over 65 had their health benefits reduced over the previous 24 months; and since 1980, the value of employee health premiums paid by employers declined from 80% to 69%, with the employee paying the difference. Surveys indicate that Americans pay 26% of their health bills out-of-pocket, with 19% paying more than 40% of their bills. None of these circumstances exist in the other nations surveyed, and it is reasonable to conclude that these are major concerns for U.S. citizens.

Based on all of the quality indicators (hospital stays, life expectancy, infant mortality) there is no evidence that the CHP delivers lower quality medical care to those who are covered in Canada—and all are covered. By any measure, the Canadians receive better care than the 41 million uninsured in the United States.

8. Canadians cross the border to the United States to receive treatments that they would either be denied or have to wait a considerable time to receive. This claim seems to be without merit. It was estimated by Congressman McDermott of Washington (who is also a practicing physician) that only 4% of the health care of Canadians is done in the United States: 2% takes place when Canadians who are visiting on vacations or business require treatment while in the United States; much of the other 2% is authorized by CHP administrators who choose to utilize and pay for expensive high-technology facilities just across the border, such as those at the University of Washington Medical School in Seattle, rather than to construct extensive facilities, purchase expensive equipment, and hire staff to utilize these facilities. Because these U.S. medical facilities are underutilized, this is a good deal for the United States and is cost-effective for Canadians.

There is some evidence that it is not the Canadians who are crossing into the United States to obtain medical treatment, but Americans who are abusing the free Canadian health-care system. It was estimated that thousands of Americans are using the CHP illegally (Farnsworth, 1993). The total number of improper claims in the province of Ontario was estimated to be 600 thousand (60 thousand suspicious claims were processed for those with American driver's licenses), costing the provincial health-care system as much as $691 million. Ontario has a population of 10 million, but 10.5 million people are registered to receive health care. The Canadians have tolerated these abuses, but are now proposing more strict penalties for fraud. Farnsworth (1993, p. A9) quoted Anne Moore, corporate services manager for the Ontario ministry, as saying, "Our intention is not to send thousands of Americans to jail, but to get them to pay their bills. Anyway, we hope it's a temporary phenomenon, until President Clinton gets his health plan through." The American public and its representatives owe the Canadians consideration, or at least an apology for the mean-spirited attributions regarding the moral character of Canadians.

9. Having federal government control of health care would violate states' rights. Under both the Clinton plan and a single-payer plan, the states would negotiate a budget with the appropriate federal agency and would be able to allocate that budget to provide for the people in that state. Any state should be permitted to establish a single-payer plan if it wished, as long as minimal national standards are maintained. A state would have more direct control than they do under the present Medicare and Medicaid programs, because they would be able to rationally budget health care with recognition of the particular needs of the citizens in the state. The proposals made to establish a single-payer plan no more violate state's rights than does a federally funded VA hospital or a federally funded military base located in a state. Before the 1994 elections, Wellstone stated that state flexibility would be a big issue in 1995, and he promised to make a major effort to permit greater flexibility to let states try a single-payer program financed with taxes. Immediately after the election, Senator Packwood predicted that it would be necessary for Congress to remodel Medicaid and Medicare along the lines of the Oregon Health Plan, and he used the "R" word. Arguments regarding states' rights seems but another red herring.

10. The Clinton Plan would result in lost jobs from the business community, and the single-payer plan would result in lost jobs in the insurance industry. The estimates of the number of jobs that would be lost under the Clinton plan range widely. The administration estimated that 150 thousand jobs would be lost through layoffs by the small businesses that do not now provide health insurance. Another estimate, based on a study subsidized by companies that do not provide health insurance for their employees, was that 3.1 million jobs would be lost. Yet another estimate, from a study sponsored by the National Federation of Independent Business, was that 12.7 million jobs would be at risk, with 1 million layoffs.

The administration attacked the latter two studies because they were not based on the final version of the Clinton Plan, which proposed subsidies for the costs to small businesses. The administration dismissed the significance of their projected 150 thousand lost jobs, because this represents only about one month's job creation which, it argued, was worth it to gain health security for all Americans and to put an end to spiraling health-care costs. There should be little in the way of job loss as a result of increased health-care costs if they are mandated industrywide. If the public needs or wants the products or services provided by businesses, then none would be competitively disadvantaged by a universal employer mandate. Richard Berman, the executive director of the Employment Policies Institute, a business-backed research organization, invoked a new biological entity when he said, "From what I'm hearing from business, my anecdotal instincts tell me that the estimate of three million job losses is a low number" (Greenhouse, 1993b).

President Clinton dismissed the single-payer plan, because it would cost too many jobs in the insurance industry. The issue of jobs should be a separate discussion from the issue of health care. Health-care reform should benefit the nation, not protect jobs in the insurance industry or keep costs low for businesses that have large numbers of employees in order for them to enjoy lower medical costs than does the general population (Mashaw & Marmor, 1993). Because health care accounts for one-seventh of the economy, any changes affect all aspects of the economy of the United States. Because of the profound economic implications, it would be best to devise an optimum health care plan and then move to the realities of the economic factors. Intermixing moral issues, medical realities, employment opportunities, naturalization, and immigration policies with the realities of the national budget brings in too many considerations all at once. Such a multitude of issues makes it likely that we will realize our objectives if the approach is to focus first on the major issue—the moral necessity of good quality universal health care—and then move to the next level, the economic issues involved to realize the goals of a moral society.

The current U.S. medical system does not function adequately, and what we have constitutes a morally indefensible state of affairs. When the alternatives of managed competition, health-care rationing, and single-source payer plans are examined, it seems that a single-source plan, similar to that used in Canada, is the most defensible economically, medically, and morally. A health-care rationing plan, such as Oregon's, would at least be a step in the right direction.

Another disturbing aspect of the process involved in the health-care debate is the political pressure being exerted by health-care industry PACs. *The Oregonian* (September 1, 1993) reported that campaign contributions from health-related business interests to members of the House Ways and Means Committee increased 46% during the first half of 1993, compared to the equivalent period in 1991—an increase from $671,742 to $981,279. The chief beneficiaries of these funds were the 17 new members of the House Ways and Means Committee which, along with the Senate Finance Committee, would have written any health-care legislation.

One HMO network announced that premiums will not rise more than 5% a year in 1994 and 1995—hardly the savings that people had in mind when considering a restructuring of the health-care system, and there is still the major issue of how health-care coverage is to be provided for the 41 million uninsured. Many members of Congress, as well as many of the middle-class taxpayers who will be covered by the new networks, seem determined to avoid any tax increases or increased cost for health coverage, and President Clinton discussed the possibility of phasing in universal coverage over a 5- to 8-year period. All in all, MC looks to be a lousy fix that would do little more than generate substantial profits for corporate entities at inordinate expense to consumers. It would solve few of the problems inherent in the U.S. health-care system, and it would be several years before universal health coverage would be available to all who need it.

Now that some of the issues have been considered, it will be of interest to consider the nature of the health-care debate that ensued in the Congress. The intent of the debate, one would have hoped, was to consider what is required to establish a just society, provide for health needs of all people in America, and do so in an equitable and economical manner.

CHAPTER 14

The Great Health-Care Debate

Little in the way of positive action or change of opinions occurred in the 103rd Congress concerning health-care reform. Issues became more and more confused, the positions frozen in place, and debates that began with political wrangling fell to even lower levels, with an attempt at the very end by a few Senators to arrive at some kind of bipartisan proposal that could be voted on. The bipartisan proposals were rebuffed by both conservatives and liberals, representatives of both poles preferred, finally, to do nothing rather than surrender any part of their position. The failure of any bipartisan coalition, plus the impending national elections (to which the members of the Senate and House wanted to devote full attention), led to an adjournment in the Senate without a vote on any finished bill, and to no debate at all in the House. The election losses suffered by the Democrats have been attributed by some analysts to the lack of success of the health-care plan, who claim that the proceedings created dissatisfaction on the part of the public regarding centralized governmental control of daily affairs.

Rather than follow the tortured path of the debate in the Senate, a chronological series of impressions will be presented to characterize the debate, identify the positions of some of the major players, and give a flavor of the statements. The process can be broken into several stages: a warm-up period; the initial Senate debate; the August–September recess; the resumption of the debate, ending in adjournment; the fallout before the elections; and the postelection retreat from any serious discussion of health-care policy at any level other than that of economics.

THE WARM-UP PERIOD

The opening shot of the campaign to reform the health-care system was Senator Harris Wofford's upset Democratic victory in Pennsylvania in November, 1991, a victory that was attributed to the promise of health-care reform. Early in the 1992 campaign for the Democratic presidential nomination, Senators Kerrey of Nebraska, Tsongas of Massachusetts, and Governor Clinton of Arkansas all emphasized the importance of health-care reform.

September 24, 1992—Candidate Clinton proposed that universal health care should be provided, and that it should involve a mandate for employers to buy insurance for their employees, for the Government to guarantee coverage for the unemployed, for small businesses to buy insurance at a discount, and for a panel to establish national limits on health spending to achieve cost-containment.

January 25, 1993—Following the inauguration, President Clinton appointed Hillary Rodham Clinton to head the Task Force on National Health-Care Reform to prepare health-care legislation by April 30.

August 16, 1993—In his first speech on health care since becoming President, Clinton proposed to the National Governors' Association that there should be an employer-mandated health-insurance program, universal coverage, with people paying in proportion to what they could afford, cooperatives in each state that would buy insurance, cost-containment to encourage competition among plans, an overall health-budget ceiling for each state, with the self-employed allowed to deduct 100% of their health-insurance costs.

September 10–11, 1993—A bootleg copy of the Clinton plan was obtained by lobbyists and lawmakers. The draft was a working copy still undergoing revisions and reflected the ideas of President Clinton and the working group convened by Hillary Rodham Clinton.

September 22, 1993—President Clinton, in a joint session of Congress, outlined a health-care proposal that included universal coverage by 1999, purchasing groups (called *health alliances*), and an 80% employer mandate. The Congress expressed varying degrees of bipartisan support.

September 23, 1993—By this time, three Republican plans had been put forward: A group of Senate moderates offered a plan developed by Senator Chafee of Rhode Island; a plan was offered by Representative Gramm of Texas; another was proposed by 106 House members. When compared to the Clinton plan, all of them relied less on the government to expand coverage or hold down costs, relying instead on free-market competition and incentives. None involved employer mandates; they would not set rapid or definite dates for universal coverage, reshape the health-care system, or impose mandatory ceilings on medical spending.

An ominous note was sounded by Krauss (1993) concerning the

mobilization of lobbyists to influence the health-care proposal. He reported the following immediate developments: Anheuser–Busch urged Bud drinkers to oppose any tax on beer; IVAC Corporation, a San Diego-based company that manufactures intravenous pumps, established a 20-employee phone bank to call other health-industry workers to oppose proposals to limit hospitals' access to medical technology; hundreds of R. J. Reynolds employees volunteered to take vacation time in the spring to travel to Washington to lobby against higher cigarette taxes; AIDS activists staged street theaters to advocate more federally funded research on the HIV virus; physicians raised concerns about health-care reform to patients during physical examinations; delegations from groups representing such professions as dance therapists, masseurs, chiropractors, and podiatrists were descending on Congress; pharmaceutical companies were writing their shareholders, warning that profits could suffer if price controls impeded drug research; Planned Parenthood encouraged members to campaign that abortion, prenatal care, and estrogen replacement be included in standard health-insurance coverage; small-business owners, who faced the prospect of having to pay for health care for their employees for the first time, flocked to Capitol Hill; the Pharmaceutical Manufacturers Association and American Academy of Family Physicians spent millions of dollars hiring lobbyists; political action committees linked to the AMA, National Association of Life Underwriters, and the American Dental Association increased campaign contributions to lawmakers; the Health Insurance Association of America started a glossy advertising and telephone bank program to insist that the Clinton plan would mean higher insurance premiums and fewer health-care choices (its first TV commercial was run before the Clinton program had been announced); the Health-Care Reform Project, which was formed by such groups as the AFL–CIO, American Airlines, and League of Women Voters, moved to build a consensus around a compromise to slow medical insurance rates and to cover the uninsured. This lobbying blow was the first that was to create a stormy sea that the reform process would have to ride out—or is the appropriate platitude "sink-or-swim"?

October 6, 1993—Representative Jim Cooper, Democrat of Tennessee, and Senator Breaux, Democrat of Louisiana, each introduced a plan without employer mandates, relying on market forces to control spending.

October 27, 1993—President Clinton delivered the MC plan to Congress. The Senate Republican Conference characterized it as "too much government, too soon." In Washington, there were an estimated 800 lobbying groups concerned with health care. Some suggested that the closer the election approached, the less likely it was that a major bill would pass.

November 4, 1993—Leon Panetta, the Administration's budget director, estimated that under the Clinton Plan, 70% of insured Americans would be paying the same or less for the benefits they receive, saving an average of about $61 per month, including copayments and deductibles. The other 30% of them would pay an average of $24 a month more. He estimated that in the first year, 15.1% of families would save more than $84 a month, while only 1 in 1,000 would pay $83 more a month.

December 16, 1993—Ten medical associations, representing 300 thousand physicians, endorsed the MC plan. Representative Jim McDermott, Democrat of Washington, defended his single-payer proposal, the AMA (representing 296 thousand physicians) urged Congress to consider alternatives to Clinton's employer mandates, and Representative Newt Gingrich, Republican of Georgia, asserted that the Clinton plan would lead to much more central planning and could bring "socialism."

January 3, 1994—*The New York Times* presented a range of forecasts by various concerned parties regarding the outlook for major health-care legislation, including universal coverage. A number of individuals were completely optimistic: Representative Rostenkowski, Democrat of Illinois; Senator Rockefeller, Democrat of West Virginia; Senator Chafee; Representative Waxman, Democrat of California; Representative Stark, Democrat of California, Representative Cardin, Democrat of Maryland; Dr. Bristow, Chairman, AMA; John Rother, Director of Legislation and Public Policy, AARP; Donna Shalala, Secretary of Health and Human Services. On the other hand, a flat "no" was predicted by some: Senator Kassebaum, Republican of Kansas; Representative Armey, Republican of Texas; Representative Bliley, Republican of Virginia.

January 25, 1994—President Clinton delivered his State of the Union address and vowed to veto any proposal that did not provide universal coverage. The Administration sent signals that compromise was possible regarding the number and structure of health alliances. Large companies might be allowed to insure themselves, and there could be flexibility in the timetable to achieve universal coverage. Senator Dole, Senate minority leader and Republican of Kansas, delivered the official Republican response to the address, stating that "the country has a health-care problem, but no health-care crisis," and Senator Moynihan, Democrat of New York, made a similar statement. *The New York Times* on that day suggested that Senator Dole might be seeking to solidify the Republican party behind him for a third run at its Presidential nomination in 1996, which Dole denied.

January 31, 1994—The nation's governors unanimously called for Congress to pass a health-care bill this year, but could not reach agreement regarding whether there should be an employer mandate. Senator Dole warned that there would be no comprehensive health-care bill this year if it was not addressed on a bipartisan basis. A media consultant for President Clinton predicted that "there will be a barrage of cynical advertising like we have never seen. I think it will be hideous." These represented the firing of the opening shots in what would be a consistent, continuing battle.

February 1–7, 1994—The Business Roundtable, representing 200 of the nation's largest companies, endorsed the Cooper plan, and the U.S. Chamber of Commerce and National Association of Manufacturers declined to support the Clinton plan. Representative Gephardt, House majority leader from Missouri, attacked the Republicans; "Every time we try to lift people up, they say it's socialism, it's big government, it's tax and spend." He quoted Newt Gingrich, House Republican whip, to have said that guaranteed coverage is "socialism, now or later, and a dictatorship on health care"; Gingrich, elsewhere, compared President Clinton's plan to Soviet Communism (Clymer, 1994a). Senator Daschle, Democrat of South Dakota, worried that, although universal coverage was essential, nothing could be done incrementally. Senator Gramm insisted that mandatory universal coverage amounted to having the government take over and run the health-care system. Twenty-seven Senate Republicans offered a bill to reach universal coverage by 2005. The tone of the rhetoric of both sides was just about set by this time.

February 11, 1994—The American College of Surgeons, with 60 thousand members, was the first large organization of physicians to endorse a single-payer system, arguing that it best preserved the rights of patients to choose their own physicians, and that it was simpler to deal with than the MC plan. They cited a CBO estimate that a single-payer plan could result in $100 billion in administrative savings.

February 16, 1994—Arguments centered on what the term *universal coverage* really meant for the different plans: single-payer plan, universal coverage by 1995; Clinton Plan by 1998; Chafee Plan by 2005; Cooper Plan, no universal coverage, but universal "access," whereby a voluntary program with insurance reform is available, with insurance purchasing cooperatives and subsidies for the poor. Cooper estimated that 80% of the uninsured would obtain coverage under his plan, leaving about 8 million people uninsured. The Christian Coalition announced a $1.4 million campaign to build grassroots opposition to the Clinton Plan among conservative Christians. Their opposition was due to their belief that the Plan "would replace the finest health-care system in the world with a bureaucratic, Byzantine, European-style syndicalist nightmare that has no precursor in the American experience." [Wow!] Dole: no employer mandate.

March 8, 1994—Representative Pete Stark proposed a plan to cover all poor and uninsured through an expansion of Medicare that would be financed by a new 0.8% payroll tax, a 75-cent increase in cigarette taxes, and an employer mandate beginning January 1, 1995 for companies with more than 100 employees.

March 15, 1994—The House Ways and Means Subcommittee on Health voted on a health-care bill, and four other committees were in various stages in their deliberations.

June 3–11, 1994—Three House committees produced bills: The Labor and Health Resources Committee, chaired by Senator Kennedy of Massachusetts, essentially adopted the Clinton Plan. The remarks on all sides were typical of the general tone that had emerged: Kennedy—"Comprehensive health reform is a defining issue for this Congress"; Dole—"It doesn't look good to me"; Breaux, Democrat of Louisiana—"Too much government and too many mandates"; Danforth, Republican of Missouri—"It's an awful lot of taxes"; Packwood, Republican of Oregon—It's a "Burger King bill, a Whopper, $190 billion in taxes ... the Democrat's way out: tax and tax and spend and spend"; unidentified Republicans regarded the House plan produced by the Education and Labor Committee, chaired by Ford, Democrat of Michigan—"ugly, irrelevant, a cruel hoax, and bad as dog's breath."

June 20, 1994—Hillary Rodham Clinton insisted that universal coverage cannot be compromised and reminded members of Congress that a veto of any health bill was possible if it does not include universal coverage. The call for universal coverage was echoed in a letter to the President from 80 organizations, including unions, physicians, and church groups. The Health Insurance Association of America also called for universal coverage, but insisted that there be no significant changes in insurance law.

July 2, 1994—Moynihan's Senate Finance Committee adopted a bill that did not have an employer mandate and aimed to reach 95% universal coverage by 2002.

July 12, 1994—The Roman Catholic Church, which sponsors more than 600 hospitals, reaffirmed its support for universal coverage and announced a national campaign against any government plan that included abortion.

July 19, 1994—President Clinton suggested that he might compromise and accept coverage for 95–98% by 2002. Polls found that 80% of the public agreed that it is "very important" for every American to receive health-insurance coverage. Dole: "The President's proposals for universal coverage are "dead" on the Senate floor, and an employer mandate is not going to happen this year."

July 22, 1994—Fifty large corporations urged Congress to pass "comprehensive health-care reform that provides universal coverage through shared employer responsibility." *The New York Times* reported that the Center for Public Integrity found that unprecedented vast sums of money were being expended to influence health-care legislation: In 1993 and 1994, hundreds of special interests cumulatively spent over $100 million to influence the outcome; at least 97 law, lobbying, or public relations firms had been hired; members of Congress had frequently traveled at the expense of the health-care industry, often with family members and often to resort areas, to meet with interest groups; at least 80 former senior members of the executive branch in the Reagan, Bush, and Clinton administrations, as well as former members of Congress, had gone to work for health-care interests; many members of Congress had investments in health-care companies, especially in the pharmaceutical industry.

July 28, 1994—The Senate majority leader, Tom Mitchell of Maine, suggested that the employer mandate should be only 50%, rather than the previously specified 80%, and should be used only if voluntary measures failed to cover a certain percentage of the uninsured. Poll results indicated that 65% of the public agreed that the president should veto any bill that did not guarantee health insurance for all Americans. Dole—"Need a week of uninterrupted reading time to study the bill"; Packwood—Republicans are in a state of "almost uniform anger, if we're going to be given a bill that's 1,000 pages long and a day to study it."

August 2, 1994—Mitchell proposed legislation to achieve 95% coverage by 2000 through voluntary measures and subsidies for the working poor (especially for their children), for pregnant women, and coverage for prescription drugs and home and community-based long-term care. If coverage did not rise to 95% by 2000, employers would pay 50% of workers' premiums "unless Congress found a better way." He proposed a change in insurance law to require companies to guarantee that policies would be renewed, that rates could not be increased if the insured became ill, that there would be no lifetime limits on reimbursement, and that there would be community rating. Gramm—"We are moving the Government into the position of actually running the health-care system."

August 8, 1994—Two conservative Senators, Gramm and Shelby, Democrat of Alabama, threatened to filibuster the Mitchell bill, which seemed to be the most likely one to be debated and voted on. Senator

Helms, Republican of North Carolina, offered an amendment to a spending bill that would force the Senate to put off the whole matter until next year. Senator Dole designated 10 of the most conservative Republican senators to take charge of the debate, including Gramm, Nickles of Oklahoma, Lott of Mississippi, and Wallop of Wyoming. Senator Mitchell planned to rely on Kennedy and Moynihan. The Republican delaying tactics were not successful, and the debate was scheduled to begin, although it was difficult to be optimistic about the outcome, given the cross-currents of opinion in the Senate and the divisive tone throughout the process thus far.

THE INITIAL SENATE DEBATE

August 9, 1994—The CBO reported that the Mitchell bill would quickly raise the percentage of insured Americans from the current 85% to a level of 95% by 1997, but would go no higher. The debate opened with Mitchell describing universal coverage as a matter of simple justice, and Dole assuring that Republicans wanted to help, although he stated that Mitchell's bill would menace the quality of American medicine by giving an excessive role to Government. In a front page article in *The New York Times* on August 10, Adam Clymer characterized the Republican intention as one of fighting not just to block government control of health care, which they saw in most of the Democratic proposals, but also to deny President Clinton the signal accomplishment that eluded Presidents Roosevelt, Truman, Nixon, and Carter, all of whom attempted sweeping health-care reforms. Hillary Rodham Clinton told reporters that the fact health-care reform was being debated on the floor of the Senate for the first time in 60 years was a historic and major accomplishment. The debate began in the Senate, but the House decided to put off any debate for the time being—which turned out to be not at all. Packwood—"Market forces work. Competition works. Price controls don't;" Kennedy—"Costs continue to escalate. No American family can feel secure if serious illness strikes"; Kassebaum—"an unprecedented forest of new government regulations"; Dole—"America has the best health-care delivery system in the world. America has the best health care delivery system in the world. America has the best health care delivery system in the world. And I repeat it three times because I'm concerned that actions we take in the chamber in the next couple of weeks or so will mean those words are no longer true."

August 10, 1994—AARP endorsed the Democratic bills in both the House and Senate, as did the AFL–CIO and National Council of Senior

Citizens. President Clinton indicated he would accept the goal of 95% coverage by 2000, which would avoid an employer mandate according to CBO estimates. The debate continued with varying degrees of hostility and cooperation.

August 11, 1994—A bipartisan group of 10 lawmakers produced their own health-care bill, which was designed by Representatives Rowland, Democrat of Georgia, Bilirakis, Republican of Florida, and Senators Cooper and Grandy, Republican of Iowa. The proposal was market-driven, relying on subsidies, had a narrower range of benefits than the Democratic bills, would restrict the size of malpractice settlements, had no employer mandates, aimed at 90% coverage by 2004, had no new taxes, and no price controls. The bill won immediate endorsements from the AMA and a coalition of five big health insurers. It was described by several as a significant compromise with Republicans and conservative Democrats. Many Democrats, however, believed it sacrificed the basic purposes of health-care reform and was lacking on every major goal of health-care restructuring. Stark—"The bipartisan bill would leave 27 million people without insurance and would lead to sharp increases in premiums for people who have coverage"; Gramm—"I can hardly believe my ears when the health-care system of the United States of America is compared unfavorably to the health-care systems of Canada, Great Britain, and Germany. Last year more people died in Canada waiting to get into the operating room than died on the operating table"; Wofford, Democrat of Pennsylvania—"Health care delayed is health care denied. I will introduce an amendment that will disqualify every member of Congress from participating in the Federal employees' health plan until we pass a bill which moves toward universal coverage, opens the Federal employee benefits plan to the American people, and ends insurance discrimination, attaining portability and no exclusions."

August 12, 1994—A moderate bipartisan group of three Republicans and three Democrats, headed by Senator Chafee, called itself the "mainstream coalition," and planned to offer a package of health-care amendments. Senator Dole blocked an effort to vote on the first amendment to the Mitchell bill, an amendment intended to speed coverage for children and pregnant women. Packwood said that 28 Republicans still had opening statements to make and that several needed three or four hours each. The Republican speeches, to this date, had averaged close to an hour, consisting mainly of assertions that the bill is bureaucratic and that the Democrats are rushing. The Democrats, in speeches of about 15–20 minutes, argued that the issues had been studied and studied and the public wanted action. Senator Wellstone, Democrat of Minnesota, said that he and five others among the most liberal Senators would consider voting against a weakened bill. Dole—"It is necessary to start over and prepare a whole new bill"; Packwood—"We will fight it out on this line if it takes all summer" (referring to General U.S. Grant's vow at the Battle of Spotsylvania Courthouse in Virginia during the Civil War); Lott—the Mitchell bill is "health scare legislation." Pear, writing in *The New York Times* on August 14, observed that the Mitchell bill made so many concessions to conservative Democrats and moderate Republicans that Mitchell undercut his base, which confused and divided his old allies.

August 16, 1994—The seventh day of debate on the Senate floor. The Senate cast its first vote on health-care legislation when the Republicans suspended their delaying tactics in the face of Mitchell's threat of 24-hour per day sessions. The 55–42 vote was to approve an amendment to require medical coverage for infants, children, and pregnant women by July 1, 1995. The voting was along straight party lines, with only two Republicans in favor, and one Democrat opposed. Chafee based his opposition on the fact that the Secretary of Health and Human Services would determine a schedule of services to be provided, and he maintained that such decisions should be made by doctors and plans and individuals, and not by the Secretary. Gramm—"Either we are going to beat this bill or I'm going to offer a hell of a lot of amendments"; D'Amato, Republican of New York—"Despite its flaws, it is still the best health-care system in the world, bar none. The best. I dare say that if the poorest of the poor in this country had a problem that they would get better medical treatment here than Boris Yeltsin in Russia"; Dorgan, Democrat of North Dakota—"Decide together that the market system doesn't work to control health-care costs"; Mitchell—"The arguments made today are almost word-for-word the arguments made against Social Security and almost word-for-word the arguments made against Medicare."

August 17, 1994—The mainstream coalition (now including 6 Republicans and 9–10 Democrats) announced it was ready to present its "bundle of amendments" to Dole and Mitchell. The bill resembled the Senate Finance Committee bill, which the CBO said would reach 92% coverage. Kennedy noted that both Dole and Packwood in the past had sponsored bills with similar benefit packages, even though now they steadfastly opposed them in the Mitchell bill.

August 18, 1994—On the floor of the Senate, the work pace was stepped up, from what one pundit described as moving from the glacial to the merely snail-like. On August 16 and 17, the Senate had managed to consider only one amendment each day. On the 18th, it approved three—one to provide more assistance for rural health care, another requiring public meetings for any board of commission created, and yet another providing a mechanism for insurance companies to drop clients who did not pay their premiums. Lott—"It's time we stop the process, that we allow the budget office to analyze what are the real costs of the various bills that are being developed ... that we come back some time in September—the 8th or 12th.... it's time to stop the secrecy—the secrecy that's been involved with this legislation from the first day."

August 19, 1994—On the 10th day of debate, the bipartisan group (now 20 Senators) offered its health plan, which proposed to reduce the Federal deficit by $100 billion over 10 years, increase coverage to 92 or 93% by 2004, included no employer mandates, did not provide prescription-drug coverage for the elderly, and provided less money for long-term care ($10 billion instead of Mitchell's $48 billion). Both the National Council of Senior Citizens and the AFL–CIO opposed the bill.

August 25, 1994—After a four day fight to pass what was expected to be a routine anticrime bill, Mitchell recessed the Senate until September 12. The issue of health-care reform had been given only 10 days of debate on the Senate floor, all taking place in August.

RECESS

The recess was intended to provide a cooling-off period. The media pundits (e.g., Clymer, 1994d) were quick to assert that if anything was to pass, it would be less than the universal-care package that President Clinton wanted. It was suggested that the only hope was to ban the most objectionable practices of the insurance industry and to scrape up more money to help insure people who most need it, if anything was to happen at all. Clymer noted that President Clinton had always proposed a set of incremental steps, moving toward universal coverage over a period of several years, and that what was now being discussed were only increments toward hope. Some argued that it was desirable to allow the states to have greater freedom to experiment with plans of their own, but Senator Durenberger, Republican of Minnesota and a prominent member of the mainstream group, opposed this proposal on the grounds that varying state rules would make it impossible for interstate corporations to provide the kind of health care he saw as the wave of the future. Reinhardt—"One should not belittle any subsidies that can be teased out of the American upper-income classes and funneled down to the lower-income classes."

Other pundits believed Dole had blocked action on comprehensive health-care legislation long enough to force its main champions to give up, at least for this year. His motivation was to deny President Clinton credit for health-care reform, even though Dole had to risk an accusation of hypocrisy, because he had shifted from his previous support of universal coverage.

The CBO issued an evaluation of a Republican plan offered by Representative Lott and Representative Michel, Republican of Illinois. This evaluation concluded that the plan would have the unintended consequence of raising the cost of standard insurance plans to the point of threatening their existence. It would reduce the deficit by $11.3 billion over 10 years but would do almost nothing to curb growing health-care expenditures and little to expand coverage, reaching about 87%—5 million poor children and 2 million poor people would acquire coverage as a result of proposed subsidies.

Toner (1994a) announced the collapse of health-care reform, asserting that people gradually turned away from the idea of a comprehensive overhaul, and that Congress responded to their will. He also noted that the system was overwhelmed by millions of dollars worth of lobbying, polling, and advertising that produced, not a new consensus, but an exhausted paralysis that fed the public's "corrosive cynicism toward their government."

Toner (1994a) quoted Peter Hart, a Democratic pollster,

> The worst of the process worked. Yes, there was a dialogue, but it was so influenced and affected by the special interest groups that the public didn't get a true and honest debate. What they learned is everything they had to fear—and very little about what they could hope for.

Toner noted that groups opposed to the Clinton program outspent those that favored it by a 4:1 ratio in advertising, and that the news media slipped into a campaign mode, focusing coverage on whether Clinton was winning or losing, as opposed to the merits of health-care reform.

The course of the debate in the Senate, and the failure of the House to even take up health-care reform was disappointing, given the grand goals of many well-intentioned people at the outset, the medical needs of over 40 million Americans, and the strong desires of the American public (a Gallup Poll in June 1994 indicated that 74% of the public still approved of universal health insurance, despite the claim by Toner that people had turned away from the idea of comprehensive overhaul). The actions of the players were discouraging displays of political posturing, distortions of facts, hopeless compromises, and deliberate attempts to see that nothing was accomplished. Within this polluted atmosphere the debate was to resume.

THE DEBATE RESUMES (SORT OF)

September 11, 1994—The Senate prepared to meet later in the week, amid a welter of optimistic statements. Chafee—talks have been going quite well; Kennedy—fewer sticking points than expected; Mitchell—"I'm hopeful we are going to get a good bill." On the other hand, a familiar note of pessimism was still there (Dole) and a new one had crept in (Wellstone). Dole—dismissed the mainstream bill as more than Congress can deal with this year; Wellstone—mainstream proposal is "an unworkable retreat."

September 12, 1994—The Senate was back in session. New poll results indicated that 76% of the American public considered universal coverage "very important."

September 14, 1994—Even more optimistic statements by those actively involved in producing a bipartisan plan. Mitchell—"It is both possible and desirable to get a good bill passed this year"; Breaux—"It's alive, it's possible, it's doable"; Chafee—"We can come up with not only a darn good bill, but a very appealing one."

September 17, 1994—With the Haiti crisis occupying Congressional attention, most observers agreed that there would not be enough time to pass a bill before adjournment. The only hope seemed to be for a series of narrow bills to accomplish specific and restricted goals. Packwood—"We've killed health-care reform. Now we've got to make sure our fingerprints are not on it"; Gramm—"I am certainly proud of my part in killing the Clinton Plan in all of its incarnations"; Specter, Republican of Pennsylvania—"Some Republicans want to do nothing. They are wrong. To avoid any risk that President Clinton is going to get credit is really contrary to what mainstream Republicans think in the country"; Danforth—"I don't see any life [in the Mainstream measure] at all"; Moynihan—"We could just say it is not achievable in this session"; Jeffords, Republican of Vermont (the only Republican who had supported the Clinton Plan)—"If a deal is not struck next week, it's too late"; Chafee—"If we don't get this thing moved out next week, it's pretty difficult."

September 19, 1994—Mitchell indicated that he was encouraged that the Senate would have time to produce a bill now that the threat of armed conflict in Haiti had evaporated. Mitchell—"My objective is to pass a bill this year," Kerrey—"We have debated this sufficiently. The task is to convince the American people to back their efforts and 'get to the obstructionists' "; Dole—"Time has almost run out. I don't see anything happening this year."

September 20, 1994—The leading Republican opponents urged President Clinton to give up on the health-care issue for the year or risk losing the rest of his agenda. Dole stated that there was little time left to pass any legislation, and that health care would take at least 7 or 8 days. Forty-five liberal organizations denounced the prospects of an agreement between the Senate mainstream group and Mr. Mitchell, whose own bill (which suffered a quiet death the previous month), they had already condemned as too limited.

September 21, 1994—Mitchell said the bipartisan group was close to giving up, because they lacked enough votes to block a filibuster threatened by the Senate Republican leadership. Chafee told Mitchell he could not count on the votes of all the 10 Republicans in the bipartisan group, and Shelby, Democrat of Alabama, promised to support a filibuster.

September 22, 1994—Senate Republicans started two simultaneous filibusters on bills that had already been approved by the House: one on campaign finance reform, and the other on California desert protection. Kerrey—"I think the forces of darkness are about to prevail."

September 23, 1994—Foley, Democrat of Washington and Speaker of the House, suggested that House Democratic and Republican leaders meet to see if there was any sort of health-insurance legislation on which they could agree. The House Republican leader, Representative Michel was receptive to the idea but Representative Gingrich scoffed at the suggestion, saying, "I don't want to be suckered. I do not trust them." He stated that he feared liberal Democrats would somehow make any measure more liberal than House leaders might agree on and "I don't want to be set up."

September 26, 1994—In a major front-page headline, the journalist Clymer declared that the NHP is dead in Congress, and that Senator Mitchell had announced on September 25 that he would not pursue the issue any further this year. Mitchell's reason was that the Republicans had threatened to oppose the trade pact that was still to be considered, and they had threatened a major filibuster of any proposed health-care legislation, with the Democrats lacking the 60 votes required to break it. Dole would not say what he thought should be done next year, but he said that health care "will be on top of the agenda." Dole—"Senator Mitchell blames Republicans for everything except the plane that crashed into the White House. The American people feared an overdose of government control. We saw Democracy in action. That's the way it is supposed to work."

Toner, in an accompanying front-page editorial, made the point that "the electorate just may not care"—a statement that is contradicted by all of the available polling data, and one which he contradicted at the end of the article. The print media pundits identified two major points toward which opposition to the proposed Clinton plan was directed:

mandatory insurance purchasing cooperatives (health alliances) and employer mandates. Clinton gave way on the first, but the second was maintained as the only way to finance the program. It was noted that, by midsummer, Mitchell and Kennedy were scrambling for compromises, and Dole was backing away from them.

Among the postmortems were the following: Republican National Committee—"We find ourselves in this position because the Clinton Administration proposed the creation of a government-run health-care system." Vice-president of the National Federation of Independent Business—"The left killed health-care reform by overreaching, by not being willing to accept a consensus approach."

A review of this scenario makes it difficult to conceive of how any consensus or compromise could have been reached, given the inflexible opposition of the Republican leadership on all key issues. The award for negative inflexibility should be given to Senator Dole; Senator Packwood should receive the colorful metaphor award; Senator Gramm, the award for finding the most demons in the closet; and Senator Mitchell, for an unflagging optimism throughout the ordeal.

POSTADJOURNMENT, THE ELECTIONS, AND AFTER

During October, the focus of the reconvened Congress was on the upcoming elections. Health care was not emphasized as a major issue in the Congressional elections, with the exception of a few candidates such as Edward Kennedy, who stressed the importance of the issue as well as emphasizing his strong and continuing role in achieving health reform. The collective wisdom seemed to be that employer mandates should be abandoned and that reliance should be placed on market resources that some experts stated were "already revolutionizing the delivery of health care in the U.S." This statement refers to the mergers and acquisitions that continue to occur at all levels of the health-care delivery system.

It was agreed that the idea of using employer mandates to achieve universal coverage would be difficult to reintroduce, and many wanted to consider the single-payer option a dead issue. California voters defeated, by a 73% to 27% margin, an initiative that would have established a single-payer plan for the state. A great deal of money was spent by insurance companies to defeat the initiative, which they "vehemently" opposed.

After the election, Michel guessed that Congress would start with employers being required to "offer" insurance for their employees, but not required to pay for it. He stated that he still expected Congress to make some modest and incremental changes. Health care was not included as an item in the House Republicans' Contract with America, developed by the House majority leader Newt Gingrich. That Contract outlined the incoming Republican majority's legislative priorities for the first 100 days of the 104th Congress. If the Republican strategy was to not have any health-care reform pass in order to produce gridlock and lead to a Democratic defeat, it was eminently successful.

Senator Dole stated that he expected early agreement on some basics that would expand access to insurance, limit restrictions based on preexisting conditions, achieve portability of insurance when the place of employment changed, introduce malpractice reform, and make reforms that would enable small business to provide employees access to health insurance. Gingrich floated a trial balloon, stating that he would consider replacing Medicare altogether (Gosselin, 1995a, p. 9): "I think we need to transform Medicare into a different system." That statement elicited such an immediate negative reaction that his aides insisted he did not mean to imply that the program needed to be cut back.

Fallows (1995) argued that the demise of health-care reform was a triumph of misinformation throughout the course of the debates and in countless newspaper and magazine postmortems. Among those counts of conventional wisdom he outlined was the false charge that the plan was hatched in secret. There was a task force of 500 members that met continuously during the Administration's first few months in office, and it went out of its way to hear a variety of views, meeting with 572 separate organizations (but not having representatives from the AMA or the Pharmaceutical Manufacturers of America—obviously a grave tactical error), as well as a couple of hundred meetings with the Congressional leadership and individual members of Congress. Fallows believes that a larger problem was that the Washington press was excluded from the deliberations, and that this secrecy toward the press was "stupid," but it did not represent closed-mindedness about ideas as the press corps portrayed it.

Another charge was that the plan was politically naive. I believe that Clinton's adamant stand against the more sensible and simpler single-payer option was a calculated political decision to strike the pose that no "new taxes" were needed, as well as a reluctance to recommend the abolition of the entire medical insurance industry. That decision was fatal, and it created political gridlock, but the strategy can be defended as a rational political choice. Ira Magaziner, the cocreator of the Clinton Plan (with Hillary Rodham Clinton), was quoted by Fallows (p. 30) to have said that they felt single-payer supporters could be bought off if a plan such as MC was offered, even though they did not really like it. The market reform people could be bought off as well, they thought. The Administration was wrong regarding both groups—both finally dug in their heels and the impasse led the Republicans to lose interest in negotiating.

Another fallacious charge is that the plan reached too far, too quickly. As Fallows noted, the reality is that the plan proposed little that was new or unprecedented, and it was not more complex than most other plans that were discussed.

There were numerous falsehoods, and these were almost gleefully exploited by media pundits, being repeated over and again by Congressional opponents to health-care reform. One of the most blatant was that the plan would prevent people from going outside the system to buy better basic coverage, even going so far as to say that there would be "doctors in jail" for providing fee-for-service. Fallows (p. 36) quoted an early portion of the bill which stated, "Nothing in the Act shall be construed as prohibiting the following: (1) An individual from purchasing any health care services." He noted that the article making the charge that people would not be permitted to select and pay for medical treatments they desired was published in *The New Republic*, which he claimed declined to publish a point-by-point rebuttal by the White House. Reinhardt told Fallows (p. 34) that "the average American patient would have had more choice under the Clinton Plan than they now will. If you work for a particular company, your choice of HMOs is whatever the company offers you."

Fallows faulted the administration for allowing untrue attacks on the health plan to go unanswered and placed major blame on the media for permitting the public to be so grossly misinformed. He concluded that the economic concerns of the various special-interest groups were what drove the process and led to the defeat of any reform.

The bottom line is that the agreed-upon moral necessity to provide adequate health care for all people gave way to the onslaught of special economic interests and fears that were fed by a well-financed campaign of misinformation and false allegations. The realities of the situation are unchanged: More and more people are being added to the ranks of the uninsured presently; more people require government support for health care under the aegis of Medicare and Medicaid, yet these programs are being targeted for reduction in the interests of achieving a balanced budget sometime in the future.

After the election, polls taken by Harvard's School of Public Health and the Kaiser Family Foundation found that 25% of the public agreed that Congress should enact a major reform bill, 41% that there should be modest changes, and only 25% agreed the system should be left alone

(Toner, 1994b). These results can be interpreted to mean that 66% of the public still express a need for some health-care reform. The poll also revealed that there were deep misconceptions by the public regarding the structure of public financing in the United States: 30% thought the largest Federal expense was military spending, 27% said foreign aid, 19% said welfare, and 15% said Social Security. In fact, military spending accounts for more than 19% of Federal spending, Social Security for nearly 22%, foreign aid for well under 2%, and the basic welfare programs for just over 1%. A responsible media might keep such facts at the ready and make sure the public is aware of them anytime policies are recommended on the basis of misinformation that led to such misunderstanding, misperceptions, and confused the public throughout the health-care debate. At the least, when front-page headlines and lead stories on TV news are misleading or false, an equally prominent statement of the accurate state of affairs should be provided. Perhaps in this way, the media could make it too costly for people to mislead, and it might even make for more interesting, entertaining, and popular news coverage.

One can only hope that the decks will be cleared and that Congress will take the lead on the health-care issue (which 66% of the public surveyed in the aforementioned poll thought it should), and President Clinton should stand aside, given the expressed lack of confidence in the Administration's previous actions regarding health-care reform (only 18% of those polled said President Clinton should take the lead). The Administration should reaffirm its commitment to the moral necessity of universal coverage, provide accurate information regarding the realities of economics and financing, and let powerful and responsible members of Congress take the lead for another go-around.

Perhaps one of the best summaries of the goings on was contained in a political cartoon by Toles that was published in *The Oregonian* on August 12, 1994 (p. 69). On a piece of paper was a paragraph containing a health-care bill that was labeled

> The Interest-Group Plan: All Americans will have free insurance, through highly profitable insurance companies, for every medical treatment regardless of cost, including drugs expensive enough to guarantee high drug company profits, and choice of doctors, whose incomes will continue to increase faster than inflation, forever, at no cost to employers, large or small, or the government or taxpayers or anyone else.

There was a notation included in the lower left-hand corner that this bill had "passed unanimously and forwarded to Santa's workshop."

CHAPTER 15

Epilogue

This book completes an examination of issues concerning the humane use and care of human beings by other human beings. This examination was begun in *Human Evolution, Reproduction, and Morality* (Petrinovich, 1995), in which argumentation was developed regarding the basic principles that should be brought to bear to understand the nature and functioning of those organic systems of which the human species is a part. These principles involve the realities of evolutionary biology, physiological substrates, and behavioral outcomes. The latter are concerned with activities that serve to form and cement social bonds between the members of human communities, especially between a mother and her newborn, and with the development of cognitive capacities. The arguments were documented through an analysis of the theoretical and research literature regarding human evolution, behavior, and psychology, with special attention devoted to the nature of developmental processes, ethological mechanisms, and the principles involved in perceptual and cognitive processing.

In that book, considerable attention was devoted to the ideas developed by moral philosophers and to some of the basic fallacies in argumentation that are often found when data and theory are brought to bear on the consideration of human social problems. It was insisted that high standards should be set for the admissible rules of argumentation. These rules should emphasize conceptual clarity, factual information, rationality, impartiality, universality, a cool emotional tone, and the conclusions reached should enjoy a reasonable level of intuitive acceptability.

Particular emphasis was placed on basic aspects of evolutionary mechanisms and the ways these mechanisms operate in relation to human behavior at several levels. The existence of evolved mechanisms was examined at the levels of human reproduction, speech and language development, perception, and cognition. The usefulness of evolutionary principles was argued whenever there was an interest to understand the human social condition, and a case was made for an evolutionary psychology adequate to consider human behavior at all levels. Objections to the use of evolutionary principles, such as genetic determinism, reductionism, and cultural variability were examined and rejected.

The results of an extensive empirical study of moral intuitions were used to support the argument that humans have a coherent system of moral beliefs that are mapped onto biological and cultural universals. These universals are found across cultures and sexes and meet the criteria that would qualify them as evolved processes, similar to those that have been found important in most sexually reproducing species. The contribution of these moral beliefs and the behaviors that result from them should be considered within the context of the conditions that likely would have existed in the environment of evolutionary adaptation, because it is within that environment that they were selected. When some understanding has been reached at this theoretical level, then it is useful to consider why these tendencies have persisted and whether they are adaptive, given the present circumstances within which the human species finds itself coping. The important dimensions involve such things as enhancing inclusive fitness and honoring social contracts, both of which are of central importance in evolutionary theory, especially when applied to humans.

Following the development of these basic principles, they were applied to issues and arguments concerning reproduction, especially abortion. The polar positions regarding abortion were considered within the frameworks of philosophy, biology, and cognition. It was argued that the status of personhood marks a critical stage determining the moral standing of organisms, and that the point of birth defines the entry into this stage. With the attainment of personhood, the organism is considered to be a moral patient, who cannot be held to standards of morality, but who is entitled to respect by moral agents, all of whom have the duties and responsibilities that characterize full moral standing. The entry into the critical stage of moral agency is defined by cognitive criteria that make it possible for an individual to understand rules and the concept of causation.

Polar positions regarding abortion—restrictive and permissive—were characterized, policy issues that society has had to face (and will continue to) were discussed, and policy recommendations were developed that flowed from the basic arguments developed to that point. The nature of the new reproductive technologies developed to assist infertile people to reproduce were examined, and the moral dilemmas they introduce were explained, especially those involved in artificial insemination, *in vitro* fertilization, the use of surrogate mothers, cloning human embryos, and research using embryos. That book, therefore, considered many of the important questions involved in the reproduction of human life and developed the basic biological and moral issues that are involved in reproduction.

The discussions in the present book turned to questions regarding living an adequate life once it has been begun, and to those regarding the end of life. The first of these questions concerned the criteria for death, which were examined both at the biological and biographical levels. It was argued that the crucial criteria should rest on an individual's ability to continue a biographical life. The implications of this position were developed in terms of taking organs from cadavers for transplantation into people who need them in order to survive. Next, the permissibility of suicide and euthanasia were considered as moral issues. A position was argued that emphasized the importance of honoring human dignity and freedom within a framework of social contracts and expectations. Although the emphasis shifted from those evolutionary principles involved in differential reproduction to principles concerning freedom, dignity, and responsibility, the underlying concerns are much the same. The analyses at both the biological and rational levels involve considerations that are phrased in terms of costs and benefits, as well as a consequentialist assessment of conflicting and competing values.

The human genome project was described, and the use of genetic screening was discussed, because these enterprises have powerful implications for the detection, prevention, and treatment of disease in developing humans—both early and late in development. The ethical objections that have been raised against pursuing these programs were examined, and it was argued that several of these objections (such as the necessity for informed consent, informed counseling, and regulation of the commercial use of genetic information) should be guaranteed so that abuses do not occur. Other objections, such as those based on genetic determinism, the specter of eugenics, and the possibility of individual stigmatization, were examined and rejected. The perils of such factors are overemphasized; these potential dangers have been acknowledged and are being avoided; and the benefits being realized justify continuation of the programs. Much of the argumentation against genetic research rests on justifications of the status quo and baseless fears of entering a slippery slope.

The issues regarding the termination of medical treatment, assisted suicide, taking organs for transplant, and the use of genetic information and material in the practice of medicine involve decisions concerning morality and responsibility. The question of morality, therefore, needs to be brought into a clear focus. All of these developments in medical practice and technology led to the emergence of a new field of philosophy, bioethics. Bioethicists have considered carefully the troublesome aspects and implications of the technological developments and have debated which moral principles should used to guide social policies when difficult decisions are to be made. These decisions involve questions such as the responsibilities of physicians to their healthy, ill, and dying patients—young and old; decisions that must be made regarding when organs can be taken from neomorts for transplantation and who should have priority to receive the scarce organs. The complex developments in technology have made it necessary to make such decisions with greater frequency, and have led to the formation of hospital review boards to set policy and assist physicians and hospital administrations to face such difficult decisions on a daily basis.

When questions of medical ethics are considered, it is necessary to establish the rules by which health care should be equitably distributed to the public. The Clinton administration faced the problems regarding a national health-care delivery system, which is a monster that sort of grew like Topsy, with the United States spending unprecedented sums of money (both in absolute amounts and proportion of GDP) on health care. Yet, there are many millions of U.S. citizens who are under- and uninsured, and the statistics regarding the adequacy of U.S. health care paint an appalling picture. Although many in government (all of whom are guaranteed the best, permanent, and affordable health care) deny there are any serious problems, others argue that they are approaching the disaster stage, and the U.S. public agrees, according to most surveys.

Realities in terms of financing and health-care delivery were discussed at length in the last few chapters of this book. The moral issues were developed, the political and social realities were considered, and various plans were proposed to solve these problems. The experiences of other countries were considered, as well as the lessons that could be learned from the actions taken in the various states that have grappled with problems in health-care financing and delivery. Given all these realities and considering the moral issues involved, it was argued that, in the interest of having a just society, it is necessary to provide universal health care to all the people living in the country, that coverage should be mandated, and that an employer mandate is the easiest way to attain that ideal, given the structure of the present insurance system. It was argued that a single-payer system modeled after the Canadian Health Plan would attain that ideal with the greatest economy.

After the dust of the health-care debate in the 103rd Congress had settled, it was clear that no solutions were reached and the issues regarding health care remain, with the problems becoming more severe.

The issues that were discussed here will undoubtedly be with us for some time to come, and it will be necessary to examine the lessons that can be learned from the experience up to this point, again and again.

Throughout this book, it was emphasized that humanists, philosophers, and social and biological scientists should abandon the position that what we strive for is knowledge only for the sake of knowledge itself. It should be accepted that scientists and philosophers have a responsibility to provide public access to what is known and guidance regarding how one can frame and argue policies based on the best available information, using the best available modes of dialogue. To this end, procedures that social psychologists have developed to facilitate rational decision making were discussed. These procedures can help reduce the confusion and obfuscation that often are introduced when various special-interest groups find their personal prerogatives or cherished beliefs threatened. Most members of society are interested in developing a just society, and one can build on this positive motivation, if the discussions are focused on goals rather than fears. The media bear a considerable share of responsibility for both the present state of confusion and the resulting inaction that occurred regarding the establishment of a national health-care plan. The media should bring to bear the marvelous technological tools they have at their disposal to enhance the quality of public debate, rather than to pander to the sensational charges that are made so often. Perhaps such a suggestion is not reasonable, given the fact that the media are part of the economic establishment that is interested in maintaining the present structure of society. If such is the case, then that reality should be acknowledge and steps taken to ensure that the public is somehow kept informed.

EVOLUTION CONSIDERED

Evolutionary principles are brought into play whenever human reproduction is involved, because the game of evolution is regulated by ultimate outcomes scored in terms of relative reproductive success. In order to succeed at this game, it is necessary to transmit one's genes to the next generation, and to have the members of that generation reproduce. This ultimate process is driven by the proximate structures and behaviors that organisms exhibit, and it is these proximate factors that undergo natural selection. Those structures that enhance reproductive success—and which are genetically transmitted—are important to the survival of the individuals possessing them. There is a continual sorting and resorting of genetic material, and one of the most effective ways to accomplish this sorting is through sexual reproduction. This mode of reproduction produces a varying set of genotypes, drawn from the existing gene pool at each generation and resulting in the necessary amount of variability for some individuals of the species to survive the vicissitudes of selection pressures in order that the species line can avoid extinction should the conditions of living change.

Evolutionary mechanisms are important not only to drive the process of reproduction, but also to develop various behavioral adaptations that enhance the likelihood of surviving to reproduce and to raise offspring successfully. Williams and Nesse (1991; Nesse & Williams, 1995) considered a variety of physiological mechanisms that have evolved to enable organisms to avoid and withstand infectious disease, to repair damages to physiological systems, to compensate for impaired functions of organs and systems, to avoid, expel, attack, and destroy pathogens, and to isolate parasites that have entered the body. They argued that, too often, medical treatments are introduced to counteract symptoms, such as fever and pain, and that these symptoms often represent useful adaptations to combat injury and disease. It is not always in the organism's best interests to circumvent useful natural adaptations merely to enhance a patient's immediate comfort.

Medical practice will be improved if there is a better understanding by the medical and public-health community of the basic principles governing competition between hosts and parasites and if this understanding is reached with an appreciation of the different vectors—cultural and ecological—that affect the mode of transmission and the virulence of diseases. Evolutionary principles can also be brought to bear to understand that it is not possible to find a perfectly natural, safe, toxin-free diet, because toxins are used by most plants to protect themselves from insects and small mammals. It would be preferable to diversify dietary regimes, which would avoid overloading any particular detoxification mechanism, rather than attempt to have a perfectly toxin-free diet.

An understanding of the forces affecting selection might also make it possible to understand *senescence*—the breakdown that occurs at the end of the life cycle. If it becomes possible to delay senescence through manipulation of the genome, there could well be deleterious effects that would be evident in earlier stages of development, where the interest is healthy development and reproduction. Gene-line manipulations should be approached cautiously, because they might negatively influence the organism during the reproductive and parental phases of life, and these costs might override the benefits of preventing degenerative changes to make it possible to live a longer time.

Williams and Nesse argued that an evolutionary perspective adds the ultimate dimension to the proximate level that is usually considered in the practice of medicine and public health. Adaptation should be viewed within the framework of a cost–benefit analysis made within the ecological context: Certain adaptations to trauma and illness produce definite benefits, but at a cost to other aspects of the organism's welfare and function.

Tooby and Cosmides (1994) discussed a serious problem that could jeopardize much of the basic research that has been done with standardized cell lines to study cancer. The cell lines used by most laboratories have a generation time of 2 to 4 days, and these lines have been used for many years. Those in the biomedical research community used these standardized cell lines to obtain a uniform culture that would enhance the comparability of results between different laboratories and the same laboratory over time. A careful examination of these cell lines, however, has shown that they have lost sex chromosomes, have undergone rapid phenotypic changes, and lost and/or duplicated some chromosomes.

There is no constancy in the cell lines, because they are evolving. This problem was discounted when the cell lines were established, on the assumption that there could not be evolution in the absence of differential selection for mortality, an assumption that is based on a misunderstanding of the evolutionary process. Evolution occurs as the result of differential reproduction, and after a large number of generations, it is expected that there would be systematic drift, even though no variable selection pressure is operating. A failure to understand the nature of basic evolutionary mechanisms has cast doubt on many years of fundamental research that was based on the assumption that cell lines were constant over time. At the same meeting of the Human Behavior and Evolution Society at which Tooby and Cosmides made their presentation, Paul Ewald argued that the use of the principles of Darwinian evolution is the 21st-century solution that will direct the progress of medical research.

At the outset of the enterprise represented by these two books, it was emphasized that it is important to develop principles and arguments that are logically consistent and can be applied universally. The arguments apply to many of the important issues in human life and death. The next step is to extend the same principles to consider the relationships between humans and other animals. There has been considerable (and heated) discussion between the animal research community and animal rights and liberation organizations. The relative value, and the moral permissibility, of various kinds of medical and behavioral research using animals have been debated, and there has been considerable social and political activism by those on different sides of the issues. The methods used to breed and care for animals used for food, hides, and medicinals have been discussed at length, and it has been argued by some that such usage of animals is not permissible morally. The arguments have been extended to consider of the permissible manipulations of the ecology within which humans exist, raising questions regarding the importance of preserving natural ecosystems, maintaining biodiversity, and whether humans have an obligation to do so.

In the next volume, I will bring the principles and arguments developed to this point to bear on this series of issues. To address these principles at an adequate level of complexity, questions of animal cognition will be considered in greater detail and concerns in the philosophy of science that bear on the question of how it can be decided whether a given line of research represents progress in science will be entertained. This will all be done to demonstrate that the evolutionary, cognitive, and philosophical principles argued in these first two books are of sufficient generality to provide a useful framework to consider those factors that influence attempts made by humans to cope with, and to exist in, the universe which we have created for ourselves.

References

Altman, S. H., & Cohen, A. B. (1993). The need for a national global budget. *Health Affairs, 12*(Suppl.), 194–203.

American Medical Association, Council on Ethical and Judicial Affairs. (1994). Ethical issues in health care system reform. *Journal of the American Medical Association, 272*, 1056–1062.

American Medical Association, Council on Ethical and Judicial Affairs. (1995). Ethical issues in managed care. *Journal of the American Medical Association, 273*, 330–335.

Anderson, J. R. (1990). *The adaptive character of thought.* Hillsdale, NJ: Erlbaum.

Angell, M. (1988). Euthanasia. *New England Journal of Medicine, 319*, 1348–1350.

Angell, M. (1993a). How much will health care reform cost? *New England Journal of Medicine, 328*, 1778–1779.

Angell, M. (1993b). Privilege and health—What is the connection? *New England Journal of Medicine, 329*, 126–127.

Angell, M. (1994). After Quinlan: The dilemma of the persistent vegetative state. *New England Journal of Medicine, 330*, 1524–1525.

Angier, N. (1992, December 4). U.S. permits use of genes in treating cystic fibrosis. *The New York Times*, A28.

Angier, N. (1994a, January 11). Biologists find key genes that shape patterning of embryos. *The New York Times*, B7.

Angier, N. (1994b, April 1). Gene experiment to reverse inherited disease is working. *The New York Times*, A1.

Annas, G. J. (1994). Asking the courts to set the standard of emergency care—The case of Baby K. *New England Journal of Medicine, 330*, 1542–1545.

Annas, G. J., & Elias, S. (1992a). *Gene mapping.* New York: Oxford University Press.

Annas, G. J., & Elias, S. (1992b). The major social policy issues raised by the human genome project. In G. J. Annas & S. Elias (Eds.), *Gene mapping* (pp. 3–17). New York: Oxford University Press.

Applebome, P. (1994, June 26). Amid cries of politicking, a widely endorsed plan dies. *The New York Times*, A11.

Asseo, L. (1995, April 25). Justices reject Kevorkian appeal. *The Boston Globe*, p. 3.

Baker, L. C., & Cantor, J. C. (1993). Physician satisfaction under managed care. *Health Affairs, 12*(Suppl.), 258–270.

Baker, R. R., & Bellis, M. A. (1989). Elaboration of the Kamikaze sperm hypothesis. *Animal Behaviour, 37*, 865–867.

Baker, R. R., & Bellis, M. A. (1993). Human sperm competition: Ejaculate manipulation by females and a function for the female orgasm. *Animal Behaviour, 46*, 887–909.

Barer, M. L., & Evans, R. G. (1992). Interpreting Canada: Models, mind-sets, and myths. *Health Affairs, 11*(1), 44–61.

Barkow, J. H., Cosmides, L., & Tooby, J. (1992). *The adapted mind.* New York: Oxford University Press.

Barnard, C. (1986). The need for euthanasia. In A. B. Downing & B. Smoker (Eds.), *Voluntary euthanasia* (pp. 173–183). Atlantic Highlands, NJ: Humanities Press International.

Barr, R. (1993, February 17). Health care abroad reaches more for less. *The Oregonian*, A3.

Barrington, M. R. (1986). The case for rational suicide. In A. B. Downing & B. Smoker (Eds.), *Voluntary euthanasia* (pp. 230–247). Atlantic Highlands, NJ: Humanities Press International.

Bartecchi, C. E., MacKenzie, T. D., & Schrier, R. W. (1994). The human costs of tobacco use. *New England Journal of Medicine, 330*, 907–912.

Bates, T., & Lane, D. (1994, November 27). Oregon's suicide statute faces tests. *The Oregonian*, A1.

Battin, M. P. (1987). Age rationing and the just distribution of health care: Is there a duty to die? *Ethics, 97*, 3;17–340.

Beardsley, T. (1993). From mice to men. *Scientific American, 268*, 18–19.

Bedau, H. A. (1982). *The death penalty in America.* Oxford, UK: Oxford University Press.

Belkin, L. (1993). *First, do no harm.* New York: Simon & Schuster.

Bellis, M. A., & Baker, R. R. (1990). Do females promote sperm competition? Data for humans. *Animal Behaviour, 40*, 997–999.

Berke, R. L. (1994, July 10). On the stump, not much talk of health care. *The New York Times*, A1.

Bickerton, D. (1990). *Language and species.* Chicago: University of Chicago Press.

Blendon, R. J., Leitman, R., Morrison, I., & Donelan, K. (1990). Satisfaction with health systems in ten nations. *Health Affairs, 9*(2), 185–192.

Blewett, L. A. (1994). Reforms in Minnesota: Forging the path. *Health Affairs, 13*(4), 200–209.

Bohlen, C. (1995, March 31). Pope offers "Gospel of Life" vs "Culture of Death." *The New York Times*, A1.

Bolles, R. C. (1970). Species-specific defense reactions and avoidance learning. *Psychological Review, 77*, 32–48.

Botkin, J. R., & Post, S. G. (1992). Confusion in the determination of death: Distinguishing philosophy from physiology. *Perspectives in Biology and Medicine, 36*, 129–138.

Brandt, R. B. (1976). The morality and rationality of suicide. In E. S. Schneidmann (Ed.), *Suicidology: Contemporary developments* (pp. 378–399). (Reprinted in R. B. Brandt (Ed.). (1992). *Morality, utilitarianism, and rights.* Cambridge, UK: Cambridge University Press, pp. 315–335.)

Brandt, R. B. (1980). *The concept of a moral right and its function.* Paper read at Virginia Polytechnic Institute and State University. (Reprinted in R. B. Brandt (Ed.). (1992). *Morality, utilitarianism, and rights.* Cambridge, UK: Cambridge University Press, pp. 179–195.)

Brandt, R. B. (1987). Public policy and life and death decisions regarding defective newborns. In R. C. McMillan, H. T. Engelhardt, & S. F. Spicker (Eds.), *Euthanasia and the newborn.* Dordrecht, Holland: D. Reidel. (Reprinted in R. B. Brandt (Ed.). (1992). *Morality, utilitarianism, and rights.* Cambridge, UK: Cambridge University Press, pp. 354–369.)

Brandt, R. B. (1992). Utilitarianism and welfare legislation. In R. B. Brandt (Ed.), *Morality, utilitarianism, and rights* (pp. 370–387). Cambridge, UK: Cambridge University Press.

Brewer, C. (1986). The hospice movement. In A. B. Downing & B. Smoker (Eds.), *Voluntary euthanasia* (pp. 203–209). Atlantic Highlands, NJ: Humanities Press International.

Brinkley, J. (1994, March 31). At Utah hospital, innovative way to track medical quality. *The New York Times*, A9.

Brinster, R. L., & Averbock, M. R. (1994). Germline transmission of donor haplotype following spermatogonial transplantation. *Proceedings of the National Academy of Sciences, USA, 91*, 11303–11307.

Brinster, R. L., & Zimmermann, J. W. (1994). Spermatogenesis following male germ-cell transplantation. *Proceedings of the National Academy of Sciences, USA, 91*, 11298–11302.

Brown, L. D. (1993). Commissions, clubs, and consensus: Florida reorganizes for health reform. *Health Affairs, 12*(2), 7–26.

Brunswik, E. (1956). *Perception and the representative design of psychological experiments.* Berkeley: University of California Press.

Burnstein, E., Crandall, C., & Kitayama, S. (1994). Some neo-Darwinian decision rules for altruism: Weighing cues for inclusive fitness as a function of the biological importance of the decision. *Journal of Personality and Social Psychology, 67*, 773–789.

Buss, D. M. (1994). *The evolution of desire.* New York: Basic Books.

Buss, D. M., & Schmitt, D. P. (1993). Sexual Strategies Theory: An evolutionary perspective on human mating. *Psychological Review, 100*, 204–232.

Butler, D. (1994). Ethics treaty to target genome implications. *Nature, 371*, 369.

Butler, S. M. (1993, September 28). Rube Goldberg, call your office. *The New York Times*, A19.

Cabrera, L. (1993, September 5). Blue Cross of Washington and Alaska releases salary figures. *The Oregonian*, 1.

Cairns, P., et al. (1994). Rates of p16 (MTS1) mutations in primary tumors with 9p loss. *Science, 265*, 415–416.

Callahan, J. C. (1987). On harming the dead. *Ethics, 97*, 341–352.

Campion, E. W. (1994). The oldest old. *New England Journal of Medicine, 330*, 1819–1820.

Cantor, C. (1992). The challenges to technology and informatics. In D. J. Kevles & L. Hood (Eds.), *The code of codes* (pp. 98–111). Cambridge, MA: Harvard University Press.

Capecchi, M. R. (1994). Targeted gene replacement. *Scientific American, 269*, 52–59.

Caplan, A. L. (1985). Ethical issues raised by research involving xenografts. *Journal of the American Medical Association, 254*, 3339–3343.

Caplan, A. L. (1986a). Requests, gifts, and obligations: The ethics of organ procurement. *Transplantation Proceedings, 18*(Suppl. 2), 49–56. (Reprinted in A. L. Caplan (Ed.). (1992). *If I were a rich man could I buy a pancreas?* Bloomington, IN: Indiana University Press, pp. 145–157.)

Caplan, A. L. (1986b). The ethics of in vitro fertilization. *Primary Care, 13*, 241–253. (Reprinted in R. T. Hull (Ed.). (1990). *Ethical issues in the new reproductive technologies*. Belmont, CA: Wadsworth, pp. 96–108.)

Caplan, A. L. (1986c). The high cost of technological development: A caveat for policymakers. In A. S. Allen (Ed.), *New options, new dilemmas: An interprofessional approach to life or death decisions*. Lexington, MA: Lexington. (Reprinted in A. L. Caplan (Ed.). (1992). *If I were a rich man could I buy a pancreas?* Bloomington, IN: Indiana University Press, pp. 285–301.)

Caplan, A. L. (1988). New technologies in reproduction—new ethical problems. *Annals New York Academy of Sciences, 530*, 73–82.

Caplan, A. L. (1989a). Hard data is the only answer to hard choices in health care. *The Mount Sinai Journal of Medicine, 56*(3), (Reprinted in A. L. Caplan (Ed.). (1992). *If I were a rich man could I buy a pancreas?* Bloomington, IN: Indiana University Press, pp. 302–314.)

Caplan, A. L. (1989b). Ethics, cost-containment, and the allocation of scarce resources. *Investigative Radiology, 24*, 918–926. (Reprinted in A. L. Caplan (Ed.). (1992). *If I were a rich man could I buy a pancreas?* Bloomington, IN: Indiana University Press, pp. 315–335.)

Caplan, A. L. (1989c). If I were a rich man could I buy a pancreas? *Transplantation Proceedings, 21*, 3381–3387. (Reprinted in A. L. Caplan (Ed.). (1992). *If I were a rich man could I buy a pancreas?* Bloomington, IN: Indiana University Press, pp. 158–177.)

Caplan, A. L. (1992). Mapping morality: Ethics and the human genome project. In A. L. Caplan (Ed.), *If I were a rich man could I buy a pancreas?* (pp. 118–142). Bloomington, IN: Indiana University Press.

Caplan, A. L. (1993, March 3). Other nations may offer ideas on health-care reforms. *The Oregonian*.

Caplan, A. L. (1994, February 10). Inmate off list as transplant candidate. *The Oregonian.*

Caprino, M. (1993, October 6). Mergers transform health care industry. *The Oregonian.*

Caskey, C. T. (1992). DNA-based medicine: Prevention and therapy. In, D. J. Kevles & L. Hood (Eds.), *The code of codes* (pp. 112–135). Cambridge, MA: Harvard University Press.

Chao, J. (1993, December 12). Dutch counseling service seeks to break taboo around suicide. *The Oregonian*, A9.

Clarke, A. L., & Low, B. S. (1992). Ecological correlates of human dispersal in 19th century Sweden. *Animal Behaviour, 44*, 677–693.

Clymer, A. (1993, September 30). First Republican supports health plan. *The New York Times*, A15.

Clymer, A. (1994a, February 5). Hype around the health care debate obscures its seriousness. *The New York Times*, A8.

Clymer, A. (1994b, April 13). Cost and access: Consensus quickly splinters. *The New York Times*, A11.

Clymer, A. (1994c, May 6). Hawaii is a health care lab as employers buy insurance. *The New York Times*, A1.

Clymer, A. (1994d, August 28). With health overhaul dead, a search for minor repairs. *The New York Times*, A1.

Cohen, J. S., Fihn, S. D., Boyko, E. F., Jonsen, A. R., & Wood, R. W. (1994). Attitudes toward assisted suicide and euthanasia among physicians in Washington state. *New England Journal of Medicine, 331*, 89–94.

Conly, S. R., & Camp, S. L. (1992). *China's family planning program: Challenging the myths.* Washington DC: Population Crisis Committee.

Cook, D. J., et al. (1995). Determinants in Canadian health care workers of the decision to withdraw life support from the critically ill. *Journal of the American Medical Association, 273*, 703–708.

Corder, E. H., et al. (1993). Gene dose of apolipoprotein E type 4 allele and the risk of Alzheimer's disease in late onset families. *Science, 261*, 921–923.

Cosmides, L., & Tooby, J. (1987). From evolution to behavior: Evolutionary psychology as the missing link. In J. Dupre (Ed.), *The latest on the best: Essays on evolution and optimality* (pp. 277–306). Cambridge, MA: MIT Press.

Crittenden, R. A. (1993, Summer). Managed competition and premium caps in Washington State. *Health Affairs, 12*, 82–88.

Daly, M., & Wilson, M. (1988). *Homicide.* New York: Aldine de Gruyter.

Davies, J. L., et al. (1994). A genome-wide search for human type 1 diabetes susceptibility genes. *Nature, 371*, 130–136.

Davies, P. D., Fetzer, J. H., & Foster, T. R. (1995). Logical reasoning and domain specificity: A critique of the social exchange theory of reasoning. *Biology and Philosophy, 10*, 1–37.

Day, K. (1995, March 22). Maryland firm wins rights to genetic therapy technique. *The Boston Globe*, p. 81.

Degler, C. N. (1991). *In search of human nature.* Oxford, UK: Oxford University Press.

De Lew, N. Greenberg, G., & Kinchen, K. (1992). A layman's guide to the U.S. health care system. *Health Care Financing Review, 14,* 151–169.

DeParle, J. (1994, October 7). Census sees falling income and more poor. *The New York Times,* A16.

Detsky, A. S. (1993). Northern exposure—Can the United States learn from Canada. *New England Journal of Medicine, 328,* 805–807.

Dick, A. W. (1994). Will employer mandates really work? Another look at Hawaii. *Health Affairs, 13*(1), 343–349.

Dickson, D. (1994). "Gene map" plan highlights dispute over public vs. private interests. *Nature, 371,* 365–366.

Dole, R. (1994, January 26). Excerpts from the Republicans' response to the President's message. *The New York Times,* A9.

Dowie, M. (1988). *We have a donor.* New York: St. Martin's Press.

Dubler, N. N. (1993). Commentary: Balancing life and death—proceed with caution. *American Journal of Public Health, 83,* 23–25.

Dworkin, R. (1993). *Life's dominion.* New York: Knopf.

Dworkin, R. (1994a, May 17). When is it right to die? *The New York Times,* A19.

Dworkin, R. (1994b, January 13). Will Clinton's plan be fair? *The New York Review of Books,* pp. 20–25.

Eckholm, E. (1993a, August 23). Double sword for president. *The New York Times,* A1.

Eckholm, E. (1993b, September 20). Less cost vs. less care. *The New York Times,* A1.

Eckholm, E. (1993c, November 7). Health plan is toughest on doctors making most. *The New York Times,* A1.

Eckholm, E. (1994, July 11). Frayed nerves of people without health coverage. *The New York Times,* A1.

Eckholm, E. (1995a, January 11). H.M.O.'s are changing the face of Medicare. *The New York Times,* A1.

Eckholm, E. (1995b, January 19). A hospital copes with the new order. *The New York Times,* sec. 3, p. 1.

Eddy, D. M. (1991). Oregon's plan: Should it be approved? *Journal of the American Medical Association, 266,* 2439–2445.

Elias, S., & Annas, G. J. (1994). Genetic consent for genetic screening. *New England Journal of Medicine, 330,* 1611–1613.

Emanuel, E. J. (1991). *The ends of human life.* Cambridge, MA: Harvard University Press.

Emanuel, E. J., & Dubler, N. N. (1995). Preserving the physician–patient relationship in the era of managed care. *Journal of the American Medical Association, 273,* 323–329.

Emanuel, E. J., & Emanuel, L. L. (1994). The economics of dying. *New England Journal of Medicine, 330,* 540–544.

Engelhardt, H. T., & Rie, M. A. (1986). Intensive care units, scarce resources, and

conflicting principles of justice. *Journal of the American Medical Association, 255*, 1159–1164.
Enthoven, A. C. (1993a). The history and principles of managed competition. *Health Affairs, 12*(Suppl.), 24–48.
Enthoven, A. C. (1993b). Why managed care has failed to contain health costs. *Health Affairs, 12*(3), 27–43.
Fallows, J. (1995). A triumph of misinformation. *Atlantic Monthly, 278*, 26–36.
Farnsworth, C. H. (1993, December 20). Americans filching free health care in Canada. *The New York Times*, A1.
Feder, B. J. (1994a, July 14). Hospital supplier in novel deal. *The New York Times*, C1.
Feder, B. J. (1994b, October 10). New hospital giant seeks savings and stream of cases. *The New York Times*, C1.
Feinberg, J. (1984). *Harm to others.* Oxford, UK: Oxford University Press.
Feinberg, J. (1985). *Offense to others.* Oxford, UK: Oxford University Press.
Feinberg, J. (1986). *Harm to self.* Oxford, UK: Oxford University Press.
Feinberg, J. (1988). *Harmless wrong-doing.* Oxford, UK: Oxford University Press.
Feinberg, J. (1989). Rawls and intuitionism. In N. Daniels (Ed.), *Reading Rawls* (pp. 108–124). Stanford, CA: Stanford University Press.
Fernald, A. (1992). Human maternal vocalizations to infants as biologically relevant signals: An evolutionary perspective. In J. H. Barkow, L. Cosmides, & J. Tooby (Eds.), *The adapted mind* (pp. 391–428). New York: Oxford University Press.
Fielding, J. E., & Lancry, P.-J. (1993). Lessons from France—"Vive la difference." *New England Journal of Medicine, 270*, 748–756.
Fisher, L. M. (1994, January 30). Profits and ethics collide in a study of genetic coding. *The New York Times*, A1.
Flew, A. (1986). The principle of euthanasia. In, A. B. Downing & B. Smoker (Eds.), *Voluntary euthanasia* (pp. 40–57). Atlantic Highlands, NJ: Humanities Press International.
Fox, R. C., & Swazey, J. P. (1992). *Spare parts.* Oxford, UK: Oxford University Press.
Freudenheim, M. (1993a, August 8). Making health plans prove their worth. *The New York Times*, F5.
Freudenheim, M. (1993b, September 1). Physicians are selling practices to companies as changes loom. *The New York Times*, A1.
Freudenheim, M. (1993c, September 19). Changing the fortunes of the medical business. *The New York Times*, sec. 3, p. 1.
Freudenheim, M. (1994a, July 12). Pharmaceutical giant is buying operator of drug-benefit plans. *The New York Times*, A1.
Freudenheim, M. (1994b, October 23). Insurance companies expect battle over health care to shift to the states. *The New York Times*, A18.
Freudenheim, M. (1995a, January 4). Hospitals are tempted but worry as for-profit chains woo them. *The New York Times*, A1.
Freudenheim, M. (1995b, January 16). Clamping the costs of job injuries. *The New York Times*, D1.

Freudenheim, M. (1995c, April 11). Penny-pinching H.M.O's showed their generosity in executive paychecks. *The New York Times*, D1.

Garcia, J., & Brett, L. P. (1977). Conditioned responses to food order and taste in rats and wild predators. In, M. Kare (Ed.), *The chemical senses and nutrition* (pp. 277–289). New York: Academic Press.

Gaylin, W. (1993). Faulty diagnosis. *Harper's Magazine*, *287*, 57–64.

Gaylin, W., Kass, L. R., Pellegrino, E. D., & Siegler, M. (1988). Doctors must not kill. *Journal of the American Medical Association*, *259*, 2139–2140.

Geary, D. (1995). Reflection of evolution and culture in children's cognition. *American Psychologist*, *50*, 24–37.

Geary, D. (in press). Sexual selection and sex differences in spatial cognition. *Learning and Individual Differences*.

Gibbs, J. P. (1978). Preventive effects of capital punishment other than deterrence. *Criminal Law Bulletin*, *14*, 34–50. (Reprinted in H. A. Bedau (Ed.), *The death penalty in America*. Oxford, UK: Oxford University Press, pp. 103–116.)

Gigerenzer, G., & Hug, K. (1992). Domain-specific reasoning: Social contracts, cheating, and perspective change. *Cognition*, *43*, 127–171.

Gilbert, W. (1992). A vision of the grail. In D. J. Kevles & L. Hood (Eds.), *The code of codes* (pp. 83–97). Cambridge, MA: Harvard University Press.

Gillon, R. (1986). Suicide and voluntary euthanasia: Historical perspective. In A. B. Downing & B. Smoker (Eds.), *Voluntary euthanasia* (pp. 210–229). Atlantic Highlands, NJ: International Press International.

Ginzberg, E. (1994). Improving health care for the poor. *Journal of the American Medical Association*, *271*, 464–467.

Glaser, W. A. (1993). The United States needs a health system like other countries. *Journal of the American Medical Association*, *270*, 980–984.

Gosselin, P. G. (1995a, January 8). Medicare, Medicaid cuts urged. *The Boston Globe*, p. 1.

Gosselin, P. G. (1995b, February 2). GOP weighs reductions in Medicaid spending. *The Boston Globe*, p. 1.

Gould, C. (1993, October 3). Health care: Loser with a future? *The New York Times*, F12.

Greely, H. T. (1992). Health insurance, employment discrimination, and the genetics revolution. In, D. J. Kevles & L. Hood (Eds.), *The code of codes* (pp. 264–280). Cambridge, MA: Harvard University Press.

Greenhouse, S. (1993a, September 17). Small-business federation looms large against Clinton. *The New York Times*, A1.

Greenhouse, S. (1993b, September 28). Job losses in health plan are denied. *The New York Times*, B10.

Greenough, W. T., Black, J. E., & Wallace, C. S. (1987). Experience and brain development. *Child Development*, *58*, 539–559.

Grumbach, K., & Bodenheimer, T. (1995). The organization of health care. *Journal of the American Medical Association*, *273*, 160–167.

Grumbach, K., Bodenheimer, T., Himmelstein, D. U., & Woolhandler, S. (1991). Liberal benefits, conservative spending: The physicians for a national health program proposal. *Journal of the American Medical Association, 265*, 2549–2554.

Guralnik, J. M., Land, K. C., Blazer, D., Fillenbaum, G. G., & Branch, L. G. (1993). Educational status and active life expectancy among older blacks and whites. *New England Journal of Medicine, 329*, 110–116.

Haig, D. (1993). Genetic conflicts in human pregnancy. *The Quarterly Review of Biology, 68*, 495–532.

Hamilton, W. D., Axelrod, R., & Tanese, R. (1990). Sexual reproduction as an adaptation to resist parasites (A review). *Proceedings National Academy of Sciences, USA, 87*, 3566–3573.

Hammond, K. R., & Adelman, L. (1976). Science, values, and human judgment. *Science, 194*, 389–396.

Hammond, K. R., & Grassia, J. (1985). The cognitive side of conflict: From theory to resolution of policy disputes. In S. Oskamp (Ed.), *Applied social psychology annual* (Vol. 6, pp. 233–254). Beverly Hills, CA: Sage.

Hammond, K. R., Mumpower, J., Dennis, R. L., Fitch, S., & Crumpacker, W. (1983). Fundamental obstacles to the use of scientific information in public policy making. *Technological Forecasting and Social Change, 24*, 287–297.

Hammond, K. R. , Stewart, T. R., Brehmer, B., & Steinmann, D. O. (1977). Social judgment theory. In M. F. Kaplan & S. Schwartz (Eds.), *Human judgment and decision processes: Applications in problem settings* (pp. 271–312). New York: Academic Press.

Hamosh, A., & Carey, M. (1993). Correlation between genotype and phenotype in patients with cystic fibrosis. *New England Journal of Medicine, 329*, 1308–1313.

Handyside, A. H., Lesko, J. G., Tarin, J. J., Winston, R. M. L., & Hughes, M. R. (1992). Birth of a normal girl after in vitro fertilization and preimplantation diagnostic testing for cystic fibrosis. *New England Journal of Medicine, 327*, 905–909.

Harris, J. (1992). *Wonderwoman and superman*. Oxford, UK: Oxford University Press.

Hilts, P. J. (1993a, March 4). A.M.A. is softening stand on changes in health system. *The New York Times*, A1.

Hilts, P. J. (1993b, March 31). Doctors' pay resented, and it's underestimated. *The New York Times*, A9.

Hilts, P. J. (1993c, November 5). Genetic test results have cost some their insurance. *The New York Times*, A20.

Hilts, P. J. (1994, July 8). Sharp rise seen in smokers' health care costs. *The New York Times*, A12.

Himmelstein, D. U., et al. (1989). A national health program for the United States. *New England Journal of Medicine, 320*, 102–108.

Hofmeister, S. (1994, October 12). $3.3 billion hospital deal is planned. *The New York Times*, C1.

Hubbard, R. (1995). Genomania and health. *American Scientist, 83*, 8–10.

Hume, D. (1784). Of suicide. In, T. H. Green & T. H. Gosse (Eds.), *The philosophi-*

cal works of David Hume. London. (Reprinted in P. Singer (Ed.), *Applied ethics*. Oxford, UK: Oxford University Press, pp. 19–27.)

Humphry, D. (1991). *Final exit*. Eugene, OR: The Hemlock Society.

Humphry, D. (1992). *Dying with dignity*. New York: Birch Lane Press.

Hutton, H., Borowitz, M., Olesky, I., & Luce, B. R. (1994). The pharmaceutical industry and health reform: Lessons from Europe. *Health Affairs, 13*(3), 98–111.

Iglehart, J. K. (1992). The American health care system: Private insurance. *New England Journal of Medicine, 326*, 1715–1720.

Iglehart, J. K. (1994). Health care reform: The states. *New England Journal of Medicine, 330*, 75–79.

Johnson, M. (1993, September 23). Conditions worsen for U.S., British children. *The Oregonian*.

Jones, S. (1993). *The language of genes*. New York: Doubleday.

Kaiser Health Reform Project. (1994). *Health reform legislation: A comparison of major proposals*. Prepared by the Kaiser Commission on the Future of Medicaid. January.

Kamb, A., et al. (1994). A cell cycle regulator potentially in many tumor types. *Science, 264*, 436–440.

Kamm, F. M. (1993). *Morality, mortality: Volume I*. Oxford, UK: Oxford University Press.

Kaplan, R. M. (1993). Application of a general health policy model in the American health care crisis. *Journal of the Royal Society of Medicine, 86*, 277–281.

Kaplan, R. M. (1994a). The Ziggy theorem: Toward an outcomes-focused health psychology. *American Psychologist, 13*, 451–460.

Kaplan, R. M. (1994b). Using quality of life information to set priorities in health policy. *Social Indicators Research, 33*, 121–163.

Kaplan, R. M., et al. (1994). Quality adjusted survival analysis: A neglected application of the quality of well-being scale. *Psychology and Health, 9*, 131–141.

Kapp, M. B. (1993). Life-sustaining technologies: Value issues. *Journal of Social Issues, 49*, 151–167.

Keil, J. E., et al. (1993). Mortality rates and risk factors for coronary disease in black as compared with white men and women. *New England Journal of Medicine, 329*, 73–78.

Kekes, J. (1993). *The morality of pluralism*. Princeton, NJ: Princeton University Press.

Keller, E. F. (1992). Nature, nurture, and the human genome project. In C. J. Kevles & L. Hood (Eds.), *The code of codes* (pp. 281–299). Cambridge, MA: Harvard University Press.

Kenrick, D. T., & Keefe, R. C. (1992). Age preferences in mates reflect sex differences in human reproductive strategies. *Behavioral and Brain Sciences, 15*, 75–91.

Kevles, D. J. (1992). Out of eugenics: The historical politics of the human genome. In, D. J. Kevles & L. Hood (Eds.), *The code of codes* (pp. 3–36). Cambridge, MA: Harvard University Press.

Kevles, D. J., & Hood, L. (1992). Reflections. In D. J. Kevles & L. Hood (Eds.), *The code of codes* (pp. 300–328). Cambridge, MA: Harvard University Press.

Kilner, J. F. (1990). *Who lives? Who dies?* New Haven: Yale University Press.

King, P. A. (1992). The past as prologue: Race, class, and gene discrimination. In G. J. Annas & S. Elias (Eds.), *Gene mapping* (pp. 94–111). New York: Oxford University Press.

Kinney, H. C., Korein, J., Panigrahy, A., Dikkes, P., & Goode, R. (1994). Neuropathological findings in the brain of Karen Ann Quinlan. *New England Journal of Medicine, 330*, 1469–1475.

Kleinig, J. (1991). *Valuing life*. Princeton: Princeton University Press.

Knox, R. A. (1995, February 18). Study finds ICU doctors withholding treatment. *The Boston Globe*, p. 1.

Kolata, G. (1992, April 29). Ethicists debating a new definition of death. *The New York Times*, B7.

Kolata, G. (1993, December 7). Nightmare or the dream of a new era in genetics? *The New York Times*, A1.

Kolata, G. (1994a, January 11). Reproductive revolution is jolting old views. *The New York Times*, A1.

Kolata, G. (1994b, November 22). Ethicists wary over new gene technique's consequences. *The New York Times*, B9.

Kolata, G. (1994c, November 22). Gene technique could shape future generations. *The New York Times*, A1.

Konner, M. (1993). *Medicine at the crossroads*. New York: Pantheon.

Krauss, C. (1993, September 24). Lobbyists of every stripe on health care proposal. *The New York Times*, A1.

Kristoff, N. D. (1993, July 21). Chinese turn to ultrasound, scorning baby girls for boys. *The New York Times*, A1.

Krueger, A. B., & Reinhardt, U. E. (1994). Economics of employer versus individual mandates. *Health Affairs, 13*(1), 34–53.

Lakshmanan, I. A. R. (1995, March 17). Citizens for life to fight suicide bill. *The Boston Globe*, p. 31.

Leary, W. E. (1994, December 3). Clinton rules out federal money for research on human embryos created for that purpose. *The New York Times*, A6.

Le Gal La Salle, G., et al. (1993). An adenovirus vector for gene transfer into neurons and glia in the brain. *Science, 259*, 988–990.

Lee, T. F. (1993). *Gene future*. New York: Plenum Press.

Lehigh, S. (1995, April 6). Patient argues for choice on how to die. *The Boston Globe*, p. 31.

Leichter, H. M. (1993a). The trip from acrimony to accommodation. *Health Affairs, 12*(2), 48–58.

Leichter, H. M. (1993b). Health care reform in Vermont: A work in progress. *Health Affairs, 12*(2), 71–81.

Leichter, H. M. (1994). Health care reform in Vermont: The next chapter. *Health Affairs, 13*, 78–103.

Lessenberry, J. (1994, November 27). Kevorkian aids in suicide as Michigan's ban expires. *The New York Times*, A12.

Letsch, S. W. (1993). National health care spending in 1991. *Health Affairs, 12*, 94–110.

Levit, K. R. Lazenby, H. C., Cowan, C. A., & Letsch, S. W. (1993). Health spending by state: New estimates for policy making. *Health Affairs, 12*(3), 7–26.

Levit, K. R., et al. (1994). National health spending trends, 1960–1993. *Health Affairs, 13*(5), 14–31.

Lewontin, R. C. (1991). *Biology as ideology*. New York: HarperCollins.

Lieberman, P. (1984). *The biology and evolution of language*. Cambridge, MA: Harvard University Press.

Locke, J. (1993). *The child's path to spoken language*. Cambridge, MA: Harvard University Press.

Luther, J. (1993, October 6). New cigarette tax increase: Lives saved, jobs lost? *The Oregonian*.

MacKenzie, T. D., Bartecchi, C. E., & Schrier, R. W. (1994). The human costs of tobacco use. *New England Journal of Medicine, 330*, 975–980.

Macklin, R. (1992). Privacy and control of genetic information. In G. J. Annas & S. Elias (Eds.), *Gene mapping* (pp. 157–172). New York: Oxford University Press.

Mahowald, M. B., Silver, J., & Ratcheson, R. A. (1987). The ethical options in transplanting fetal tissue. *Hastings Center Report, 17*, 9–15. (Reprinted in R. T. Hull (Ed.) (1990). *Ethical issues in the new reproductive technologies*. Belmont, CA: Wadsworth, pp. 259–271.)

Maier, S. F., Watkins, L. R., & Fleshner, M. (1994). Psychonueroimmunology: The interface between behavior, brain, and immunity. *American Psychologist, 49*, 1004–1017.

Marshall, E. (1993). A tough line on genetic screening. *Science, 262*, 984–985.

Mashaw, J. L., & Marmor, T. R. (1993, August 18). The states as health care labs. *The New York Times*, A19.

McKusick, V. A. (1992). The human genome project: Plans, status, and applications in biology and medicine. In G. J. Annas & S. Elias (Eds.), *Gene mapping* (pp. 18–42). New York: Oxford University Press.

McMahan, J. (1993). Killing, letting die, and withdrawing aid. *Ethics, 103*, 250–279.

Meier, B. (1993a, February 14). Effective? Maybe. Profitable? Clearly. *The New York Times*, sec. 3, p. 1.

Meier, B. (1993b, August 17). Blue Cross tries to push plans for managed care. *The New York Times*, B1.

Meier, B. (1994, March 31). Health plans promote choice, but decisions may not be easy. *The New York Times*, A1.

Metz, R. (1995, February 21). HMO profits may spur drive to force them to cut rates. *The Boston Globe*, p. 50.

Meyers, D. (1993, September 17). Critics of health care plan take aim at "bureaucracy." *The Oregonian*, A15.

Miki, Y., et al. (1994). A strong candidate for the breast and ovarian cancer susceptibility gene BRCA1. *Science, 266*, 66–71.

Morse, L. J. (1993). A declaration of independence for health system reform. *New England Journal of Medicine, 329*, 804–805.
Muller, I. (1991). *Hitler's justice.* Cambridge, MA: Harvard University Press.
Multi-Society Task Force on Persistent Vegetative States. (1994). Medical aspects of the persistent vegetative state. *New England Journal of Medicine, 330*, 1499–1508, 1572–1579.
Murphy, D. J., et al. (1994). The influence of the probability of survival on patients' preferences regarding cardiopulmonary resuscitation. *New England Journal of Medicine, 330*, 545–549.
Murray, C. J. L. (1990). Mortality among black men. *New England Journal of Medicine, 322*, 205–206.
Murray, T. H. (1993). Moral reasoning in social context. *Journal of Social Issues, 49*, 185–200.
Myerson, A. R. (1994, June 29). Hospital chain sets guilty plea. *The New York Times*, C1.
Myerson, A. R. (1995, March 20). Helping health insurers say no. *The New York Times*, D1.
Nabel, G. J., et al. (1993) Direct gene transfer with DNA—liposome complexes in melanoma: Expression, biologic activity, and lack of toxicity in humans. *Proceedings of the National Academy of Sciences, USA, 90*, 11307–11311.
Nagel, T. (1979). Death. In T. Nagel (Ed.), *Mortal questions* (pp. 1–10). Cambridge, UK: Cambridge University Press.
Neergaard, L. (1995, February 23). AIDS testing is urged for pregnancies. *The Boston Globe*, p. 3.
Nelkin, D. (1992). The social power of genetic information. In, D. J. Kevles & L. Hood (Eds.), *The code of codes* (pp. 177–190). Cambridge, MA: Harvard University Press.
Nesse, R. M., & Williams, G. C. (1995). *Why we get sick.* New York: Times Books.
Neubauer, D. (1993). A pioneer in health system reform. *Health Affairs, 12*(2), 31–39.
O'Keefe, M. (1994a, November 10). Assisted-suicide measure survives heavy opposition. *The Oregonian*, A1.
O'Keefe, M. (1994b, November 15). Bishops label measure 16 "cancer." *The Oregonian*, A1.
Olshansky, S. J., Carnes, B. A., & Cassel, C. K. (1993). The aging of the human species. *Scientific American, 268*, 46–52.
O'Neill, P. (1994, November 14). Is the law liberation or Pandora's box? *The Oregonian*, B1.
Pappas, G., Queen, S., Hadden, W., & Fisher, G. (1993). The increasing disparity in mortality between socioeconomic groups in the United States, 1960 and 1986. *New England Journal of Medicine, 329*, 103–109.
Pascal, L. (1980). Judgement day or the handwriting on the wall. *Inquiry, 23* 242–251.
Passell, P. (1993, June 22). The unknown devil. *The New York Times*, D1.
Payami, H., et al. (1994). Alzheimer's disease, apolipoprotein E4, and Gender. *Journal of the American Medical Association, 271*, 1316–1317.

Pear, R. (1993a, January 5). Health-care costs up sharply again, posing new threat. *The New York Times*, A1.
Pear, R. (1993b, July 7). Drug industry gathers a mix of voices to bolster its case. *The New York Times*, A1.
Pear, R. (1993c, August 16). U.S. to guarantee free immunization for poor children. *The New York Times*, A1.
Pear, R. (1993d, August 21). Health industry is moving to form service networks. *The New York Times*, A1.
Pear, R. (1993e, September 5). Health aides see a tax on benefits beyond basic plan. *The New York Times*, A1.
Pear, R. (1994a, March 29). Gaps in coverage for health care. *The New York Times*, C18.
Pear, R. (1994b, April 13). Inquiry challenges doctors on ordering diagnostic tests. *The New York Times*, A11.
Pear, R. (1994c, June 8). Vermont shows how a health bill can fail. *The New York Times*, A15.
Pear, R. (1994d, October 7). Health care insurance percentage is lowest in 4 Sun Belt states. *The New York Times*, A16.
Petrinovich, L. (1989). Representative design and the quality of generalization. In L. W. Poon, D. C. Rubin & B. A. Wilson (Eds.), *Everyday cognition in adulthood and late life* (pp. 11–24). Cambridge, UK: Cambridge University Press.
Petrinovich, L. (1990). Avian song development: Methodological and conceptual issues. In D. A. Dewsbury (Ed.), *Contemporary issues in comparative psychology* (pp. 340–359). Sunderland, MA: Sinauer.
Petrinovich, L. (1995). *Human evolution, reproduction, and morality*. New York: Plenum Press.
Petrinovich, L., & O'Neill, P. (in press). The influence of wording and framing effects on moral intuitions. *Ethology and Sociobiology*.
Petrinovich, L., O'Neill, P. & Jorgensen, M. (1993). An empirical study of moral intuitions: Toward an evolutionary ethics. *Journal of Personality and Social Psychology*, *64*, 467–478.
Pinker, S. (1994). *The language instinct*. New York: Morrow.
Pinker, S., & Bloom, P. (1990). Natural language and natural selection. *Behavioral and Brain Sciences*, *13*, 707–727.
Popper, K. R. (1957). *The poverty of historicism*. New York: Harper & Row.
Proctor, R. N. (1992). Genomics and eugenics: How fair is the comparison? In G. J. Annas & S. Elias (Eds.), *Gene mapping* (pp. 57–93). New York: Oxford University Press.
Profet, M. (1992). Pregnancy sickness as adaptation: A deterrent to maternal ingestion of teratogens. In J. H. Barkow, L. Cosmides, & J. Tooby (Eds.), *The adapted mind* (pp. 327–365). New York: Oxford University Press.
Prottas, J. M. (1993). Altruism, motivation, and allocation: Giving and using human organs. *Journal of Social Issues*, *49*, 137–150.

Quinn, M. (1994, June 11). California's health pool: Limits, but lower rates. *The New York Times*, A1.
Quinn, M. (1995, February 9). Health plans force changes in the way doctors are paid. *The New York Times*, A1.
Rachels, J. (1979). Euthanasia, killing, and letting die. In J. Ladd (Ed.), *Ethical issues relating to life and death* (pp. 146–161). Oxford, UK: Oxford University Press.
Rachels, J. (1986). *The end of life*. Oxford, UK: Oxford University Press.
Rawls, J. (1971). *A theory of justice*. Cambridge, MA: Harvard University Press.
Redelmeier, D. A., & Fuchs, V. R. (1993). Hospital expenditures in the United States and Canada. *New England Journal of Medicine, 328*, 772–778.
Regan, T. (1983). *The case for animal rights*. Berkeley: University of California Press.
Reich, W. (1993, February 27). Shame on the Dutch. *The New York Times*, A19.
Reinhardt, U. E. (1992, December 14). Health insurance for all—now. *The New York Times*, A11.
Reinhardt, U. E. (1993). Reorganizing the financial flows in American health care. *Health Affairs, 12*(Suppl.), 172–193.
Reinhardt, U. E. (1994a). The Clinton Plan: A salute to American pluralism. *Health Affairs, 13*(1), 161–177.
Reinhardt, U. E. (1994b). Germany's health care system: It's not the American Way. *Health Affairs, 13*(4), 22–24.
Richards, R. J. (1987). *Darwin and the emergence of evolutionary theories of mind and behavior*. Chicago: University of Chicago Press.
Ringer, R. (1993, September 24). Value Health plans to buy competitor. *The New York Times*, D3.
Robertson, J. A. (1992). The potential impact of the Human Genome Project on procreative liberty. In G. J. Annas & S. Elias (Eds.), *Gene mapping* (pp. 215–225). New York: Oxford University Press.
Rogal, D. L., & Helms, W. D. (1993). Tracking states' efforts to reform their health systems. *Health Affairs, 12*(2), 27–30.
Rohter, L. (1993, April 4). Florida blazes trail to a new health-care system. *The New York Times*, A1.
Rosenthal, E. (1993, January 24). Insurers second-guess doctors, provoking debate over savings. *The New York Times*, A1.
Rosenthal, E. (1994, May 12). Patients share bigger burden of rising health care costs. *The New York Times*, A1.
Rosenthal, E. (1995, February 13). Elite hospitals in New York City facing a crunch. *The New York Times*, A1.
Rubin, H. A. et al. (1993). Patients' ratings of outpatient visits in different practice settings. *Journal of the American Medical Association, 270*, 835–840.
Rublee, D. A. (1994). Medical technology in Canada, Germany, and the United States: An update. *Health Affairs, 13*(4), 113–117.

Saltman, R. B. (1992). Single-source financing systems. *Journal of the American Medical Association, 268*, 774–779.

Saltus, R. (1994, December 30). Heart transplant demand is far outpacing supply. *The Boston Globe*, p. 1.

Saltus, R. (1995a, March 20). Scientists develop new gene gun to assists in fight against cancer. *The Boston Globe*, p. 3.

Saltus, R. (1995b, March 21). Tumor-blockers near human tests. *The Boston Globe*, p. 72.

Saltus, R. (1995c, April 11). US ruling bars discrimination based on genes. *The Boston Globe*, p. 5.

Schellenberg, G. D., et al. (1992). Genetic linkage evidence for a familial Alzheimer's disease locus on chromosome 14. *Science, 258*, 668–671.

Schieber, G. J., Poullier, J.-P., & Greenwald, L. M. (1993). Health spending, delivery, and outcomes in OECD countries. *Health Affairs, 12*(2), 120–129.

Schieber, G. J., Poullier, J.-P., & Greenwald, L. M. (1994). Health system performance in OECD countries, 1980–1992. *Health Affairs, 13*(4), 100–112.

Schweitzer, A. (1962). The ethics of reverence for life. *Christendon, 1*, 225–239. (Reprinted in H. Clark, *The ethical mysticism of Albert Schweitzer*. Boston: Beacon Press, pp. 180–194.)

Shafer, T., et al. (1994). Impact of medical examiner/coroner practices on organ recovery in the United States. *Journal of the American Medical Association, 272*, 1607–1613.

Simons, M. (1993a, February 9). Dutch move to enact law making euthanasia easier. *The New York Times*, A1.

Simons, M. (1993b, February 10). Dutch parliament approves law permitting euthanasia. *The New York Times*, A5.

Simpson, J. L., & Carson, S. A. (1992). Preimplantation genetic diagnosis. *New England Journal of Medicine, 327*, 951–953.

Singer, P. A., & Siegler, M. (1990). Euthanasia—a critique. *New England Journal of Medicine, 322*, 1881–1883.

Solomon, M. Z., et al. (1993). Decisions near the end of life: Professional views on life-sustaining treatments. *American Journal of Public Health, 83*, 14–23.

Sorenson, J. (1992). What we still don't know about genetic screening and counseling. In G. J. Annas & S. Elias (Eds.), *Gene mapping* (pp. 203–212). New York: Oxford University Press.

Spector, P. (1994, September 24). Failure, by the numbers. *The New York Times*, A19.

Stein, C. (1995, January 8). Still uninsured. *The Boston Globe*, p. 70.

Steinbock, B. (1979). The intentional termination of life. In *Ethics in science and medicine* (pp. 59–64). New York: Pergamon Press. (Reprinted in J. P. Sterba (Ed.) (1991). *Morality in practice*. Belmont, CA: Wadsworth, pp. 187–192.)

Steinbock, B. (1992). *Life before birth*. Oxford, UK: Oxford University Press.

Steinbrook, R., & Lo, B. (1992). The Oregon medicaid demonstration project—Will it provide adequate care? *New England Journal of Medicine, 326*, 340–344.

Steinwachs, D. M., Wu, A. W., & Skinner, E. A. (1994). How will outcomes management work? *Health Affairs, 13*(4), 153–162.

Stevens, C. M. (1993). Health care cost containment: Some implications of global budgets. *Science, 259*, 16–17, 105.

Stroebe, M., Gergen, M. M., Gergen, K. J., & Stroebe, W. (1992). Broken hearts or broken bonds: Love and death in historical perspective. *American Psychologist, 47*, 1205–1212.

Suzuki, D., & Knudtson, P. (1990). *Genetics*. Cambridge, MA: Harvard University Press.

Swartz, K. (1994). Dynamics of people without health insurance. *Journal of the American Medical Association, 271*, 64–66.

Temkin, L. S. (1993). *Inequality*. New York: Oxford University Press.

Terry, D. (1993, August 5). Kevorkian aids in suicide, No. 17, near police station. *The New York Times*, A8.

Theodore, J., & Lewiston, N. (1990). Lung transplantation comes of age. *New England Journal of Medicine, 322*, 772–774.

Thompson, D. F. (1993). Understanding financial conflicts of interest. *New England Journal of Medicine, 329*, 573–576.

Toner, R. (1993, August 26). House Republican leaders assert health plan will push taxes up. *The New York Times*, A1.

Toner, R. (1994a, September 4). Making sausage: The art of reprocessing the democratic process. *The New York Times*, sec. 4, p. 1.

Toner, R. (1994b, November 16). Pollsters see a silent storm that swept away Democrats. *The New York Times*, A14.

Toner, R. (1995, January 31). Gingrich promises medicare tough look, bottom to top. *The New York Times*, A1.

Tooby, J., & Cosmides, L. (1992). The psychological foundations of culture. In J. H. Barlow, L. Cosmides, & J. Tooby (Eds.), *The adapted mind* (pp. 19–136). New York: Oxford University Press.

Tooby, J., & Cosmides, L. (1994, June). *Universal Darwinism and the perennially elusive cancer cure*. Paper delivered at the Human Behavior and Evolution Society Meetings, University of Michigan, Ann Arbor.

Tooby, J., & Cosmides, L. (1995). Mapping the evolved functional organization of mind and brain. In M. Gazzaniga (Ed.), *Cognitive neuroscience* (pp. 1185–1197). Cambridge, MA: MIT Press.

Torrey, B. B., & Jacobs, E. (1993). More than loose change: Household health spending in the United States and Canada. *Health Affairs, 12*(1), 126–131.

Travis, J. (1993). New piece in Alzheimer's puzzle. *Science, 261*, 828–829.

Trivers, R. L. (1972). Parental investment and sexual selection. In B. Campbell (Ed.), *Sexual selection and the descent of man* (pp. 136–179). Chicago: Aldine.

Trojan, J., et al. (1993). Treatment and prevention of rat glioblastoma by immunogenic C6 cells expressing antisense insulin-like growth factor I RNA. *Science, 259*, 94–97.

Truog, R. D., Berde, C. B., Mitchell, C., & Grier, H. E. (1992). Barbiturates in the care of the terminally ill. *New England Journal of Medicine, 327*, 1678–1682.

Truog, R. D., & Brennan, T. A. (1993). Participation of physicians in capital punishment. *New England Journal of Medicine, 329*, 1346–1350.

Ubel, P. A., Arnold, R. M., & Caplan, A. L. (1993). Rationing failure. *Journal of the American Medical Association, 270*, 2469–2474.

Vaux, K. L. (1988). Debbie's dying: Mercy killing and the good death. *Journal of the American Medical Association, 259*, 2140–2141.

Vicedo, M. (1992). The human genome project: Towards an analysis of the empirical, ethical, and conceptual issues involved. *Biology and Philosophy, 7*, 255–278.

Vidmar, N., & Ellsworth, P. C. (1974). Research on attitudes toward capital punishment. *Stanford Law Review, 26*, 1245–1270. (Reprinted in H. A. Bedau (Ed.) (1982). *The death penalty in America.* Oxford, UK: Oxford University Press, pp. 68–84.)

Wade, N. (1994, February 22). A short cut to human genes. *The New York Times*, B5.

Walton, D. (1992). *Slippery slope arguments.* Oxford, UK: Oxford University Press.

Wang, X. T. (1992). *Multivariate framing effects on human choice behavior: A search for rationality behind "irrational" decision making biases.* Paper presented at Human Behavior and Evolution Society Meetings, Albuquerque, NM. July.

Watson, J. D. (1992). A personal of the project. In, D. J. Kevles & L. Hood (Eds.), *The code of codes* (pp. 164–173). Cambridge, MA: Harvard University Press.

Wexler, N. (1992). Clairvoyance and caution: Repercussions from the human genome project. In D. J. Kevles & L. Hood (Eds.), *The code of codes* (pp. 211–243). Cambridge, MA: Harvard University Press.

White, J. (1993). Markets, budgets, and health care cost control. *Health Affairs, 12*(3), 44–57.

Wiggins, S., et al. (1992). The psychological consequences of predictive testing for Huntington's disease. *New England Journal of Medicine, 327*, 1401–1405.

Wilcox, A., Skjaerven, R., Buekens, P., & Kiely, J. (1995). Birth weight and perinatal mortality. *Journal of the American Medical Association, 273*, 709–711.

Wilfond, B. S., & Nolan, K. (1993). National policy development for the clinical application of genetic diagnostic technologies. *Journal of the American Medical Association, 270*, 2948–2954.

Williams, G. C., & Nesse, R. M. (1991). The dawn of Darwinian medicine. *The Quarterly Review of Biology, 66*, 1–22.

Wills, C. (1991). *Exons, introns, and talking genes.* New York: Basic Books.

Wilson, M. (1987). Impacts of the uncertainty of paternity on family law. *University of Toronto Faculty Law Review, 45*, 216–242.

Wilson, M., & Daly, M. (1992). The man who mistook his wife for a chattel. In J. Barkow, L. Cosmides, & J. Tooby (Eds.), *The adapted mind* (pp. 243–276). London: Oxford University Press.

Woolhandler, S., & Himmelstein, D. U. (1991). The deteriorating administrative efficiency of the U.S. health care system. *New England Journal of Medicine, 324*, 1253–1258.

Woolhandler, S., Himmelstein, D. U., & Lewontin, J. P. (1993). Administrative costs in U.S. hospitals. *New England Journal of Medicine, 329*, 400–403.

World Bank. (1993). *World development report 1993: Investing in health.* New York: Oxford University Press.

Wright, R. (1995, March 13). The biology of violence. *New Yorker*, pp. 68–77.

Young, P. S. (1994, July 8). Moving to compensate families in human-organ market. *The New York Times*, B7.

Youngner, S. J., Landefeld, C. S., Coulton, C. J., Juknialis, B. W., & Leary, M. (1989). "Brain death" and organ retrieval. *Journal of the American Medical Association, 261*, 2205–2210.

Author Index

Subject Index